PHOTOSYNTHESIS

PHOTOSYNTHESIS

Third Edition

David W. Lawlor
IACR-Rothamsted, Harpenden, Hertfordshire, UK

Springer

David W. Lawlor
IACR-Rothamstead, Harpenden, Hertfordshire, UK

Published in the United States of America, its dependent territories and Canada by arrangement with BIOS Scientific Publishers Ltd, 9 Newtec Place, Magdalen Road, Oxford OX4 1RE, UK

© **BIOS Scientific Publishers Limited 2001**

First published © Longman Group UK Limited 1987
Reprinted © Longman Group UK Limited 1990
Second Edition © Longman Group UK Limited 1993
Third Edition © BIOS Scientific Publishers Limited 2001

A CIP catalogue record for this book is available from the British Library.

ISBN 0–387–91607–5 Springer-Verlag New York Berlin Heidelberg SPIN 10760262

Springer-Verlag New York Inc.
175 Fifth Avenue, New York
NY 10010–7858, USA

Production Editor: Paul Barlass
Typeset by J&L Composition Ltd., N Yorks, UK
Printed by TJ International Ltd., Padstow, Cornwall.

Contents

Abbreviations

ABA	abscisic acid
ACP	acyl carrier protein
AEC	adenylate energy charge
ALA	δ-aminolaevulinic acid
APS	adenosine phosphosulfate
ATP	adenosine triphosphate
BCCP	biotin carrier protein
bchl	bacteriochlorophyll
bpheo	bacteriopheophytin
CA	carbonic anhydrase
CA1P	2'-carboxy-D-arabinitol 1-phosphate
CABP	2-carboxyarabinitol-1,5-bisphosphate
CAM	crassulacean acid metabolism
CF	coupling factor
chl	chlorophyll
cpDNA	chloroplastic DNA
DCCD	dicyclohexylcarboiimide
DGDG	digalactosyl-diglyceride
DHA	dehydroascorbate
DHAP	dihydroxyacetone phosphate
DNP	dinitrophenol
DOXP	1-deoxy-D-xylulose-5-phosphate
DTT	dithiothreeitol
E4P	erythrose-4-phosphate
EF	endoplasmic fracture surface
ENDOR	electron nuclear double resonance
EP	extrinsic proteins
ES	endoplasmic surface
ESR	electron spin resonance
F1,6BP	fructose-1,6-bisphosphate
F2,6BP	fructose-2,6-bisphosphate
F6P	fructose-6-phosphate

FBPase	fructose-1,6-bisphosphatase
FCCP	carbonyl ρ-trifluoromethoxyphenyl hydrazone
Fd	ferredoxin
FRS	ferredoxin-reducing substance
fru-1,6-P_2	fructose-1,6-bisphosphate
G1P	glucose-1-phosphate
G6PDH	glucose-6-phosphate dehydrogenase
GAP	glyceraldehyde-3-phosphate
GAPDH	glyceraldehyde-3-phosphate dehydrogenase
GDC	glycine decarboxylate
GOGAT	glutamate-glyoxylate aminotransferase
GGPP	geranylgeranyl diphosphate
GS	glutamine synthetase
GSH	glutathione (γ-glutanyl-cysteinyl-glycine)
HI	harvest index
HPLC	high-pressure liquid chromatography
INH	isonicotinyl hydrazide
IR	inverted repeat
LCA	limit of compensatory ability
LHC	light-harvesting chlorophyll
LHCP	light-harvesting chlorophyll–protein
LSC	large single copy region
LSU	large subunit
MDA	monodehydroascorbate
MGDG	monogalactosyl diacylglyceride
MSO	methionine sulfoxine
mtDNA	mitochondrial DNA
$NADP^+$	oxidized pyridine nucleotide
NADP-MDH	NADP-dependent malate dehydrogenase
NADP-ME	NADP-specific malic enzyme
NADPH	nicotinamide adenine dinucleotide phosphate
nDNA	nuclear DNA
NiR	nitrite reductase
NR	nitrate reductase
OD	optical density
PAGE	polyacrylamide gel electrophoresis
PAR	photosynthetically active radiation
PARAP	polarized absorption recovery after photobleaching
PC	plastocyanin
PCR	photosynthetic carbon reduction
PEP	phosphoenol pyruvic acid
PEPc	PEP carboxylase
PEPCK	PEP carboxykinase
PF	protoplasmic fracture surface
PFD	photon flux density
3PGA	3-phosphoglyceric acid
PGK	3-phosphoglycerate kinase

pmf	proton motive force
PN	pyridine nucleotides
PQ	plastoquinone
PRK	phosphoribulokinase
PS	protoplasmic surface
RC	reaction center
RFLP	restriction fragment
RGR	relative growth rate
Ru5P	ribose-5-phosphate
RuBP	ribulose bisphosphate
RUE	radiation use efficiency
S7P	sedoheptulose-7-phosphate
SBPase	sedoheptulose-1,7-bisphosphatase
SDS	sodium dodecyl sulfate
SGAT	serine-glyoxylate aminotransferase
SHMT	serine hydroxymethyltransferase
SOD	superoxide dismutase
SPS	sucrose phosphate synthase
SSC	small single copy region
SSU	small subunit
TK	transketolase
TPI	triose phosphate isomerase
TPP	thiamine pyrophosphate
TPT	triose phosphate/inorganic phosphate translocator

Preface

The assessments of the second edition of *Photosynthesis: molecular, physiological and environmental processes*, which appeared in 1993, encouraged the idea of upating for a third edition. This was justified by the pace of developments in all core aspects of the topic and in the wider subject area in which photosynthesis plays a role. Significant new information, approaches, concepts and evaluations have emerged in the eight years since the last revision was completed. Much new material has therefore been incorporated, and many alterations made whilst retaining the essential structure and subject matter of earlier editions. The aim has been to provide within one – relatively concise – book an overview and balanced assessment of the different components which constitute photosynthesis, in detail appropriate to advanced undergraduate courses in plant sciences. The book should also be of use in orienting postgraduates, and indeed non-specialists from different disciplines, in the general field of photo-synthesis. The text is, I hope, comprehensive but accessible to a wide readership who may develop enthusiasm for this most fundamental of biological processes. Also, I would feel the work justified if it helps to show the 'seamlessness' of the process and its ramifications, to balance the extreme 'reductionism' required of practitioners where focus is essential.

I am grateful to BIOS Scientific Publishers Ltd for the encouragement to pursue a third edition and to the enthusiasm and patience which Dr Paul Barlass, Production Editor, employed to see the project to completion.

I acknowledge, sincerely, IACR-Rothamsted for allowing me to undertake this project, and also many colleagues there for their support and generosity with ideas. They include Dimah Habash, Alfred Keys, Martin Parry, Rowan Mitchell and Matthew Paul; also students including James Jacob, Janaki Mahotti, Wilmer Tezara, Alejandro Pieters and Leni Demmers-Derks, and collaborating scientists, particularly Oula Ghannoum and Jann Conroy. I thank the many other scientists and unknown reviewers who made useful comments, constructive criticisms and provided opportunities for me to explore the topic with them: Bert Drake and Gabriel Cornic especially. I have also gathered information from a wide range of often unacknowledged sources. Thanks go to all, but the responsibility for the weaknesses and errors of omission and interpretation is mine. My family deserve very special thanks for tolerating me during the writing: Kirsten was responsible not only for so much of the technical support but also for 'kick-starting' the writing, Kurt provided unstinting help and Gudrun cannot be thanked sufficiently.

David W Lawlor
Harpenden

Introduction to the photosynthetic process

'Life is woven out of air by light', I. Moleschott

1.1 Introduction

Photosynthesis is the process by which many types of bacteria, algae and vascular plants convert the energy of light into the chemical energy of organic molecules. This process consists of many physical and chemical reactions involving biochemical components which enable organisms to exploit solar energy (inexhaustible even on the scale of evolutionary processes). It results in the accumulation of energy for the complex, non-photosynthetic physico-chemical reactions of virtually all living organisms, including those without the capacity to exploit solar energy by photosynthesis. Thus, photosynthesis provides the energy for the whole of the living world. In a simplified form the role of photosynthesis may be written:

$$\text{Low-energy inorganic chemical state} + \text{light energy} \xrightarrow{\text{Photosynthetic organism}} \text{High-energy chemical state}$$

Sunlight is the ultimate energy source for all biological processes on earth, with the exception of those relatively few organisms which live on chemical energy derived from geochemical reactions. For example, other sources of energy such as inorganic molecules, like hydrogen sulfide (H_2S), may be broken down and the energy exploited; these sources were important, perhaps dominant, in earth's early history and continue to be so in particular ecological conditions, e.g. deep-sea volcanic vents. However, such sources are finite and perhaps of limited long-term importance, although renewable by geochemical processes. For much of biological time, the exploitation of light to change matter from a lower to a higher energy state has been essential for life. In addition organisms, plants included, have used light as a signal, conveying information about the state of the environment to regulate a host of processes in metabolism, morphology, reproductive biology and so on. This is a most important aspect of the role of light. However, this book is concerned primarily with light as a source of energy for photosynthesis. Energy is needed to rearrange electrons in molecules and to synthesize chemical bonds, but a complex process may not take place spontaneously and a mechanism is required; this book examines the mechanisms and how they function to capture energy and transduce it to form complex biochemical products from simple inorganic molecules.

According to the laws of thermodynamics, biological processes will tend to go from a high-energy to a low-energy condition, losing energy in the process, until equilibrium between the organism and the environment is achieved – the state of death and decay – unless energy is available to drive and maintain the reaction in the reverse direction. Living organisms are in an unstable thermodynamic state and require energy to keep chemical constituents in a highly ordered condition and to do work against the thermochemical energy gradient. This is needed to accumulate matter, chemical constituents such as ions or gases from the environment which are required for metabolism, growth and development, or to move, *et cetera*, all of which characterize the living state in contrast to the world of inanimate matter.

The movement of matter, chemical interconversion or changes in energy state cannot proceed with absolute efficiency, and involve the loss of some of the energy, usually as heat, at a temperature very close to that of the environment, so that it cannot be used to do other work. Finally, the lost energy is radiated to the cold of space. Once a biological system has accumulated free energy it can convert it to different chemical forms or into physical energy or exchange it between organisms, but with time the useful energy will be lost and thermodynamic equilibrium (i.e. death) will be attained.

Without continuous supply of high-energy 'food' living organisms cannot survive. All nonphotosynthetic organisms, such as animals, fungi and bacteria, are dependent upon preformed materials. Some organisms, principally bacteria, can utilize the energy of bonds in inorganic molecules as an energy source. However only those organisms able to use the supply of energy from the sun can increase the total free energy of the earth and of living organisms which cannot use either such chemical sources or solar energy. Solar energy has enabled the larger part of the biosphere to become independent of the limitations imposed by other energy sources.

Photosynthesis is achieved by some bacteria, blue-green algae (also called cyanobacteria), algae and higher plants by a mechanism able to capture the fleeting energy of a light particle and make it available to biochemistry. Given an abundant supply of sunlight they can survive, grow and multiply using only inorganic forms of matter readily available in their environment. Sunlight, in addition to being the only form of energy which adds to the total energy supply of the earth, drives the weather and geochemical events, and also the biological cycles. Solar energy dominates the earth, although geochemical processes also contribute to the energy balance. The earth is bathed in a sea of energy in the form of electromagnetic waves, differing in wavelength and energy, derived from the thermonuclear reactions in the sun. Short-wavelength radiation, such as X-rays, is highly energetic and may destroy complex molecules by ionization. Ultraviolet light, of wavelengths greater than X-rays but shorter than visible light, breaks bonds within organic molecules and destroys many biological tissues. Infrared radiation has a longer wavelength than visible light and lower energy. It causes chemical bonds to stretch and vibrate but is not very active in biological processes, although it is important in the energy (heat) balance of the biosphere. The energy of visible light is sufficient to cause changes in the energy states of the valency electrons of many molecules and can be used by living organisms to effect the transition from a low- to a high-energy state. Molecules which absorb visible light and are relatively stable function to transduce physical energy to a chemical form, allowing the evolution of complex organic molecular 'living' systems, using light as their ultimate energy source.

1.2 **Solar radiation**

Early biological evolution may have occurred under conditions dominated by geochemical processes, with respiratory processes consuming organic molecules from the environment (the 'primeval soup') providing energy for biochemical processes. However, light was exploited early in evolution as a source of energy to drive biological processes. Considering the need for a continuous supply of energy, it is perhaps not surprising that solar radiation provides the energy used in the biosphere and is the basis of life. Light also serves to control many biological processes, for example day length regulates the development of many plants, but it does not add significantly to the global energy and does not drive the major energy fluxes in the biosphere. The physical characteristics of light and of the molecules with which it reacts, are crucial to the process of capturing energy and will be considered in Chapter 2.

1.3 **Concepts of photosynthesis**

Photosynthesis may be generalized as a process by which the physical energy of light is used to convert chemical substances to a more energetic state. In all organisms it involves: capture of the energy of a photon of light by a substance pigment; formation of an electronic excited state and use of this 'excited electron' to reduce an acceptor substance and to form 'energy-rich' molecules. These reduced substances and energy-rich molecules are used to form other, complex organic molecules.

1.3.1 *Energy and oxidation and reduction*

The electron lost from the excited pigment is replaced by an electron from another source in the environment.

$$\text{Pigment (P)} + \text{light} \longrightarrow \text{excited pigment (P*)} \tag{1.1}$$

$$\begin{aligned} \text{P*} + \text{acceptor molecule (A)} &\longrightarrow \text{P minus electron (P}^+\text{)} + \\ \text{A plus electron (A}^-\text{)} &\longrightarrow \text{P}^+ \text{ addition of electron} \longrightarrow \text{P} \end{aligned} \tag{1.2}$$

In photosynthetic bacteria many compounds may donate electrons to the oxidized pigment P^+, for example the sulfur bacteria use H_2S and liberate elemental sulfur (S):

$$H_2S + light + bacteriochlorophyll \longrightarrow S + 2\,H^+ + 2\,e^- \tag{1.3}$$

Cyanobacteria (blue-green algae), algae, bryophytes, ferns, gymnosperms and angiosperms use water as the source of electrons and oxygen (O_2) is released:

$$2\,H_2O + light + chlorophyll \longrightarrow O_2 + 4\,H^+ + 4\,e^- \tag{1.4}$$

1.3.2 Adenylate energy and pyridene nucleotide reductant

In both types of photosynthetic system, e^- is used to form pyridine nucleotide, either nicotinamide adenine dinucleotide (NADH) or more generally in photosynthetic systems nicotinamide adenine dinucleotide phosphate (NADPH). In addition, H^+ accumulates in the photosynthetic mechanism forming a proton concentration gradient, the energy of which is used to produce phosphorylated compound adenosine triphosphate (ATP). NADPH provides e^- and H^+ and ATP, the energy and phosphate group required in a large number of biochemical reactions. Both NADPH and ATP are synthesized by complex biochemical processes driven by light energy. The transformations of energy and material require many individual chemical steps (perhaps thousands), if the processes required to form the whole organism are counted, as they must be in a complex system.

1.3.3 Reduction processes

The energized electron may be used to reduce, ultimately, a range of inorganic substances. One of the most important is carbon dioxide (CO_2), leading to the synthesis of carbohydrates (CH_2O, e.g. sugars) and other organic molecules based on carbon; the basic process is $A^- + H^+ + CO_2 \longrightarrow (CH_2O)$, where the H^+ 'follows' the electronically negative state to balance the charges and comes from the environment. In plants using water to supply electrons, the reaction is:

$$CO_2 + H_2O + Light\ energy \xrightarrow{\substack{Chlorophyll\text{-}\\containing\ plants}} (CH_2O) + O_2 + chemical\ energy \tag{1.5}$$

The energy required for the formation of glucose ($C_6H_{12}O_6$) is 2879 kJ mol^{-1}.

ATP and reductant are also used to assimilate other inorganic compounds. Nitrate ions (NO_3^-) are reduced to ammonia, which is then consumed in the synthesis of amino acids:

$$NO_3^- + 9\,H^+ + 8\,e^- \xrightarrow{\substack{Light,\ chlorophyll,\\enzymes\ (nitrate\\and\ nitrite\ reductase)}} NH_3 + 3\,H_2O \tag{1.6}$$

Photosynthetic bacteria and blue-green algae, but not higher algae or plants, assimilate atmospheric nitrogen to form ammonia:

$$N_2 + 6\,H^+ + 6\,e^- \xrightarrow{Light,\ chlorophyll,\ enzymes\ (nitrogenase)} 2\,NH_3 \tag{1.7}$$

Sulfate ions are also reduced before entering metabolism:

$$SO_4^{2-} + 9\,H^+ + 8\,e^- \xrightarrow{\text{Light, chlorophyll, enzymes}} HS^- + 4\,H_2O \qquad (1.8)$$

Many algae, in the absence of oxygen, are able to produce hydrogen gas from water using light energy captured by chlorophyll; the enzyme hydrogenase catalyzes the reaction.

$$H_2O \xrightarrow{\text{Light, chlorophyll, hydrogenase}} H_2 + \tfrac{1}{2}O_2 \qquad (1.9)$$

These examples (considered in more detail later) show that photosynthesis is more than the assimilation of CO_2 with the production of oxygen, but is a process with many possible products and capable of being used biologically in many ways.

1.4 Photosynthesis as an oxidation–reduction process

All the photosynthetic processes summarized in equations (1.2)–(1.9) involve oxidation and reduction. Reduction is the transfer of an electron (e^-) or electron plus proton (H^+) from a donor (D) molecule to an acceptor (A); the donor is oxidized and the acceptor reduced. An electrically neutral compound becomes negatively charged and may accept a proton from water to restore electrical neutrality:

$$\text{Donor D} + \text{acceptor A} \rightarrow D^+ + A^-; \quad A^- + H^+ \rightarrow AH; \quad D^+ + e^- \rightarrow D \qquad (1.10)$$

Van Niel established that photosynthesis in all organisms conforms to donor + acceptor reacting under the influence of light in the presence of pigment and the necessary mechanism. The oxidized donor molecule may accumulate in the environment, for example oxygenic photosynthesis is responsible for oxygen in the atmosphere. Biological reduction–oxidation (redox) reactions are usually catalyzed by enzymes. Examples in plant metabolism are the reduction of oxaloacetic acid to malic acid by malate dehydrogenase (equation 9.2).

Reduction and oxidation reactions are of fundamental importance to our understanding of the mechanisms of photosynthesis. The primary reaction of photosynthesis, linking the physical energy of chlorophyll molecules excited by light with biochemical processes, is the transfer of electrons from a special form of chlorophyll to an acceptor molecule driven by the energy captured. The acceptor is reduced and the special form of chlorophyll oxidized. Electrons are then donated from different sources to the oxidized chlorophyll, reducing it and allowing the process to be repeated (*Figure 1.1*). The generation of the reduced acceptor and energy-rich compounds, NADPH and ATP respectively, requires light; these reactions are therefore called 'light reactions'. NADPH and ATP are consumed to reduce other inorganic substances and to convert them into organic molecules. Given a supply of these compounds the chemical reactions of photosynthesis may proceed without light and are thus called 'dark reactions', but they can and do proceed in both light and darkness. The term 'non-light requiring reactions' expresses this important point.

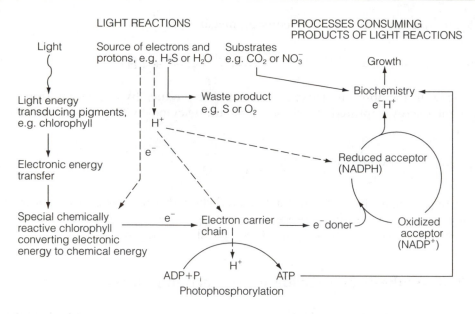

Figure 1.1. Essential features of photosynthesis in all organisms

1.5 Energy and electron transport

In photosynthesis, electrons pass from a donor to an acceptor, the two forming a redox pair or couple. The ability of electrons to transfer is determined by the energy required, called the redox potential. The redox potential is determined by comparison with a standard hydrogen electrode which has, by definition, a voltage (E_0) of 0 V under standard conditions and pH of 0. Biological reactions take place in solution close to pH 7, so that the redox potential of biological redox substances is measured at pH 7 and is called E'; E' is –0.42 V compared with E_0. The relation between redox potential and concentration of oxidized and reduced substances, [ox] and [red], is given by:

$$E' = E_0 + \frac{RT}{nF} \ln \frac{[ox]}{[red]} \tag{1.11}$$

when n is the number of reducing equivalents (e^-) transferred, R is the gas constant (8.314 J mol^{-1} K^{-1}), T the absolute temperature (K) and F the Faraday constant (F = 96 485 C mol^{-1}). Redox potentials depend on concentration and are given as a midpoint potential when the two forms are equal in concentration.

Substances with more negative redox potential are energetically able to reduce (donate e^- to) others of more positive potential; it is impossible in the reverse direction unless energy and a mechanism are available. The maximum useful energy from a reaction is given by the difference in redox potential ΔE; the free energy of a reaction, ΔG, is related to redox potential by $\Delta E = -\Delta G/nF$.

1.6 Nature of the light-gathering process and generation of reductant and ATP

Equation (1.2) includes a pigment; in equations (1.3) and (1.4) this is identified as bacteriochlorophyll (bchl) and chlorophyll (chl), respectively, but this ignores the many complex oxidation–reduction processes. Absorption of light by bchl or chl leads to ejection of an electron from a special form of the pigment; the electron is then captured in a chemical form as a reduced acceptor (a quinone molecule) in an electron transport chain. Electrons pass from a higher to lower energy state along a chain of electron acceptors and donors, which are alternatively reduced and oxidized in the process, until a stable reduced compound is formed. In higher plants this is ferredoxin, which passes electrons to inorganic compounds, such as nitrate ions, or to secondary reductants, such as oxidized pyridine nucleotide ($NADP^+$). Energy transformations in living organisms take place at nearly isothermal conditions and involve multiple steps with rather small energy changes between them, which are rather easier to control than large 'jumps', and contribute to achieving optimum rates of processes and efficiency. It is not possible to obtain both maximum rates and efficiency in a system; organisms may have evolved to maximize the energy output at rather lower efficiency.

Energy is required to carry out the biological catalytic reactions leading to the reorganization of the primary reactants into complex products and is provided by the hydrolysis of ATP or other related phosphorylated adenylate compounds. ATP has three phosphate groups. When the bond joining the terminal group is hydrolyzed it provides energy and releases phosphate as a charged phosphate group, which may be donated to other compounds, activating them; phosphorylation is an essential step in many biochemical reactions.

$$ATP + H_2O \longrightarrow ADP + P_i + energy \ (-31 \ kJ \ mol^{-1}) \tag{1.12}$$

ATP is resynthesized by photophosphorylation, a process in which the movement of electrons along the chain of electron carriers also pumps H^+ across the cell membrane (in bacteria) or the chloroplast thylakoid membranes (in other photosynthetic organisms), producing a concentration gradient of H^+ which is coupled to ATP synthesis (Chapter 6). The essential nature of the photosynthetic process is therefore a chemical oxidation–reduction reaction driven against the thermodynamic energy gradient by the energy of light captured by pigments. As already mentioned, the light-transducing mechanism generates reductant in the form of organic molecules which pass it to biological reactions, themselves independent of the direct effect of light. In addition, light energy drives the synthesis of ATP, which is also essential for biochemical reactions. The study of photosynthesis considers the atomic and molecular processes underlying the capture of light energy and its conservation, and the relation between the production of energetically favorable compounds and the assimilation of inorganic molecules. Ultimately, the fundamental processes of photosynthesis are related to the performance of the organism at an ecological level.

1.7 Occurrence of photosynthesis

Photosynthesis is the only biological process that accumulates energy. It occurs in organisms as diverse as bacteria and trees (*Table 1.1*) and is similar in all, with light capture by a pigment and conversion of the energy to chemical form. Differences between organisms reflect their evolutionary history; the greatest difference is between the photosynthetic bacteria, a very diverse group unable to oxidize water (i.e. they are anoxygenic), and blue-green algae, algae and higher plants which oxidize water and evolve O_2 (oxygenic).

The physical nature of light, but not its intensity or spectral distribution, has been a constant feature of the environment since the origin of the earth, and it is to be expected that the different pigment systems for light capture share many features. Bacteriochlorophyll and chl occur in several forms but are basically similar in all organisms and are related biosynthetically, with light absorption and transport of electrons taking place in membranes. In photosynthetic bacteria there is only a single light-driven process, whereas in

Table 1.1. Main groups of photosynthetic organisms, their structure and photosynthetic characteristics

Prokaryotes	Eukaryotes
Single cell or little complexity	Mainly multicellular, complex intercellular interactions
Cell nucleus without membrane	Cell nucleus with membrane
Photosynthesis in vesicular membranes not in discrete compartment	Photosynthesis in vesicular membranes in discrete compartment – chloroplast
Anoxygenic forms	*Anoxygenic forms*
Examples:	None
Photosynthetic bacteria (purple sulfur – Rhodospirillaceae, green sulfur – Chromatiaceae)	
Processes:	
Do not evolve O_2 – many obligate anaerobes. Source of reductant hydrogen sulfide, sulfur, thiosulfite, hydrogen, organic compounds, never water. Can reduce gaseous nitrogen	
Oxygenic forms	*Oxygenic forms*
Examples:	*Examples:*
Blue-green algae	Algae (green – Chlorophyceae, red – Rhodophyceae)
	Higher plants (Bryophyta, Angiospermae)
Processes:	*Processes:*
Evolve O_2. Source of reductant water	Evolve O_2 – source of reductant water
Reduce gaseous nitrogen	Do not reduce gaseous nitrogen

all other organisms there are two, one of which oxidizes water. Bacteria use a variety of substrates to provide reductant, suggesting that many substrates were available at early stages of evolution but later the abundance of water, and perhaps shortage of other sources of reductant, made it the preferred source. Electron transport coupled to ATP synthesis was an early feature of photosynthetic systems, requiring a closed membrane vesicle separating regions of high and low proton concentration which are needed for ATP generation. The gradient is from high proton concentration, [H$^+$], outside to low [H$^+$] inside the photosynthetic bacterial cell (light energy is used to pump H$^+$ out of the cell) and from high [H$^+$] in the thylakoids to low [H$^+$] in the cytosol of blue-green algae or in the chloroplast stroma of higher plants. Thylakoid membranes, although internal, are equivalent to the external membrane of bacterial cells. The space within the thylakoid is equivalent to the medium surrounding a photosynthetic bacterium. Thus, the mechanisms driving ATP synthesis and the structure of ATP synthase (Coupling Factor (CF)), the enzyme responsible, are comparable, despite the great structural differences between cells of different groups. Photosynthetic bacteria are important as relatively simple organisms for analysis of photosynthesis and for understanding the evolutionary process.

1.8 Types and structures of photosynthetic organisms

As *Table 1.1* indicates, there is a fundamental distinction between prokaryotes and eukaryotes with photosynthesis occurring in both, but with anoxygenic and oxygenic forms occurring in the prokaryotes and oxygenic only in the eukaryotes. There are also many structural and functional differences within these main classifications.

1.8.1 Photosynthetic bacteria

The simplest known form of 'photosynthesis' is that of the halophilic bacterium *Halobacterium halobium*, which has patches of purple pigment-protein molecules, bacteriorhodopsin, in its outer membrane. This pigment captures light and the energy provides the potential for transport of H$^+$ from inside the cell (*Figure 1.2a*) to outside. The pH gradient so formed provides (together with the electrical charge on the membrane) the energy for synthesis of ATP when the H$^+$ diffuses back into the cell through the ATP synthesizing enzyme. There is no electron transport associated with this type of photosynthesis.

Other photosynthetic bacteria (*Figure 1.2b*) have a rather more complex form of photosynthesis in which electrons are transported and used to drive the formation of the pH gradient for ATP synthesis. The outer cell membrane is invaginated, forming extensive multilayered membrane systems containing the photosynthetic pigments. These membranes are only indirectly in contact with the external medium. The bchl pigment and associated proteins of the photosystems capture light energy which leads to the formation of a reduced acceptor. At the same time the electrons reduce quinones in the membrane which carry H$^+$ from the cytosol to the external medium, in this case the space bounded by the membranes; this is equivalent to the cell exterior. A pH gradient results and the diffusion of H$^+$ back into the cell through the CF drives ATP synthesis within the cell.

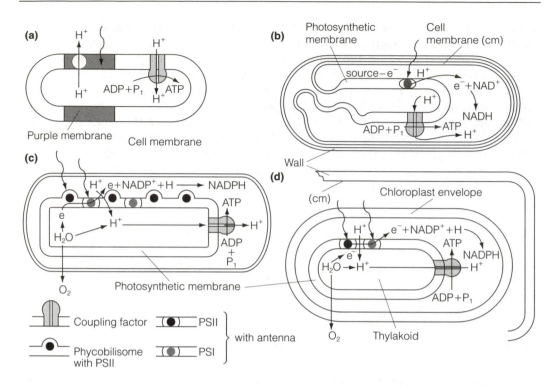

Figure 1.2. Photosynthesis in (a) *Halobacterium halobium* represents a very simple light-driven pump of H$^+$, coupled to ATP synthesis by CF enzyme. In the photosynthetic bacterial system (b), a photosystem (resembling photosystem 1 (PSI) of higher plants) in the photosynthetic membrane captures light and drives the extraction of electrons from organic or inorganic substances to reduce NAD$^+$ and also pumps H$^+$ to the equivalent of the outside of the cell; H$^+$ is coupled to ATP synthesis as in (a). A further development in cyanobacteria (c) is the presence of two photosystems, one of which has an arrangement of light-harvesting pigments in distinctive phycobilisomes on the membrane. It is coupled to water oxidation, using e$^-$ for reduction of NADP$^+$, H$^+$ for ATP synthesis and releasing O$_2$. The higher plant system (d) of the thylakoids has photosystems I and II embedded in the photosynthetic membrane, with oxidation of water, H$^+$ pumping and ATP synthesis and e$^-$ reducing NADP$^+$

The prokaryotic cyanobacteria (blue-green algae) have oxygenic photosynthesis and a photosynthetic membrane structure resembling the more complex bacteria, without chloroplasts but with an extensive membrane system arranged around the cell periphery in parallel sheets. The red algae (Rhodophyta) are similar. The membranes have regularly arranged particles, called phycobilisomes (*Figure 1.2c*) alternating on each side of the membranes, which form a continuous system the inside of which is equivalent to the external medium. The phycobilisomes contain a complex of pigments linked to the reaction center. There are two types of photosystems in the membranes, one associated with electron transport (PSI), and the other with water oxidation and electron transport (PSII). As with the bacteria, transport of electrons leads to synthesis of reductant and also to formation of a pH gradient, which drives the synthesis of ATP.

1.8.2 *Higher plants*

In the eukaryotic algae, bryophytes, ferns, gymnosperms and angiosperms, all with oxygenic photosynthesis, the photosynthetic membranes (thylakoids) are enclosed within an envelope consisting of a double membrane, forming a structure called the chloroplast. The inside of the thylakoid membranes (lumen) is equivalent to the outside of the cell (*Figure 1.2d*) and the membranes are differentiated into stacked (granal) and unstacked (stromal) forms in many of these plants. There are two photosystems, as in the cyanobacteria, but of different form, particularly the light gathering pigments which are joined with proteins forming antennae buried in the membrane, not on its surface. However, as in the cyanobacteria, electron transport leads to the reduction of an acceptor, the formation of the pH gradient and synthesis of ATP via the coupling factor in the chloroplast.

1.9 Photosynthesis in relation to other plant functions

Respiration of the assimilates produced in the photosynthetic reactions provides a way of using the energy of light and matter assimilated in periods of darkness or in parts of the organism not exposed to light. It is also exploited by heterotrophic organisms which cannot feed themselves but rely on photosynthetically competent ones to provide energy and material.

If the respiratory and photosynthetic processes are compared with the starting materials on the left-hand side of equation (1.13) and the products on the right-hand side, the net result is energy conversion and loss, and closed cycles of carbon and water.

$$\text{Photosynthesis: } CO_2 + H_2O + \text{light energy} \xrightarrow{\text{biological catalysts}} (CH_2O) + O_2$$

$$\text{Respiration: } (CH_2O) + O_2 \xrightarrow{\text{biological catalysts}} CO_2 + H_2O + \text{heat energy} \qquad (1.13)$$

The net result of respiration and photosynthesis is: light energy $\longrightarrow$ heat energy. Thus, photosynthesis and respiration work in opposition, forming a cyclic, closed system for matter. The physical energy of light is converted into chemical energy and ultimately heat. Photosynthesis and respiratory processes are very carefully regulated by complex biochemical mechanisms which link the processes with the result that the products of photosynthesis are not consumed in a futile cycle of assimilation/respiration and so that growth of cells, organs and organisms is achieved in such a way that the organism is capable of continued growth and reproduction. Plants are composed of individual subsystems (cells, organs and processes), which are controlled by their internal conditions and outside factors but are highly integrated. Photosynthesis is the driving factor for all processes, yet it is not autonomous but intimately related to the other cellular functions in complex ways which are not yet well understood. On the ecological scale, the food pyramid is based on the primary photosynthetic producers, with primary consumers of living plants, such as herbivores (e.g. phytophagous insects and grazing mammals) and consumers of dead plant matter (e.g. bacteria, fungi and insects) in the lower part of the pyramid, leading to the apex of end consumers, with humans taking a large part of the primary and secondary production and eliminating much of the competing biological world. Indeed, the continued increase in global human population must place additional demands on the primary photosynthetic productive capacity of the earth, particularly those parts most capable of producing the food, fiber and fuel required. Therefore, understanding of photosynthetic processes is a necessary facet of the long-term need to maintain or possibly to improve photosynthetic processes to serve human requirements. By increasing productivity per unit of land employed in agriculture, it may be possible to reduce the impact of human populations on the biosphere in general. A study of photosynthesis requires a global, as well as atomic, view.

1.10 Evolution of photosynthesis

As a biochemical process based on proteins and organic molecules which are rapidly decomposed, photosynthesis has left little direct trace in the geological record. Yet enough is now known of the comparative biochemistry to suggest a plausible hypothesis of the development of this essential process. Here only the barest outline is attempted, and articles by Schopf (1978) and Bendall (1986) should be consulted. The earth was

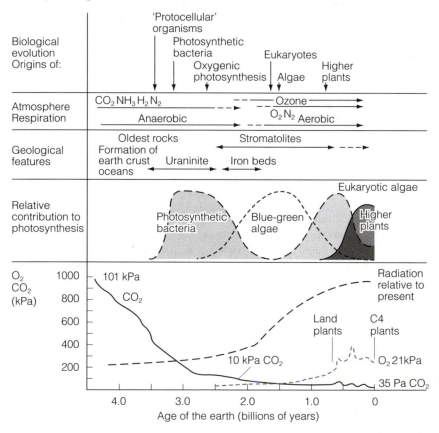

Figure 1.3. Probable sequence of events in evolution of photosynthesis and the relation to some other geological and biological processes and CO_2 + O_2 partial processes and solar radiation; highly schematic

formed (*Figure 1.3*) some 4.6×10^9 years ago, and for the first 0.5×10^9 years cooled and solidified. Because of the earth's distance from the sun and its size, which determined the heat received and the force holding gases on the surface, both liquid water and an atmosphere were retained. The primitive atmosphere was highly reducing, with methane (CH_4), H_2, H_2S, CO_2, CO, NH_3, *et cetera*, but there was no free O_2; it was anoxic. This is considered crucial to the evolution of life because oxygen destroys organic molecules. Also there was no ozone, which today forms a thin layer in the upper atmosphere, absorbing ultraviolet (UV) rays. This energetic form of radiation, together with high temperatures and abundant gases evolved by volcanic activity, provided conditions for synthesis of organic molecules such that the primitive oceans resembled 'hot dilute soup' to quote Haldane. How self-replicating biological systems were formed and evolved under these conditions is beyond the scope of this book and is still unresolved. However it occurred, organisms resembling present-day bacteria in size and cell structure are present in rocks 3.5×10^9 years old. A form of light-driven metabolism probably developed early, as derivatives of carotenoids have been detected in rocks of that age, although contamination by later material is possible. A primitive organism might have synthesized ATP by a light-driven proton pump such as the bacterium *Halobacterium halobium* (Sections 1.8.1 and 11.6). Organic carbon from deposits of this age show discrimination against the heavy isotope of carbon (^{13}C), suggesting that some form of photosynthesis occurred very early in evolution. Such organisms would have been independent of the supply of preformed ATP and would have had an evolutionary advantage by exploiting an almost limitless source of energy.

Many processes were linked to the light reactions and electron flow and ATP, including N_2, CO_2 and S assimilation. However, water was not split in early photosynthesis until 3.5×10^9 years ago so no photosynthetic O_2 was formed and the atmosphere was reducing. It was only after evolution of the water-splitting process, driven by light through two linked photosystems, that water could be oxidized and O_2 (the waste product) escaped to the atmosphere. This time scale is suggested by much evidence (Schopf, 1978). Extensive beds of fossil limestone, called stromatolites, were formed 3×10^9 years ago and contain blue-green algae, as do present-day stromatolites. However, oxygen production probably started before that time. Many geochemical processes would have consumed O_2; for example, iron (Fe^{2+}) reacts to form insoluble $Fe_3O_4^-$. Oceans would have been slowly depleted of Fe^{2+} with deposition of iron ores. This may be the origin of the 'red beds' formed 2.2×10^9–1.7×10^9 years ago. To borrow Schopf's expression 'the world's oceans rusted'. A non-photosynthetic origin of O_2 by UV-splitting of water was too slow for such a massive chemical process. Uraninite (UO_2) is a uranium ore, insoluble at O_2 concentrations above 1%; no deposits younger than 2×10^9 years have been found. Thus, between 3.5×10^9 and 3.0×10^9 years ago photosynthesis developed using H_2O as source of reductant and atmospheric O_2 increased. CO_2 concentration was many hundred-fold greater than today and deposition of reduced carbon (deduced from the change in $^{13}C/^{12}C$ ratio of ancient carbon deposits) may also have contributed to the removal of C and the increase in O_2 in the atmosphere. By 1.5×10^9 to 1.0×10^9 years ago aerobic conditions were established as the chemical buffers were exhausted and O_2 exceeded 1 kPa. Oxygen in the upper atmosphere formed an ozone layer, which absorbed UV radiation, allowing the evolution of higher organisms and invasion of the land. Oxygen, by acting as a terminal receptor for respiratory processes, greatly (10-fold) increased the amount of energy to be obtained by respiration of organic substances. Most of the present-day biota, including man, owe their existence not only to the photo-assimilates retained by plants, but to the waste product of photosynthesis!

Eukaryotes, with large internally compartmented nucleated cells, may have evolved early in earth's history, although evidence, such as the presence of steranes (molecules derived from sterols thought only to be made by nucleated, that is eukaryotic, cells) in rocks, only supports an age of 1.7×10^9 years and multicellular eukaryotic seaweeds occur in strata 1.4×10^9 years old. They evolved rapidly from about 1×10^9 years ago, forming multinucleate, macroscopic organisms (both plants and animals) perhaps linked to changes in the climate about 900–600 million years ago associated with tectonic and volcanic activity and large loss of C by burial in sediments and the onset of global climate changes (including glaciation). Evidence from the structure and function of the nucleic acids of chloroplasts and mitochondria of higher plants suggests that these organelles arose from the invasion of non-photosynthetic eukaryotic cells by bacteria and blue-green algae. Photosynthesis evolved more complex biochemistry, separation of respiration and photosynthesis and their regulation. Photosynthesis shaped the biosphere both directly and through its effects on the earth's climate and geology. Carbon from photosynthesis was sequestered in oil, coal and gas, decreasing the atmospheric CO_2 and increasing the ratio of O_2 to CO_2. This may have been unfavorable for photosynthesis as the CO_2-fixing enzyme ribulose bisphosphate carboxylase is less efficient under these conditions. On land, prevention of water loss from the plant by a thick cuticle also reduced supply of CO_2 for photosynthesis. Evolution of quantitatively different types of photosynthesis and particularly qualitatively different mechanisms, C4 and CAM (Chapter 9), based on the earlier metabolic forms are probably a response to an environment of decreasing CO_2/O_2 ratio and drier atmosphere, with intense radiation where they are more efficient. Present human activity is increasing the CO_2 concentration of the atmosphere by burning fossil fuels. This may improve plant growth in the short term but it will also affect the world's climate (Chapter 11). Humankind's future is closely linked to the photosynthetic production of food, fuel and fiber, and increasing population will lead to greater demands on the efficiency of the process.

1.11 **Photosynthesis in complex living organisms**

During the course of this book, with so much emphasis on the details of separate parts of the mechanisms by which photosynthesis is achieved, for example the atomic structure of enzymes and of protein–pigment complexes that absorb light, it should be kept in mind that all the processes occur within complex 'systems' made up of the interactions of the many smaller parts. These interactions produce the characteristics of the complete system, which are often very different from what might be expected from the individual parts. From the atomic level of organization and the physical and chemical events, the analysis moves into the realm of the physics and chemistry of biochemical and eventually physiological processes. Indeed, the photosynthetic 'realm' extends even further into ecology and global chemistry. Analyzing the aggregated effect of multiple interactions in 'whole systems' is as fascinating, challenging and important as 'reductionist' probing of smaller and more intricate sub-structures. Analysis of photosynthesis has been achieved by application of the widest range of techniques to understand the details of the mechanism. That need continues. However, there is renewed urgency to understand complex systems if the photosynthetic process is to be altered for specific ends, for example to increase dry matter production of crop plants so that yields can be improved, and to ensure that food production will keep pace with the increase in the human population of the world. Analyses of plants with altered photosynthetic biochemistry by genetic engineering has shown clearly that the expectations based on understanding of biochemistry *in vitro* were not justified – the systems behaved in complex ways – output rarely matching expectation. Also, the long known but often ignored interaction between the individual parts of photosynthesis, the whole system and the environment is highly 'nonlinear' with complex responses. This book therefore addresses the basic sub-system processes of photosynthesis, with some emphasis on the atomic, molecular events, and proceeds to more complex systems and their organization and function. The hope is that by taking this approach the scale of events from the very small and fast to the large and slow may be seen, allowing the reader to put into context the processes and systems which occur as part of the 'photosynthetic whole'.

References and further reading

Atwell, B.J., Kriedemann, P.E. and Turnbull, C.G.N. (1999) Plants in action: adaptation in nature, performance in cultivation. Macmillan Education, South Yarra, Victoria, Australia.

Arnon, D.I. (1977) Photosynthesis 1950–75: changing concepts and perspectives. In *Encyclopedia of Plant Physiology (ICS), Vol. 5, Photosynthesis I* (eds. A. Trebst and M. Avron). Springer, Berlin, pp. 7–56.

Bendall, D.S. (ed.) (1986) *Evolution from Molecules to Men.* Cambridge University Press, Cambridge.

Benedict, C.R. (1978) Nature of obligate photoautotrophy. *Annu. Rev. Plant Physiol.* **29**: 67–93.

Briggs, W.R. (ed.) (1989) *Photosynthesis.* Allan R. Liss, New York.

Budyko, M.I. (1974) *Climate and Life* (English Edn, ed. D.H. Miller). Academic Press, London.

Clayton, R.K. (1980) *Photosynthesis: Physical Mechanisms and Chemical Patterns.* IUPAB Biophysics Series, Cambridge University Press, London.

Cogdell, R. and Malkin, R. (1992) An introduction to plant and bacterial photosystems. In *Topics in Photosynthesis, Vol. 11, The Photosystems: Structure, Function and Molecular Biology* (ed. J. Barber). Elsevier, Amsterdam, pp. 1–15.

Davis, G.R. (1990) Energy for planet earth. *Sci. Am.,* **263**: 20–27.

Dennis, D.T. and Turpin, D.H. (1990) *Plant Physiology, Biochemistry and Molecular Biology.* Longman, Harlow.

Dose, K. (1983) Chemical evolution and the origin of living systems. In *Biophysics* (eds. W. Hoppe, W. Lohmann, H. Markl and H. Ziegler). Springer, Berlin, pp. 912–924.

Giese, A. (1964) *Photophysiology, 1, General Principles: Action of Light on Plants.* Academic Press, New York.

Hall, D.O. and Rao, K.K. (1999) *Photosynthesis* (6th Edn), Cambridge University Press, Cambridge.

Krauss, N. et al. (1993) Three-dimensional structure in system I of photosynthesis at 6 Å resolution. *Nature* **361**: 326–331.

Olsen, J.M. and Pierson, B.K. (1986) Photosynthesis 3.5 hundred million years ago. *Photosynthesis Res.* **9**: 251–259.

Papiz, M.Z. Prince, S.M., Hawthornthwaite-Lawless, A.M. McDermott, G. Freer, A.A., Isaacs, N.W. and Cogdell, R.J. (1996) A model for the photosynthetic apparatus of purple bacteria. *Trends Plant Sci.* **1**: 198–206.

Porter, G. (1989) Solar energy from photochemistry. In *Techniques and New Developments in Photosynthesis Research* (eds. J. Barber and R. Malkin). Plenum Press, New York, pp. 3–7.

Rabinowitch, E.I. (1945) *Photosynthesis.* Interscience, New York.

Schopf, J.W. (1978) The evolution of the earliest cells. *Sci. Am.* **239**: 84–103.

Sundstrom, V. and van Grondelle, R. (1991) Dynamics of excitation energy transfer in photosynthetic bacteria, in photosynthetic bacteria. In *The Chlorophylls* (ed. H. Scheer). CRC Press, Boca Raton, FL, pp. 1097–1124.

Van Gorkom, H.T. (1987) Evolution of photosynthesis. In *Photosynthesis. New Comprehensive Biology,* Vol. 15 (ed. J. Amesz). Elsevier, Amsterdam, pp. 343–350.

Weng, G., Bhalla, U.S. and Iyengar, R. (1999) Complexity in biological signalling systems. *Science* **284**: 92–96.

Woese, C.R. (1987) Bacterial evolution. *Microbiol. Rev.* **51**: 221–171.

Light – the driving force of photosynthesis

2.1   Characteristics of light

Photosynthetic mechanisms use the energy of photons to produce the high-energy organic compounds of carbon, nitrogen and sulfur, which are consumed in metabolism. Therefore, the physical nature of light and its interaction with matter are at the core of photosynthesis, and are described in a much simplified, qualitative way in order to understand the biological processes. The quantitative characteristics of light in relation to photochemistry are discussed in the books by Clayton (1980) and Hoppe *et al.* (1983). Light is electromagnetic radiation, emitted when an electrical dipole (a paired positive and negative charge, separated by a small distance) in an atom oscillates and causes a change in the field of force. The dipole produces an electrical and a magnetic vector, which are in phase but at right angles. Fluctuations in the field strength of these vectors are perpendicular to the direction of travel of the wave and, hence, light is a transverse wave. The electromagnetic wave is characterized by both wavelength, λ (in meters), which is the distance between successive positive or negative maxima on the sine wave, and by frequency, ν, the number of oscillations per unit time (s^{-1}).

Frequency is determined by the oscillations of the dipole. Wavelength and frequency are related by the velocity of propagation of the wave, υ (ms^{-1}):

$$\upsilon = \lambda\nu \qquad\qquad (2.1)$$

The velocity of light (c) is 3×10^8 ms^{-1} *in vacuo*. *Table 2.1* gives the approximate wavelengths and frequencies of the main groups of electromagnetic waves. Photosynthesis uses radiation of 400–800 nm wavelength and the term 'photosynthetically active radiation' (PAR) is applied to that spectral distribution. Of course, the radiation has no 'photosynthetic activity'; a better term might be 'photosynthetically effective radiation' (see Monteith and Unsworth, 1990, Chapter 13). Frequencies between 7.5 and 3.8×10^{15} s^{-1} (wavelengths 400–700 nm) are visible to the human eye and are called light. However, the sensitivity (response) of the human eye to different wavelengths within the 400–700 nm band is not the same as the sensitivity (response) of photosynthesis. The simplicity of the word and similarity between the wavelengths of visible light and those used in photosynthesis are the reasons why the word 'light' is applied to radiation

Table 2.1. Electromagnetic radiation, wavelength, frequency and energy

Type of radiation	Wavelength	Frequency[a] (s^{-1})	Energy per photon[b] (eV)	(J)	Energy 1 mol photons[c] (J)
Infra-red	800 nm	3.8×10^{14}	1.55	25.16×10^{-20}	15.2×10^4
Visible red light	680 nm	4.4×10^{14}	1.82	29.13×10^{-20}	17.5×10^4
Visible green	500 nm	6.0×10^{14}	2.50	39.72×10^{-20}	23.9×10^4
Visible violet-blue	400 nm	7.5×10^{14}	3.12	49.65×10^{-20}	29.6×10^4
Near ultraviolet	200 nm	1.5×10^{15}	6.25	9.93×10^{-19}	59.5×10^4
Ultraviolet	100 nm	3.0×10^{15}	123.0	19.86×10^{-19}	119.6×10^5
X-rays	0.01 nm	3.0×10^{19}	1.24×10^5	19.86×10^{-15}	119.6×10^8

[a] Calculated from eqn 2.1. [b] Calculated from eqn 2.2. [c] Energy of 1 photon multiplied by Avogadro's number of photons. Note: 1 nm = 10^{-9} m = 10 ångström = 1 mμ.

effective in photosynthesis. However, the word 'light' should be used carefully when applied to photosynthesis.

Electromagnetic radiation passes through space without matter to transmit it, in contrast to wave propagation in solids, liquids or gases. The wave form of light is shown by interference phenomena and the slower transmission of light in dense media. Light has, however, the characteristics of both a wave and a particle. Emission of electrons from metal surfaces caused by light, called the photoelectric effect, and radiation of energy from atoms at distinct frequencies rather than as a continuous spectrum show that light is particulate.

In 1900, Planck resolved the conflict between the wave and particle concepts by describing light as discrete particles of energy, called quanta, which can only be absorbed or emitted by matter in indivisible units or quanta. Thus, processes involving light are quantized, that is, 'all or nothing'. The quantum of energy is carried as the oscillating force field of the electromagnetic wave. The particle carrying a quantum of energy is called a photon, which has no rest mass. Quantum and photon are distinct concepts; the former is the energy carried by the photon. Changes in the state of atoms or molecules caused by light involve a transition in the energy state of electrons within the substance. This transition can only take place if all the energy of a photon is transferred to the electron. If the quantum is larger or smaller than the energy required for the transition, then the photon will not be 'captured', that is the energy of the photon will not excite the electronic state of the molecule, and light will not be absorbed. When the energy of photons of a particular wavelength is absorbed, that wavelength of light is removed. Hence, if a broad band of wavelengths is passed through the absorbing material, then the emitted wavelengths are depleted (to a smaller or greater degree depending on the amount of material and its characteristics) in the one absorbed, and the spectrum of the light passing through is altered.

The energy of a photon (ε) depends on the frequency of the electromagnetic wave, which is related to the wavelength, and is given by:

$$\varepsilon = h\upsilon = hc/\lambda \qquad (2.2)$$

where h is Planck's constant (6.62×10^{-34} J s), which has the units of energy $\times$ time or 'action'. The greater the frequency and, from equation (2.1), the smaller the wavelength, the larger the energy of the photon. Characteristics of selected wavelengths are given in *Table 2.1*. Where the energy of light is to be related to a photochemical effect, as in spectroscopy, the wave number ($\bar{\upsilon} = 1/\lambda$; with units of cm^{-1} by convention) is employed as it is directly proportional to the energy from Planck's law (equation (2.2)) with $\varepsilon = hc\bar{\upsilon}$. The Système International (SI) unit of energy is the joule (symbol J), but much of the older literature uses the calorie. Also, energy levels of molecular orbitals, ionization potentials, *et cetera*, are often expressed as electron volts (eV), which is the energy acquired by an electron falling through a potential difference of 1 V. Conversion factors for different units of energy, used in the literature on radiation biology are shown in *Table 2.2*.

A single photon is a small unit in biological terms; at noon on a bright day the earth's surface receives a maximum of about 1.3×10^{21} photons $m^{-2}\,s^{-1}$ (400–700 nm), so a larger unit of radiation, the mole of photons (more usually referred to as a mole of quanta) is used. It is the number of photons corresponding to Avogadro's number of particles (6.023×10^{23}), and may be thought of as the number of photons required to convert a mole of a substance

Table 2.2. Conversion factors for energy units used in the photosynthetic literature

1 electron volt (eV)	$= 1.602 \times 10^{-19}$ J
1 watt	$= 1$ J s^{-1}
1 kWh	$= 3.6 \times 10^6$ J
1 joule	$= 0.239$ calories
	$= 6.242 \times 10^{18}$ eV
1 calorie	$= 4.184$ J
1 kJ mol quantum^{-1}	$= 1.036 \times 10^{-2}$ eV
Planck's constant	$= 6.62 \times 10^{-34}$ J s
	$= 4.136 \times 10^{-15}$ eV s

1 mol contains as many elementary particles as there are carbon atoms in 0.012 kg of ^{12}C, or Avogadro's number, 6.023×10^{23}, of particles. 1 mol of photons

Energy of 1 photon (J)	$= 1.986 \times 10^{-16}/\lambda$
With λ in nm 1 mol quanta	$= \dfrac{1.986 \times 10^{-16} \times 6.023 \times 10^{23}}{\lambda}$
	$= \dfrac{119.616 \times 10^6 \text{ J}}{\lambda}$
1 mol quanta $m^{-2}\,s^{-1}$	$= \dfrac{1.2 \times 10^8}{\lambda}$ J $m^{-2}\,s^{-1}$

to another form with 100% efficiency, if captured in a single discrete step. A mole quanta is often called an 'einstein' although the unit is not permitted in the Système International, and should not be used. The relationship between wavelength, frequency and energy of individual photons and a mole quantum of photons is given in *Table 2.1*.

2.1.1 *Measuring and expressing light*

Confusion may arise over the many ways of measuring light and the units of expression because either the number of quanta, or their energy (or both) may be determined. There are also measures of the illuminance, that is, the visual impression to the human eye. To study quantitative and kinetic aspects of the response of chemical and biological processes to light only photon number or energy in defined spectral regions are useful. Photon number incident on a surface normal to the beam is given by photon flux (mol m^{-2} s^{-1}). Energy is given by the radiant flux (J m^{-2} s^{-1}; as 1 J s^{-1} = 1 watt (W) this is equivalent to W m^{-2}). Illuminance is given by the luminous flux, measured in lux (= lx = lumen m^{-2}). In the older literature the foot candle (1 fc = 1 lumen per square foot = 10 764 lx) was used extensively but is no longer acceptable. The biological literature uses the term photon flux density (PFD) as equivalent to photon flux; PFD is not used in the Commission International de l'Eclairage which recommends use of photon flux and photon irradiance. Photon flux alone may avoid confusion caused by 'density'. Photon number (mol), photon flow (mol s^{-1}) and photon flux or photon irradiance (mol m^{-2} s^{-1}) may be qualified as normal or as a spherical flux (depending on the spatial distribution of the light). As mentioned earlier, radiation from about 400 to 700 nm is used in higher plant photosynthesis and is called photosynthetically active radiation or PAR.

PAR is measured by quantum sensors, including germanium diodes and lead sulfide resistors; their sensitivity changes with wavelength so that at different parts of the spectrum a true photon number is counted. Energy detectors or radiometers respond to energy independent of wavelength; the instruments include thermopiles and bolometers which can measure mono- or polychromatic light. Detectors must have correct geometrical characteristics to detect light coming from different directions; most used in photosynthesis studies are cosine corrected (see Hall *et al.*, 1993; Chapter 13).

Spectroradiometers are instruments that measure the energy of light as a function of wavelength over the whole spectrum and are used to determine spectra of light sources. Spectrophotometers are used to determine the response of photochemical and biological processes to wavelength. Illuminance is measured by light meters, which respond to light with a similar spectral response to the human eye. As the human eye is most sensitive to green light (*ca* 550 nm wavelength) measurements of illuminance should not be used in studies of photosynthesis (see Hall *et al.*, 1993; Chapter 13).

2.1.2 *Electronic states of matter, photon emission and photon capture*

Atoms consist of the central nucleus surrounded by the electrons, in particular energy levels. Molecules have interactions between electrons of the constituent atoms, which results in complex energy levels. Light is emitted by atoms or molecules undergoing changes in energy state, e.g. as a result of heating, which excites electrons from the ground state to higher excited states. Radiation is emitted when the electrons drop back to the lower energy state (*Figure 2.1*), at discrete wavelengths corresponding to the energy difference between the ground and excited states. Atoms give rise to line spectra because the energy levels are distinct. Excited molecules, particularly the more complex ones, may emit radiation at several wavelengths, often close together, giving broader emission spectra. This is due to more closely spaced molecular orbitals from which electronic transitions occur.

The wavelength of the radiation emitted depends on the energetic characteristics of the states, thus, the temperature of the material determines the wavelength of light emitted. Lamps, for example, have a color temperature at which a given spectrum of light is produced according to the particular material emitting it.

Absorption of radiation in the visible spectrum depends on the electronic states of the excited atoms and molecules in the absorbing substance. Substances absorbing visible light are called pigments and absorb particular wavelengths of light, the light not so absorbed being reflected or transmitted. Thus chlorophyll absorbs light at about 450 and 650 nm, corresponding to the light detected by the average human eye as the colors blue and red, respectively. However, the light between those two wavelengths is relatively poorly absorbed, so the color of vegetation appears green to the eye. As photon capture in photosynthesis is predominantly by large complex molecules, brief consideration will be given to molecular orbitals and how electrons are arranged in them.

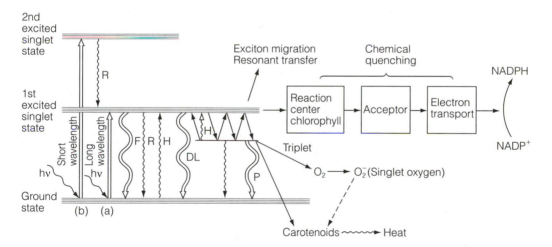

Figure 2.1. Concept of absorption of photons (hν) by an atom, energizing an electron to an excited state (a) and its subsequent decay with release of energy. Capture of a more energetic photon (b) results in higher energy level orbitals being filled and then decay by radiationless transition (R). Heat (H) may also raise an electron to higher energy level and the energy is emitted when the electron drops back to the ground state. The main energy-dissipating processes are by radiationless transition (R), prompt fluorescence (F), delayed light emission (DL), phosphorescence (P), and by transfer of excitation linked to chemical reactions. In photosynthetic organisms these include assimilation of CO_2, and transfer of triplet energy to oxygen or carotenoids or of excitation energy to other chlorophyll and pigment molecules

To capture a photon of visible light, the energy levels of the atomic or molecular orbitals must have a difference in energy corresponding to the absorbed quantum. The electronic orbitals of atoms are analogous to molecular orbitals. The main energy levels are referred to as the ground state (lowest energy), and the first and second (or higher) excited states. The energy levels are designated by a total quantum number. Within each level the number of possible orbitals depends on the magnetic motion and orientation of the electrons in relation to the nuclei, as described by quantum mechanics. Electrons spin within an orbital according to the prevailing magnetic field. Spin may be parallel to the field or antiparallel to it, and is designated by a spin quantum number, S, which has values of $+\frac{1}{2}$ or $-\frac{1}{2}$. Two electrons can only occupy the same orbital if their spins differ. When two electrons of opposite spin occupy the lowest energy orbital the configuration is a stable electronic ground state, S_o: adding the spins of electrons together gives the total spin S and the spin multiplicity of the electrons is given by $2S + 1$. When all spins balance out $S = 0$ and the spin multiplicity is 1, which is called a singlet state. If there is spin reversal, then S equals 1 and $2S + 1$ becomes 3, giving a triplet. For triplet states, spin reversal must occur. This is a relatively infrequent event and has a low probability, thus the formation of a triplet is uncommon and the lifetime may be long. Oxygen is an example of a ground-state triplet. With an electron absent from an orbital, S equals $\frac{1}{2}$, and the spin multiplicity is 2 or a doublet. Singlet and triplet states are important in photochemistry, and the doublet in free radicals, which may cause photochemical damage in photosynthesis. The chemical and physical texts listed provide quantitative descriptions of molecular structure and quantum mechanics.

With a given electronic configuration, a molecule absorbs light of particular wavelength, smaller or larger quanta are not absorbed and they cannot, by the Grotthus–Draper Law, cause a physical or chemical change. If the oscillating electronic vector of a photon causes an electron in a molecule to resonate (i.e. to vibrate at the same frequency) then the energy of the photon will be captured. The direction of the electrical and magnetic fields of the photon must be in correct orientation to the electronic oscillations to cause resonance. Capture of a quantum is rapid by all criteria of 'normal' biological interactions; an electronic transition from one energy level to another occurs in a time inversely proportional to the frequency of the wave ($1/v$). For red light (*Table 2.1*) this is about 2.3×10^{-15} s, which is much faster than nuclear vibrations (10^{-13} s) and so does not influence the nuclear configuration of the molecule.

In complex molecules, with many nuclei and electrons, the possible combination of orbitals is greatly increased, and the energy levels are split into vibrational and rotational levels as electrons within the molecule are influenced by magnetic and electrical forces. Molecules absorbing visible light exhibit differences in energy of about 1–5 eV between ground and excited states corresponding to wavelengths between 1000 and 200 nm and vibrational energy differences of approximately 0.1 eV; rotational energy levels are not observed at these wavelengths in solid state, only in gases.

Larger molecules usually have more complex energy levels because the outer electrons which provide the bonding are 'delocalized' over the whole molecule. These π electrons travel in the extended π orbitals, even in the ground state and may undergo transition to an excited orbital, π*, in the same way that electrons in other, more 'rigid' orbitals can.

The π^* orbitals are delocalized; the electrons are not orientated in the same way in π and π^* states; they are, respectively, bonding and antibonding. As π electrons are free to move in large volume, the energy levels are smaller and, therefore, the binding energy is greater than for a system of double and single bonds and the molecule is more stable. Also, the energy required for the ground to excited state transition is low and the capacity to absorb light of longer wavelength is increased.

Delocalization is of great importance in organic molecules, including the photosynthetic pigments which have extensive orbitals over large molecules. Extensive systems confer high efficiency of energy capture, together with stability under normal conditions *in vivo*. Also, orbitals of different energies provide for absorption of different wavelengths of light. Groups of atoms in a molecule which are responsible for absorbing light energy are called chromophores. In the 200–800 nm range chromophores always have loosely bound electrons, in π orbitals, the size of which determines wavelengths absorbed. For example, the peptide bond in proteins absorbs short energetic (ultraviolet) wavelengths (190 nm), and the nucleic acid bases, which have a larger π system, absorb at 260 nm. The photosynthetic pigments β-carotene (absorbing at 400–500 nm) and chlorophyll (absorbing between 400 and 700 nm) have increasingly larger π systems. The absorption maxima become more dependent on the environment of the molecule as the size of the π system increases because the probability of interaction between the molecule and environment is increased and the small changes in energy can affect the properties of the electronic system substantially. Energy levels of a molecule depend on the environment, on intramolecular rearrangement, binding to other compounds, and so on, which alter the absorption of particular wavelengths. Chlorophyll *in vivo* is associated with protein, which 'tunes' the absorption of light over a range of wavelengths.

Chemical reactions involve reorganization of the outer shell or valency electrons, which form chemical bonds. For an excited electron to be used chemically it must be held in a configuration which is stable for long enough to transfer to a chemical acceptor directly or via an exchange of energy (not electrons). In photosynthesis the light-harvesting pigments, which capture the energy, transfer excitation via other pigment molecules to special reaction centers, composed of a form of chlorophyll, which convert the excitation energy to chemical energy.

The wavelength and number of photons captured determines the maximum energy available for biological processes and therefore governs the overall energetics. However, the efficiency of energy capture and all the linked conversion processes limits the actual energy which can be obtained. It is possible to overcome the inefficiency of a photochemical process if the energy from the capture of several photons can be 'stored' or gathered in some way and then used, by multistep processes, for photochemical conversions. This occurs in photosynthetic organisms.

2.2 Dissipation of excitation energy

Electrons remain in the excited state for a period called the 'lifetime', dropping back to the stable ground state in an exponential decay. Processes dissipating the energy of the excited state are illustrated in *Figure 2.1*. Thermal relaxation (also called radiationless transition) between the closely spaced vibrational energy levels leads to loss of excitation as heat within 10^{-12} s and is the normal pathway of energy loss from the second to first excited states. Excitation energy may be transferred to a chemical acceptor; this is the central event in photosynthesis. Electrons also decay from the excited singlet state to the ground state emitting 'prompt' fluorescence in 10^{-9} s. Electrons in a lower ground state at normal temperatures reach the center of the excited state when excited, but decay rapidly by thermal relaxation to a lower level of that excited state, before finally dropping to an energy level in the ground state of different energy from which they started. The excitation energy that is required to excite a molecule is therefore greater than the energy emitted, that is the wavelength of the photon absorbed is shorter than that of the photon emitted and the fluorescence spectrum is shifted (Stoke's shift) towards longer, red wavelengths. If the molecule is excited to the second excited state then the absorption spectrum shows two bands. However, radiationless transition from the second to the first excited state allows fluorescence only from the first excited state, and only one red-shifted fluorescence band is detected (*Figure 2.2*).

When electrons decay from the excited triplet before returning to the ground state and releasing a photon, the process is called phosphorescence. The photon is of longer wavelength than fluorescence due to a smaller energy difference between the triplet and ground states. Phosphorescence may be much delayed as triplet states have a very long lifetime, from milliseconds to tens of seconds, due to the change in spin which accompanies the triplet transition and occurs with low probability. Electrons in the triplet, when energized to the excited singlet by thermal energy, decay to ground state releasing a photon in delayed light emission (also called luminescence, delayed fluorescence and 'after glow'); the wavelength of the photon is shifted like fluorescence as both come from the same energy level. Delayed light emission is strongly dependent on temperature and on a small energy gap between the triplet and excited singlet. Fluorescence of

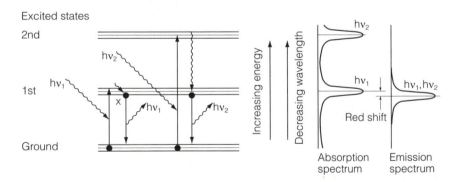

Figure 2.2. Capture of photons of low (hv_1) or high (hv_2) energy, shown by the absorption spectrum, raises an electron to the first and second excited states. Electrons drop from the second to first excited state by radiationless decay, and photons of similar wavelength are emitted from the first excited state, giving a single peak in the emission spectrum. Radiationless decay from the higher to lower energy levels of the first excited state (at X) redshifts the emitted photons

chlorophyll in the thylakoid membrane is an important characteristic. Prompt fluorescence comes from chlorophyll which cannot pass excitation onto another molecule, either because it is not joined to it or because the system is already overexcited by light. This fluorescence may arise because the reduced acceptor of energy in the reaction centers (Section 3.2) passes energy back to the chlorophyll which then fluoresces. However, energy is lost from the chlorophyll matrix over a long period by delayed fluorescence and the rate depends on the state of the complete system. Fluorescence from chlorophyll *in vivo* is a very sensitive indicator of the energy status of the photosynthetic system, indicating how the multiple processes of energy absorption and utilization interact.

The rate of decay of the excited state depends on radiationless transitions, fluorescence and photochemistry. The processes have rate constants for de-excitation, respectively, Kd, Kf and Kp so that the overall rate constant is $K = (Kd + Kf + Kp)$. With n_o excited states initially, the decrease to n excited states in time, t, is given by $n = n_o e^{-Kt}$ where e is the base of natural logarithms. The number of excited states decreases exponentially to $(1 - 1/e)$ of the initial number of excited states in a time $= 1/K$, which is the lifetime, the time required for 63% of the excited electrons to decay. The exponential decay process can also be characterized by the 'half-life', the time needed for half the original population of excited electrons to decay. Single excited states are relatively short lived, that of chlorophyll being about 5×10^{-9} s.

2.2.1 Transfer of excited states

Decay of excited electrons to the ground state, either by radiationless decay or fluorescence, wastes the energy for photosynthesis and *in vivo* the main mechanism which captures the energy of the excited pigment is transfer to a special form of chlorophyll at a 'reaction center' (RC); this is a chemically reactive form of bacteriochlorophyll or chlorophyll *a* which can pass an excited electron to an acceptor molecule and thus start the chemical reactions of photosynthesis. It is a gate between the world of physical energy and biological chemical energy and thus the heart of the whole process of life.

Evolution of photosynthetic organisms was based on utilizing the energy of photons which provided the energy for other evolutionary developments, including the formation of complex biophysical–biochemical and chemical structures for energy capture and transduction into the biological realm. The systems have multistep approaches to energy capture and transduction. Thus, energy capture by a chlorophyll molecule generally does not pass directly from an excited pigment molecule to the RC but the energy is transferred via several intermediate molecules (which transfer energy but do not react chemically) before reaching the RC. In addition, as shown by Emerson and Arnold in the 1930s, the capture of photons by a single chlorophyll molecule under even large solar photon fluxes on earth (and particularly under small fluxes in many environments, including water) is rare, of the order of 2–3 per second, so a large 'antenna' (a group of pigment molecules) for capturing photons coupled to an RC increases efficiency and capacity greatly. Current living organisms could not function in the absence of the antenna transferring energy to the RC. However, these are quantitative aspects of energy transfer: the essential feature of the mechanism is shown in equation (2.3). The pigment (P) is excited by light (P*) and donates the energy to an acceptor A, which is thus excited and after a number of such transfers excites the RC:

$$P + h\nu \rightarrow P^*; \qquad P^* + A_1 \rightarrow P + A_1^*; \qquad A_1^* + A_2 \rightarrow A_1 + A_2^* \rightarrow A_2^* + A_n \text{ repeated;}$$
$$A_n^* + RC \rightarrow A_n + RC^*; \qquad RC^* + \text{chemical acceptor} \rightarrow RC + \text{reduced}$$
chemical acceptor $\hspace{10cm}$ (2.3)

The pigments D and A may be of the same type, for example chl *a*, or different, e.g. chl *a* and chl *b* (see Figure 3.2) and may be aggregated to form large energy collecting pigment groups or 'antennae' to maximize light capture.

The electromagnetic mechanisms of energy transfer depend on the types of molecules, their size, energy levels of the electronic orbitals, *et cetera*, and on concentration (i.e. on distance between molecules) and orientation. The inverse of the mean period of time for energy transfer (also called the 'pair jump time') is proportional to the interaction energy of the two molecules. In solids or concentrated solutions, orbitals 'overlap' and form extensive 'superorbitals'. The concepts were developed particularly by Duysens, who showed that the energy migrates largely via the lowest singlet excited state. When excited, an electron enters the delocalized conduction bands and leaves a 'hole' (a positive charge) in the ground state; electrons migrate through the pigment matrix. This semiconductor type of mechanism leads to photoconductivity, which is observed in dried chloroplasts but is not thought to play an important role *in vivo*. However, it has been suggested to occur in closely bound chlorophyll groups. Another 'strong' interaction occurs between closely packed molecules at 1–2 nm spacing; it is rapid (10^{-12} s) and depends on (1/distance) and there is no radiationless decay. None of these mechanisms is thought to be important in photosynthesis. More important is repeated energy transfer between donor and acceptor leading to energy migration between groups of the same and different molecules at distant spacing. Such mechanisms are characteristic of transfer between pigment (dye) molecules which have smooth, broad and overlapping absorption and fluorescence spectra and small interaction energy. This incoherent transfer with weak coupling is the only one relevant to photobiology. It means that, after the system is excited by a very short pulse of light to excite all the pigments, there is no uniform propagation of the energy through the system, rather there is rapid randomization of the excitation and there are only a few steps in the transfer. Dipole interaction causes inductive resonance in the acceptor, so that the excitation is passed to it. There is no mass or electron transfer; the excitation migrates as a spincoupled electron-hole pair and is localized on a definite molecule. This 'weak interaction' involves rates of transfer of 10^{-12}–10^{-14} s and is called a Förster mechanism, after the discoverer. The interaction decreases as R_o^{-6}, where R_o is the Förster distance, defined as the critical distance between the dipole oscillating centers in each donor and acceptor molecule at which energy transfer rate equals the losses due to fluorescence and radiationless decay. The pair jump time equals the reciprocal of the first order rate constant for fluorescence for each participating molecule multiplied by the ratio of the distance between dipole centers (R) to the critical distance R_o. Not only does distance between molecules affect exciton transfer but also their mutual orientation (which is determined by the protein matrix to which chlorophylls are bound, for example, and also on the lipids of the membranes) and the nature (dielectric parameters) of the solvent or medium. For photosynthetic systems *in vivo*, R_o is between 4 and 10 nm for chlorophyll with an average of about 6 nm for efficient transfer. It minimizes the number of vibrations undergone by the donor and is temperature dependent.

Transfer is primarily via singlet excited states. Orientation between dipoles is important; transfer is zero with perpendicular orientation and maximal with parallel orientation. Energy levels are most critical; transfer only occurs if the fluorescence bands overlap because the electron drops by radiationless decay to the lowest vibrational level of the excited state from which it decays as a fluorescence photon if its energy is not transferred. The acceptor absorption band is not at the lowest vibrational level but in the center of the energy band and will take excitation of the correct energy, similar to fluorescence.

As energy is lost at each transfer, a donor–acceptor chain of decreasing ground state to first excited state energy difference favors funneling of energy in a particular direction and increases the rate of exciton transfer. There may be transfer between similar molecules (homogeneous) by random walk with excitation passing in the pigment matrix at random but, as this leads to energy loss, transfer between different types of molecules (heterogeneous) is more likely. It is faster, requiring fewer steps (perhaps 10–100), and is very efficient. However *in vivo* there may be different transfer mechanisms in different parts of the pigment antenna within the many different types of antenna systems in photosynthetic bacteria, cyanobacteria and higher plants.

2.2.2 Light absorption and absorption spectra

When light passes through a solution of a compound, some wavelengths are transmitted, that is, passed through without alteration, some scattered and the remainder absorbed in proportion to the concentration of absorbing substance. The same processes occur in gases and solids and in the complex state of living organisms such as the green leaves of plants. Absorbed light determines the rate of photosynthesis. The number of photons of a given wavelength absorbed is the difference between the incident and transmitted and scattered (reflected) photons. The number of photons absorbed at different wavelengths constitutes an absorption spectrum, which gives much information on molecular configuration, electronic transition and energy levels. Substances have characteristic spectra from which they may be identified and quantified. Absorption spectra are measured with a spectrophotometer; radiation of the required wavelength band or monochromatic light is passed through a known thickness of solvent (in which the substance to be studied will be dissolved) contained in a cuvette of material transparent to the wavelength to be measured (e.g. glass for visible light, quartz for UV light). This is a reference or standard solution to correct for reflectance and transmission characteristics of the cuvette. A solution of the substance is then substituted for the solvent and the decrease in transmitted light is measured (see Hoppe *et al.* (1983) and Amesz and Hoff (1996) for techniques).

With I_o photons per unit area and time passing into the cuvette of thickness l (cm) and a fraction I not absorbed and passing through it, the fraction absorbed is $I_a = I_o - I$. Thus, dI photons are absorbed in a thin layer dl of area A by n molecules, uniformly distributed per unit volume. The average cross section, σ, of each molecule is the area of the molecule for photon capture. With a probability, p, that a photon will be captured by a molecule, the absorption cross section, k, is equal to $p\sigma$ (units, cm^{-2}). Thus, the absorption of photons is given by:

$$-dI = kn\ dl\ I \tag{2.4}$$

which integrates to give the Beer–Lambert law:

$$I = I_o e^{-knl} \tag{2.5}$$

where knl is the absorbance. This may be expressed as:

$$I = I_o 10^{-\varepsilon cl} \tag{2.6}$$

where c (units, molar) is the molar concentration of absorber and ε is the molar extinction coefficient or molar absorptivity (unit, $M^{-1} cm^{-1}$), characteristic of the absorber in a particular solvent at a given wavelength. Transmittance, I/I_o, is related to optical density (OD), also called absorbance or extinction, as:

$$OD = \log I_o/I = \varepsilon cl \tag{2.7}$$

Absorbance increases linearly with concentration in dilute solution and is often quoted as $A^{1\%}_{1\,cm}$ or absorbance of a 1% concentration in a 1 cm layer.

It is possible to measure the concentration of several components (which do not react) in a solution if their extinction coefficients are known at different wavelengths, as at any one wavelength:

$$OD = \varepsilon^A c^A l + \varepsilon^B c^B l \tag{2.8}$$

and by measuring at different wavelengths (at least as many as there are components) the equations can be combined to give the concentrations. This is the basis for Arnon's much used method of measuring chlorophyll a and b in the same extract (see Neubacher and Lohmann, 1983).

2.3 Global photosynthesis

All the energy for global photosynthesis comes from the sun (although light of suitable wavelength from any source, for example electric lamps, may be used by plants). Radiation is lost from the sun's corona as if from a 'black body' (a perfect radiator and absorber of energy) with a temperature of about 5800K, in agreement with the Stefan–Boltzmann law. The distribution of the wavelengths of the solar spectrum is concentrated between 400 and 1200 nm with the peak around 600 nm (*Figure 2.3*) in the yellow-orange. At the earth's surface the spectrum is modified by selective absorption of wavelengths by constituents of the atmosphere such as carbon dioxide and water vapor.

The earth's surface receives about 5.2×10^{21} kJ year^{-1} of solar radiation, about 50% of which is of wavelengths used in photosynthesis. Only about 0.05% of this energy (3.8×10^{18} kJ year^{-1}) is captured in organic molecules and the rest is re-radiated into space as heat. The turnover of CO_2 is about 3×10^{12} tonne year^{-1} and of nitrogen about 5% of this. Almost 50% of total photosynthesis is by marine organisms. The enormous scale of this most important photochemical reaction is comparable to geological processes such as

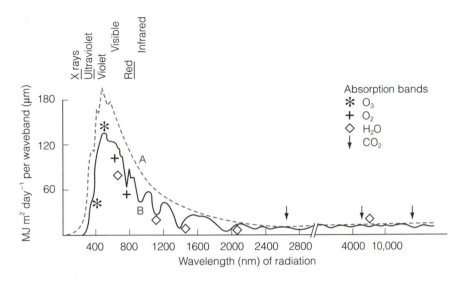

Figure 2.3. Solar radiation above the earth's atmosphere (A) and at the earth's surface (B). The energy (in megajoules per day) for each wavelength is shown. Symbols on curve B show absorption of radiation by components of the earth's atmosphere

mineral weathering. Photosynthesis produces 5×10^{11} tonnes of organic matter per year and the earth's standing organic matter is estimated at 10^{12}–10^{13} tonnes of dry matter. At present, the rate of destruction of forests and burning of fossil fuels is about 15% of the rate of photosynthesis so that the CO_2 content of the atmosphere, which is small (approximately 340 μl l^{-1} or 34 Pa partial pressure), turns over rapidly, with a half-life of 10 years. Oxygen turnover is slower, 6500 years, as the concentration is much greater (21 volumes per 100 volumes) and water turns over in 3 million years, owing to the huge global water reserves.

References and further reading

Amesz, J. and Hoff, A.J. (eds.) (1996) *Advances in Photosynthesis*, Vol. 3, Biophysical Techniques in Photosynthesis. Kluwer Academic, Dordrecht.

Borisov, A.Y. (1989) Transfer of excitation energy in photosynthesis: some thoughts. *Photosynthesis Res* **20**: 35–58.

Broda, E. (1978) *The Evolution of the Bioenergetic Processes.* Pergamon Press, Oxford.

Clayton, R.K. (1980) *Photosynthesis: Physical Mechanisms and Chemical Patterns.* IUPAB. Biophysics Series, Cambridge University Press, London.

Gates, D.M. (1962) *Energy Exchange in the Biosphere.* Harper and Row, New York.

Giese, A. (1964*) Photophysiology, 1, General Principles: Action of Light on Plants.* Academic Press, New York.

Hames, B.D. and Rickwood, D. (eds.) (1987) *Spectrophotometry and Spectrofluorimetry, A Practical Approach.* IRL Press, Oxford.

Hoppe, W., Lohmann, W., Markl, H. and Ziegler, H. (1983) *Biophysics.* SpringerVerlag, Berlin.

Lichtenthaler, H.K. (1992) The Kautsky effect: 60 years of chlorophyll fluorescence induction kinetics. *Photosynthetica* **27**: 45–55.

Neubacher, H. and Lohmann, W. (1983) Applications of spectrophotometry in the ultraviolet and visible spectral regions. In *Biophysics* (eds. W. Hoppe, W. Lohmann, H. Markl, and H. Ziegler). SpringerVerlag, Berlin, pp. 100–109.

Nobel, P.S. (1991) *Physiochemical and Environmental Plant Physiology.* Academic Press, San Diego, CA.

Owens, T.G. (1994) Excitation energy transfer between chlorophylls and carotenoids. A proposed molecular mechanism for non-photochemical quenching. In: *Photoinhibition of Photosynthesis* (eds. N.R. Baker and J.R. Bowyer). BIOS Scientific Publishers, Oxford, pp. 95–110.

Parson, W.W. and Ke, B. (1982) Primary photochemical reactions. In: *Photosynthesis,* Vol. I, *Energy Conversion by Plants and Bacteria* (ed. Govindjee). Academic Press, New York, pp. 331–385.

Salisbury, F.B. (1991) Système International: the use of SI units in plant physiology. *J. Plant Physiol.* **139**: 1–7.

Light harvesting and energy capture in photosynthesis

Photosynthesis is the use of energy from photons, derived from the nuclear reactions of the sun, to produce excited electrons which can be used for reduction reactions in controlled biochemical processes. Light harvesting pigments capture the energy of photons, forming excited pigments. This exciton energy is used to energize electrons in a 'reaction center (RC)', which provides the physico-chemical environment that enables the electrons to reduce acceptor molecules. Essentially, RC acts as a catalyst, altering the energetics of the reactions, enabling the reaction to take place and speeding it up. The reduced acceptors then undergo a series of reactions, ultimately reducing carbon dioxide, nitrate and sulfate ions and forming biochemical products. Events in capture of photon energy and at the RC are crucial to the entire photosynthetic process. Although the fundamental mechanism is similar in all photosynthetic organisms, from anoxygenic bacteria through to higher plants, differences in molecular mechanisms at the RC and associated with it have substantial biochemical consequences. Here the basic features of the mechanism are described. Details are provided in Ort and Yocum (1996) and other references.

Three functions of the light-harvesting apparatus contribute to the utilization of light quanta to produce a chemical intermediate of higher energy state:

- light absorption – the energy of a photon is captured by an antenna pigment molecule and an electron is excited;
- energy transfer – excitation energy moves through the antenna to an RC, a special form of pigment, in which it excites an electron; and
- electron transfer – an energized electron from the RC passes to a chemical acceptor and the oxidized RC is reduced by an electron from organic or inorganic molecules (anoxygenic) or water (oxygenic).

The processes are controlled by the physical and chemical characteristics of the pigments, particularly chlorophyll (chl) in higher plants and bacteriochlorophyll (bchl) in photosynthetic bacteria. However, there are many other pigments involved in photosynthesis in different organisms.

3.1 **Light-harvesting pigments**

Several types of pigments harvest the energy of light (*Table 3.1*). Only special forms of chl *a* and bchl *a* form RCs (see Section 3.2 and Chapter 5) in higher plants and photosynthetic bacteria, respectively. All other pigments are therefore accessory pigments, forming groups arranged to extend the size of the light-capturing unit, thus acting as antennae (like a radiotelescope dish) for capturing photons. They cannot initiate reduction (electron donating) reactions and only transfer excitation to RCs. A group of many antenna molecules of different types donates energy to a single RC complex, forming a very efficient 'energy-concentrating' mechanism. As mentioned in Chapter 2, the capture of energy of many photons by the antenna is crucial for photosynthesis, enabling coordinated activities such as water-splitting (see Section 5.2.2) to occur and giving much faster rates of electron transport than would be possible with the reaction center alone capturing photons. The increased excitation to the RC is several hundred times that of a single RC alone. Pigments absorb photons in different parts of the spectrum (i.e. of different energy) and, in combination, enable an organism to absorb light over a range of different wavelengths. This is of great importance ecologically, as organisms may then exploit a greater energy supply or grow in different radiation environments. For example, phycobilins of red algae absorb blue light, which predominates in deep water where the plants grow. *Figure 3.1* illustrates the pigments, wavelengths absorbed and composition

Table 3.1. Pigments of photosynthetic organisms: only some of the more important pigments and groups of plants in which they occur are given. Primary pigments are those involved in the photochemical process; accessory pigments function only in light harvesting

Organism	Primary pigment	Accessory pigment(s)	Wavelengths absorbed (nm) (approximate)
Prokaryotes			
Purple bacteria	bchl *a*	bchl *a*	Blue–violet–red, 470–750
		bchl *b*	Blue–violet–red, 400–1020
Green sulfur bacteria	bchl *a*	bchl *c* = *Chlorobium* chl	Blue–red, 470–750
Blue-green algae	chl *a*	Phycocyanin	Orange, 630
		Phycoerythrin	Green, 570
		Allophycocyanin	Red, 650
Eukaryotes			
Red algae	chl *a*	Phycocyanin	Orange, 630
		Phycoerythrin	Green, 570
		Allophycocyanin	Red, 650
Brown algae	chl *a*	chl *c*	Violet–blue–red
Higher plants	chl *a*	chl *b*	Violet–blue–orange–red, 454–670
Most higher plants, algae and bacteria		Carotenoids and xanthophylls	Blue–green, 450

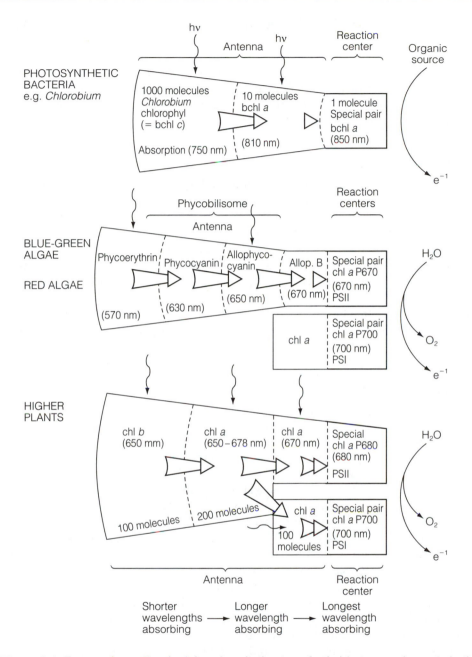

Figure 3.1. Energy absorption (〜▶) and excitation transfer (⇒) between pigments in the light-harvesting antenna and to the photochemical reaction centers of different photosynthetic organisms

of the light-capturing system for different organisms. Details are provided by Scheer (1991).

Photosynthetic bacteria have several forms of pigments donating energy to the RC bacteriochlorophyll. In higher plants chl *b* is an auxiliary pigment passing excitation to chl *a*. In all O_2-evolving organisms excitation moves to RCs composed of forms of chl *a* which pass the energy on as an excited electron to chemical reactions.

3.1.1 Chlorophylls

Chlorophylls are probably the most abundant biological pigments and vegetation appears green to the human eye due to their absorption of blue and red light. Leaves may contain up to 1 g of chlorophyll per square meter of surface area, but this varies with species, nutrition (particularly nitrogen fertilization), age, *et cetera*. Chlorophyll is extracted with fat-solubilizing solvents (e.g. ether or acetone) as it is a lipophilic molecule only found in membranes containing lipid, where it is bound to hydrophobic protein. Chl *a* and chl *b* and other pigments may be separated by chromatography.

Chemically chlorophylls are chlorin macrocycles with four-fold symmetry, derived from porphyrin. Chlorophyll *a* (*Figure 3.2*) is a conjugated macrocyclic molecule (mass 894) with a planar 'head' of four pyrrole rings; it is about 1.5×1.5 nm, but the overall size is much greater due to a phytol group, a terpene alcohol chain some 2 nm long, which positions the molecule in membranes. The chemical groups and H^+ on the outer edge of the pyrrole unit confine the electrons to a single plane, increasing absorption of red wavelengths, which is an advantage for plants on land or in surface water. A nonionic magnesium atom, bound by two covalent and two coordinate bonds in the center of the molecule, coordinates the rings. Magnesium is crucial to the capture of light energy – an iron atom (as in related pigments) cannot do this. Magnesium is a 'close shell' divalent cation which changes the electron distribution and produces powerful excited states. Chlorophyll *a* is a good donor of electrons (at a large negative potential), and oxidized chl *a* at the RC is a very strong oxidizing substance able to accept electrons (indirectly) from water. The large size and extensive ring structures of the chlorophylls, with 10 double bonds, allows electrons to delocalize in the π orbitals over the 'head' of the molecule, increasing the area for capture of a photon, giving many redox energy levels which are important for efficient energy capture and transfer, and producing a complex absorption spectrum. In chl *b* a formyl (-CHO) group replaces the methyl ($-CH_3$) group on ring II (*Figure 3.2*), which increases the blue and decreases the red absorption maxima and alters solubility. Chlorophyll *b is* less soluble in petroleum ether than chl *a*, but more soluble in methyl alcohol, for example. Bacteriochlorophyll *a* (*Figure 3.2*) is similar to chl *a* but has an acetyl instead of a vinyl group on ring I, and ring II is saturated with hydrogen instead of unsaturated. This loss of the double bond alters the π system. Such differences, which are probably later evolutionary developments, increase absorption at longer wavelengths, but bchl cannot generate a sufficiently strong oxidant to remove electrons from water. Bacteriochlorophyll *g*, which is very similar to the chl *a* of higher plants, has been described from *Heliobacterium chlorum* and may represent a stage in the evolution of chl *a*. Possibly the evolution of chl *a*, essential for the development of the oxygenic photosystem and water splitting, took place in bacteria together with the development of the

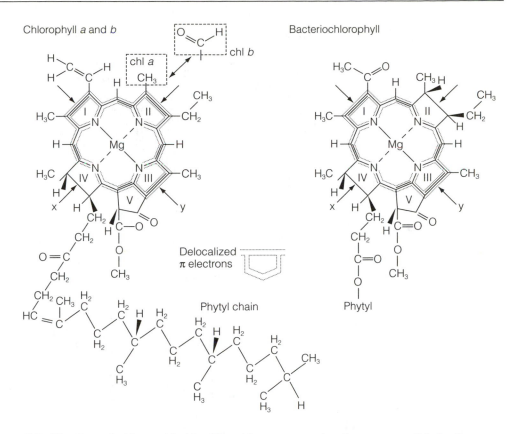

Figure 3.2. Structure of chl *a* and bchl *a*. The side groups on ring II are responsible for the difference between chl *a* and *b*. Axes *x* and *y* are the principal electronic transitions

special binding to the structural proteins which is essential for the modification of the electronic energy states to achieve particular absorption characteristics and sufficient oxidizing power to reduce (indirectly) water.

3.1.2 Absorption spectra of chlorophyll

Measured in organic solvents after extraction from the plant (*Figure 3.3*), chl *a* absorbs most strongly at 430 (Soret band) and 660 nm and chl *b* at 450 and 640 nm. The absorption maxima 'shift' with the solvent: in 40 different solvents the red absorption of chl *a* is between 660 and 675 nm. Polar solvents, such as acetone, cause strong dipole–dipole interactions, weaken London dispersion forces and change hydrogen bonding, altering the electronic configuration of the molecule and hence absorption. Aggregation of chlorophyll also causes a shift: crystalline chl *a* has its long-wave absorption minimum at 740 nm.

Differences in absorption spectra and molar extinction coefficients enable chlorophylls to be distinguished and measured spectrometrically in unpurified solutions. Chlorophyll *a*

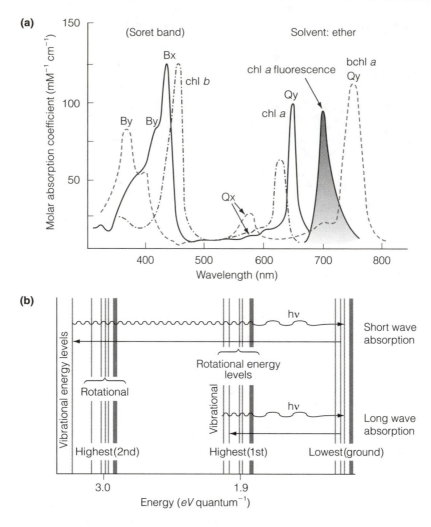

Figure 3.3. (a) Absorption spectra of chl *a* and *b* and bchl *a* in ether showing principal electronic transitions in blue (Soret) and red wavelengths. An energy level diagram (b) for chl *a* is related to the main absorption peaks

has a molar extinction coefficient of 1.2×10^5 M^{-1} cm^{-1} at 430 nm; a 10^{-5} M solution is intensely colored and absorbs some 80% of the incident light. Chlorophyll is a very efficient pigment with a cross-section absorbing area per molecule of 3.8×10^{-16} cm^{-2}. On a bright noon day at the earth's surface (a PAR photon flux of 2×10^{-3} mol quanta m^{-2} s^{-1}) a chlorophyll molecule will capture about 45 photons s^{-1}.

The maxima in light absorption correspond to different energy levels in the molecule (*Figure 3.3*). The highest energy level is the second (or higher) singlet, excited at 430 nm. The excited electrons make the transition in 10^{-12} s from the second to the first excited singlet state, to which electrons are excited by red light. Thus, the energy of blue light is dissipated and a blue photon is no more effective in photosynthesis than a red.

The lifetime of the lowest excited singlet state is about 5×10^{-9} s before it decays to the singlet ground state. In a solution of monomeric chlorophyll there is little radiationless dissipation to the ground state, but in aggregated chlorophyll it dominates. Absorption spectra of chlorophylls show the electronic transitions along axes of the molecule. Polarized light and paramagnetic or electron spin resonance (ESR) are used to analyze the transitions, which are very important in understanding the mechanism of energy transfer. The x-axis of chlorophyll is through the nitrogen (N) atoms of rings II and IV and the y-axis through the N–N atoms of I and III (*Figure 3.2*). The two main absorption bands in the blue and two in the red are called B and Q, respectively, due to $\pi-\pi^*$ transitions. The polarizations of the transitions along the axes are called x and y, and may be from the lowest vibrational energy state (called 0) or the next higher energy state (called 1). Thus, absorption at 430 nm is a Bx (0,0) transition and at 660 nm it is a Qx (0,0). The Qy transition is most altered by solvent (e.g. water bound to Mg) and association in membranes, which increases the red absorption.

3.1.3   Chlorophyll fluorescence

Electrons in the higher levels of the first excited singlet, S, state (energy E_a) decay by radiationless transition (R) to the lower levels and, if not used in photochemistry or transferred to other molecules, decay to the singlet ground state, S_o, by emission of 'prompt' fluorescence of lower energy (E_f) than the exciting light, as $E_f = E_a - R$. Thus, a solution of chlorophyll irradiated with blue light emits red fluorescence. In chl a of the thylakoids, prompt fluorescence is emitted at a peak of 685 nm. It shows the accumulation of excitation energy in the antenna and is inversely related to the use of electrons; it indicates the state of electron transport and biochemical processes relative to energy capture (Chapter 9).

3.1.4   Absorption spectra of chlorophylls in membranes

The absorption of light by chl a and chl b has been discussed (Section 3.1.2) for the pigments in solution. However, the situation in the thylakoid membranes, which are hydrophobic lipid-protein environments with respect to the pigments, is very different with the pigment bound in specific positions and orientations to proteins, and subject to the influence of the different types of lipids (and thus their proportion). In all photosynthetic organisms, the pigments are arranged in antennal complexes, closely linked to reaction centers. In some bacteria, e.g. purple bacteria such as *Rhodopseudomonas viridis*, the light harvesting antenna complex is very modular in form. The bchl a, bchl b, and carotenoids are non-covalently attached to two types of very hydrophobic proteins of 5 and 7 kDa molecular mass (α and β respectively) in the lipid membrane. The α and β subunits form heterodimers (α_2 and β_2) spanning the membrane, and a bchl a and bchl b are attached very precisely, within 1–1.5 nm of each other, so they are coupled with respect to excitation energy. About six of these heterodimers then aggregate into beautifully symmetrical rings and groups of these rings surround a reaction center, to which they deliver excitation energy.

In blue-green algae (cyanobacteria) the absorbing pigments are arranged in phycobilisomes attached to the reaction center in a symetrical way, with those with the shortest wave-length absorption at the periphery and the longest near the RC (see *Figure 3.6*). Higher plants also have close association of highly structured pigment-protein complexes with the RCs, again with the shorter wavelength absorbing pigments delivering excitation energy to the longer and then to the RC.

Aggregation of chlorophylls in membranes is analyzed mathematically (deconvoluted) by fitting curves of a normal (Gaussian) distribution to absorption spectra of thylakoids. Chlorophyll *a* shows up to 10 absorption maxima, the main ones absorbing at about 660, 670, 678, 685 and 689 nm. A 650 nm absorption is due to chl *b*. Light-harvesting chlorophyll–protein (LHCP) complex (see Section 4.5) has six bands, including that at 678 nm. Chlorophyll *b* in LHCP may be in groups exchanging by the strong excitation mechanism and linked to chl *a* by the Förster mechanism. The chl *b* antenna shows strong interaction, but not chl *a*.

3.2 Chlorophylls at the reaction center

Normal absorption spectra cannot measure small changes in absorption occurring in a few molecules if the bulk of the pigments have overlapping absorption bands. Most bchl in bacteria and chl *a* and *b* molecules in higher plants form the antenna, and only a few bchl *a* or chl *a* form RCs which pass electrons to acceptor molecules. Antenna chlorophyll 'swamps' the absorption even in preparations enriched in RC chlorophyll. To overcome this, difference spectra compare absorption between light and dark with repetitive flashes or with chemical oxidation and reduction (see Satoh, 1996), enabling the chemically reactive form of chlorophyll to be studied. ESR and electron nuclear double resonance (ENDOR) techniques are also used to detect changes in paramagnetism caused by the transfer of electrons to an electron acceptor, independent of changes in the nonparamagnetic bulk chlorophylls and other pigments.

In isolated spinach chloroplasts from which most of the chlorophyll had been removed, difference spectra showed a decrease in absorption (called photobleaching) near 700 nm following illumination (*Figure 3.4*). The signal was altered by the redox state of the chlorophyll; when oxidized there was a loss of absorption at 680 and 690 nm and an increase at 686. The form of chl *a* responsible is called P700 (P for pigment and the wavelength of the absorption change) and is the RC for photosystem I (PSI) (Section 5.3). Only one in 300–400 chlorophyll molecules is in the special form, P700. Difference spectra also show fast change (60 ps) in absorption at 680 nm, separate from, but equivalent to the P700; it is associated with the RC chl *a* of photosystem II (PSII) and is called P680.

In bacterial chromatophores, a small (2%) decrease in absorption at about 870 nm wavelength is caused by illumination or chemical oxidation; the absorption then increases slowly in the dark. By removing the major part of the bchl with detergent, the signal from P870, the bacterial RC, is enhanced. Changes in absorption with illumination or

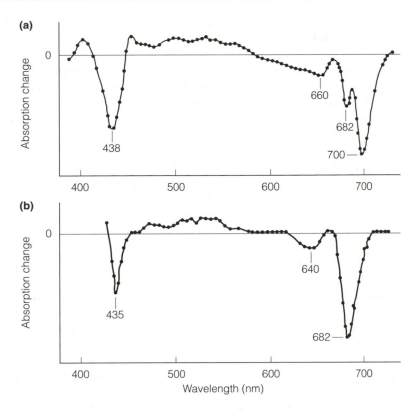

Figure 3.4. Difference spectrum of spinach chloroplasts between light and dark: (a) the change at 700 nm is due to absorption by PSI reaction center chl *a* (after Kok, 1961). (b) The reaction center chl *a* absorption of PSII (P680) (after Döring *et al.*, 1969)

chemical modification show charge separation during transfer of electrons from RC chl *a* (or bchl) to an acceptor, A.

$$\text{chl } a + A \xrightarrow{\text{light}} \text{chl } a^+ + A^- \tag{3.1}$$

P700, P680 and P870 convert light energy to chemical form. This step is the true 'photochemical act' in photosynthesis and thus is unique. The organization and function of higher plant RCs and comparison with bacterial RCs, which are similar to PSI, is discussed in Ort and Yocum (1996). The structure of the higher plant PSII RC is also considered by Satoh (1996) and Noguchi *et al.* (1993). Many of the concepts developed in the earlier work on the RCs have been substantiated by the detailed structural analysis of the bacterial RC (Section 5.1.2). The molecular structure and arrangement of RCs are now known in some detail from X-ray analysis of isolated bacterial and higher plant complexes, although details which are essential for understanding the energetics and efficiency of the mechanism are still lacking, particularly in PSI.

Photo-oxidation in PSI RCs is accompanied by an ESR signal identified as an unpaired electron delocalized over a large π system. The ESR signal and photobleaching of P700 is

1:1, that is, an excitation produces one free radical and ejects one electron from chl *a*. The structure of P700 is, on evidence from ESR and ENDOR, called a 'special pair' of chl *a* molecules joined covalently as a dimer in a specific way, symmetrically arranged in parallel and overlapping on ring I of bchl. A strong exciton coupling results and alteration of the absorption bands, and the energy difference is +0.5 V. This structure resembles the special pair bchl dimer complex (linked to protein and electron acceptors) in photosynthetic purple bacteria. The electron is distributed over the π system of both molecules of the dimer (see Malkin, 1996).

Several models of the PSI RC have been proposed. One model is based on the X-ray structural analysis and on calculations of the molecular orbitals and electron spin densities of the dimer components. It resembles the bacterial RC (*Figure 3.5*). The methyl group (CH3-) and the -C-CH$_3$ = O group on ring I of the bchl are linked to the L subunit of the proteins forming the 'framework' of the RC, and interact with the other bchl of the dimer bchl of the RC on the M protein. There is a strong influence of the electronic state of the atoms in each half of the dimer on the other half. The distant methyl groups on rings III in each bchl affect the spin density of the molecule and cause an imbalance so that the dimer is not symmetrical. Possibly this is necessary to achieve a directional effect, enabling the RC to eject an electron on excitation, as required for the chemical reaction with the acceptor molecule. This could give rise to signals which appear to be from a monomeric chlorophyll. Clearly the orientation of the molecules, their *x–y* transitions and close association (0.4 nm) is determined by the supporting protein. The spin densities of the dimer components will be influenced by the environment as well as in the molecule itself. However, the electronic state transition involves only the bchl dimer and not the proteins. The arrangement of the dimer makes a very extensive and stable but finely tuned state that is possible with extensive π orbitals. In membranes the Q*y* transition of special pair chl *a* is less closely orientated and chl *b* is possibly at angles greater than 35°.

The characteristics of bchl or chl *a* as a special pair dimer would be that the Q*y* transition is shifted to the required longer wavelength absorbing form, the energy levels are correct for sharing the unpaired electron, and it is stable, allowing an excited electron to transfer to an acceptor (with which it forms a radical pair) whilst holding the electron for sufficient time for reaction to occur. Also the cation free radical formed on oxidation should not as reactive as the monomer, thus reducing the chances of back reactions with the acceptor. The type of model suggested for the bacterial RC is possibly applicable to the PSI but not to PSII RCs. There is no generally accepted model of PSI RC structure or function and there is still much to be done in establishing the features of this most important part of the photosynthetic process. The dimer model may not be fully acceptable for the PSI RC, as more refined ESR and ENDOR spectroscopy favors a monomer of chl *a*, which has a more realistic redox potential. The molecular arrangement of the PSII RC, P680, has been characterized by several groups but with differences in interpretation of the structure. The models include two chl molecules, with exciton interaction, which form the RC, but the geometry of the arrangement is not as in the bacterial RC – the chlorin rings are further apart or rotated. This minimizes the interactions. Another suggestion is that P680 is a single chl molecule, in different orientation with respect to the membrane plane compared to the bacterial RC, and with a different mechanism of transferring energy.

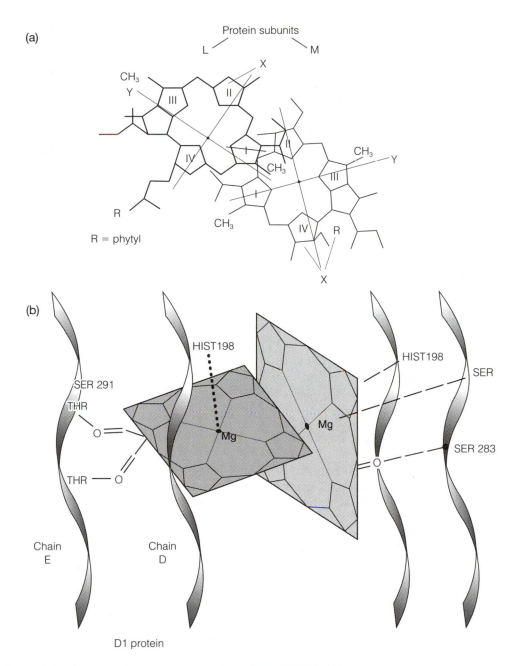

Figure 3.5. Structure of the reaction centers of (a) bacteria with the bchl *a* dimer P865 based on X-ray structure analysis, ENDOR and other techniques and calculated molecular orbitals. The bacteriochlorophylls are arranged in approximate parallel orientation (after Lendzian *et al.* in Michel-Beyerle, 1990). (b) Structure of higher plant photosystem II reaction center, with the chl *a* molecules arranged in close proximity but with the π orbitals at approximately right angles (simplified after Satoh, 1996). This may allow development of the larger energy difference in photosystem II than in the bacterial reaction center

Noguchi *et al.* (1993) suggested that P680 is composed of two molecules of chl, but not oriented in parallel. One (P1) is almost at right angles to the other (P2), thus having very little overlap in energy levels (*Figure 3.5*). P1 is bound in position by serine and threonine residues in the E-loop of the PSII D1 protein and coordination of the Mg with a histidine (198) on the D-loop. The P2 chl is linked to the D2 protein D and E segments. Thus, P680 differs considerably from the bacterial RC, and the arrangement gives only weak exciton coupling and is probably responsible for the very much larger redox potential, although details about how this might be achieved are unclear. The rate of primary charge separation in PSII is not established: the RC itself is probably of the order of 3–25 ps, but in the whole complex it is slower (100 ps), because of the exciton transfer to the RC from the antenna.

Events at the RC are of great importance and interest. The mechanism of the oxidation processes is one of exciton dissociation, with the dissociated negative charge transferred (via electron tunneling) over distances of 2–3 nm, into an acceptor pool. This pool must be reduced at a rate of two orders of magnitude greater than the rate of photon absorption and the back-reactions must be of the same order as the photon capture. These conditions give an efficiency of the PSI RC of about 20% of the solar radiation energy, although losses outside the RC, in the electron transport and related processes give an efficiency of 5%. Such efficiency has been measured for rapidly growing plants under optimal conditions. Detailed understanding of the mechanism may help to improve the efficiency of energy transduction in photochemical processes of many types. Detailed description of the bacterial photosystem complex is given in Chapter 5 together with discussion of the PSII of oxygen-producing organisms.

3.3 The antenna, light harvesting and energy use

The pigment antenna is a mechanism for accumulating quanta in dim light to drive a process requiring several quanta. The effective cross-sectional area of the RC is increased more than 100-fold and, by combining pigments, different wavelengths of light are used. In bright light, excited chlorophyll can 'hold' energy for the next RC containing an electron when acceptors are open. However, there are many factors *in vivo* which are not well understood. Energy movement between closely aggregated chlorophylls is possibly by the 'rapid' (exciton) mechanism in homogeneous transfer (Section 2.2.1). These groups interact with less-organized pigments by the slow Förster mechanism. Fluorescence occurs in 10^{-9} s and the Förster mechanism in 10^{-10} or 10^{-11} s and radiationless relaxation in 10^{-12} s. Therefore, excitation may pass by random walk between many molecules before it reaches an RC, which can accept an electron. Theoretically, many steps (perhaps up to 1000) may occur between antenna and RCs, but heterogeneous transfer directs excitation movement so there are many fewer, perhaps 200–300 steps. This is shown by, for example, loss of fluorescence polarization induced by polarized light. In dim light practically all photons are tunneled to RCs and transfer is about 100% efficient. To achieve this the chlorophyll molecules are arranged at 4–10 nm distances. Reaction centers of PSI trap the excitons from the antenna in 60–70 ps but PSII is slower, 200–500 ps; these rates are derived from fluorescence lifetime measurements. Models of energy transfer between groups of chlorophylls are discussed in relation to the light-harvesting chlorophyll–protein complexes in thylakoids. Formation of a reduced acceptor is the most important way of dissipating chlorophyll excited states and requires a particular structure of the reaction center–acceptor complex, correct molecular orientation and overlap of molecular orbitals in order that net transfer is faster than back-reactions caused by thermal excitation. In photosynthesis excitation drives an electron from the RC over an energy barrier to an acceptor, the pigment acting as catalyst or sensitizer.

3.4 Accessory light-harvesting pigments

Pigments (*Table 3.1*) other than bchl or chl *a* form a major part of the light-gathering antenna. In organisms such as algae, accessory pigments are important as they capture light of the wavelengths not absorbed by chlorophyll and pass the energy to the RC.

3.4.1 Bacteriorhodopsin and bacteriochlorophyll

The simplest form of transduction of light energy, that of *Halobacterium halobium* (see Section 3.2 and Chapter 5) involves only one pigment, bacteriorhodopsin. However, the 'true' photosynthetic bacteria contain a wide range of accessory pigments, some of which are mentioned in *Table 3.1*. Bacteriochlorophyll *a* and bacteriopheophytin (bpheo) absorb light at significantly longer wavelengths than chl *a* and chl *b*. The RC absorbs in the infrared region between 870 and 960 nm with the Soret bands below 400 nm in the UV-A region (*Figure 3.3a*). Such a shift in light absorption is very important to the eco-logical performance of these bacteria as they inhabit oxygen-poor environments, often in the shade of green algae and plants which cannot use the wavelength, thus enabling the bacteria to fill a special niche.

The green bacteria (Chlorobiaceae and Chloroflexaceae) contain bchl *b*, *d* or *a* in addition to bchl *a* and carotenoids. The pigments are often arranged into antennae structures called chlorosomes, which are bound to the cytoplasmic membranes and contain four types of polypeptides. One of these is probably a dimer of molecular mass 3.7 kDa and 10–16 bchl molecules are bound to it. In the Chloroflexaceae bchl *a* is bound to a 5.8 kDa hydrophobic protein which absorbs at 792 nm and donates the energy to the RC.

3.4.2 Phycobiliproteins and phycobilisomes

Phycobiliproteins of the red algae and blue-green algae are composed of a chromophore, a bilin pigment (phycocyanin, phycoerythrin or allophycocyanin) attached to a protein, characteristic of the organism. The chromophores are straight chain tetrapyrroles, related to porphyrins and therefore chlorophyll, but are water-soluble. Phycocyanin and phyco-erythrin absorb mainly at 630 and 550 nm, respectively and allophycocyanin at 650 nm; they have high molecular extinction coefficients. Phycobiliproteins transfer energy to RCs with almost 100% efficiency. The chromophore groups are bound to the proteins with thioether bridges and form a variety of different wavelength-absorbing com-plexes, phycoerythrobilin, phycobilin, phycourobilin and phycobiliviolin. Energy absorbed by the shorter-wavelength phycoerythrin is transferred to the phycocyanin and allophycocyanin, which absorb at progressively longer wavelengths. This enables the energy between 480 and 630 nm to be exploited. The energy is then transferred to chl *a*, probably by a Förster mechanism with efficiency approaching 100%.

In the cyanobacteria and red algae, the phycobilin pigment–protein complexes are organized into phycobilisomes. These are arranged very regularly on the thylakoid membranes and are anchored to PSII particles (*Figure 3.6*). The center of the phycobilisome is a nucleus of allophycocyanin organized in three cylindrical complexes of stacked subunits. Onto this nucleus, radially arranged, are three or four rods, also made up of segments of pigment–protein complexes. Next to the allophycocyanin is phycocyanin and towards the extremity of the rods phycoerythrin or phycoerythrocyanin. The individual segments are joined by linker protein (L) of molecular mass 25–30 kDa. The subunits are α–β heteromers, organized into cyclic trimers, thus producing a hexamer, $(α–β)_6L$. The α subunit of 17 kDa binds one or two chromophores and the β subunit one to four chromophores. The α and β subunits are always in a 1:1 ratio. In addition, large subunits occur. The composition of the phycobilisomes of some algae is strongly influenced by the wavelengths of light they experience during development, a phenomenon called chromatic adaptation. This allows an additional and more effective exploitation of light with ecological benefits for survival and reproduction of the organism.

3.4.3 Carotenoids

Carotenoids, that is the carotenes and xanthophylls, are terpenoids, synthesized by the same synthetic pathway from isoprene units (five carbons), except for the final steps. They have several functions in photosynthetic systems – collecting light energy for photosynthesis, removing high energy states (triplets) of chlorophyll, detoxifying singlet oxygen, removing excitation energy from the thylakoid as heat, and contributing to the structure of thylakoid membranes. The importance of the carotenoids has long been

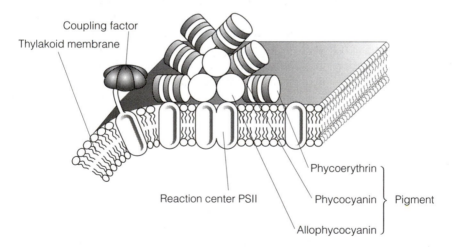

Figure 3.6. The phycobilisome antenna complexes of a red alga on the thylakoid membrane (a) and details of its subunit structure and attachment to the membrane in the cyanobacterium *Mastigiocladus laminosus*, with three central cylinders to which rods of pigment–protein complexes are attached

recognized and their roles are seen as an integral part of the regulation of all aspects of photosynthesis, especially in a world dominated by the very large concentration of oxygen in the atmosphere. The main carotene of leaves is β-carotene and lutein is the principal xanthophyll, although violaxanthin and neoxanthin are also very important. Carotenes are fat-soluble orange pigments (maximum absorption at 530 nm). They are long- (3 nm) chain hydrocarbons of isoprene with alternating double bonds, nine or more in different positions in the different photosynthetic carotenoids. Xanthophylls contain oxygen and also hydroxyl groups and are therefore not hydrocarbons; they differ in the position of the oxygen in the terminal ring structure. The most important function of carotenes in higher plants is not as accessory pigments, for they are only 30–40% efficient in transferring energy to RCs, but in dissipation of the excess energy of chlorophyll and of 'detoxifying' reactive forms of O_2. Carotenoids occur in all photosynthetic organisms that evolve O_2, and are closely associated with the RCs. Plants lacking carotenoids, because of chemical inhibition of synthesis or mutation, are damaged during photosynthesis. Photosynthetic bacteria which live in environments with very little oxygen are damaged if the concentration increases, because the reactive oxygen species formed in the light cannot be removed as the cells lack carotenoids. Oxygen-tolerant photosynthetic bacteria, such as *Rhodopseudomonas sphaeroides*, are destroyed in the light in the presence of oxygen when a mutation prevents production of carotenoids.

In the light, ground-state oxygen reacts with excited-state chlorophyll, giving singlet oxygen which is very reactive and oxidizes (bleaches) chlorophyll, purines in nucleic acids and polyunsaturated fatty acids, which form lipid peroxides. Superoxide ($O_2^{\cdot-}$) with a free radical (an unpaired electron) is produced in photosynthesis; it is both a reducing and an oxidizing agent. Hydrogen peroxide, which is formed when O_2 reacts with electrons from photosynthesis, reacts with $O_2^{\cdot-}$ to form the hydroxyl (OH) radical, which is very reactive indeed and damages most biological material. It is therefore essential to prevent formation of these compounds or destroy them quickly, before the photosynthetic system is damaged. In light, oxygen becomes toxic and photosynthesis self-destructive! Peroxide and $O_2^{\cdot-}$ are destroyed by the enzymes catalase and superoxide dismutase (SOD), respectively. Carotenoids quench excited chlorophyll, preventing it from reacting with O_2, thus decreasing the probability of formation of toxic oxygen species. Carotenoids also dissipate $O_2^{\cdot-}$, thus providing an important and efficient protection. It is a measure of the importance of detoxification mechanisms for preventing the formation of active oxygen species that several different systems operate within photosynthesis. Carotenoids prevent the formation of active oxygen by taking the energy from the chlorophyll triplet state by energy migration. This requires correct differences in energy and electronic configurations. The carotenoid and chlorophyll molecules must be at the required distance and in the proper orientation in order to form the carotenoid excited triplet state. For the triplet energy migration the distance between carotenoid and chlorophyll must be very small, one order of magnitude smaller than the Förster distance, so close association between pigments is essential for π electron systems to overlap. The triplet excited state is then dissipated by radiationless decay (*Figure 3.7a*), transferring energy to the medium as heat.

Carotenoids are an integral part of the photosystems. Separation of light-harvesting complexes (LHCII, for example) by nondenaturing electrophoresis shows that two lutein

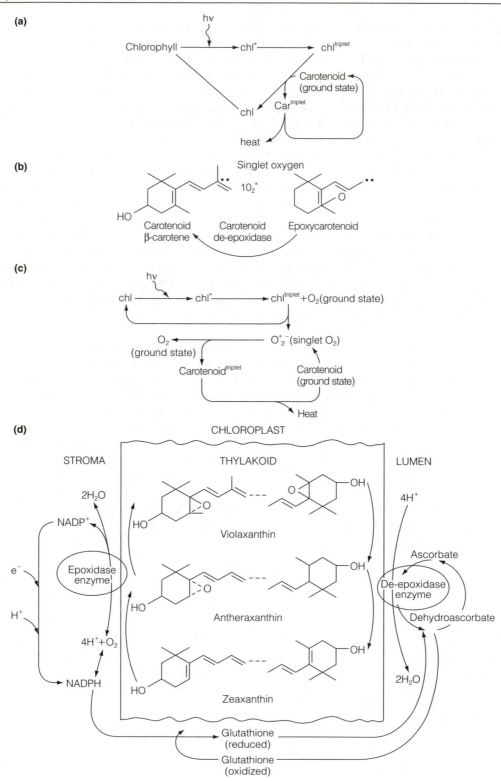

(a)

(b)

(c)

(d)

molecules are arranged as a cross in the middle of the structure. Violaxanthin forms a large proportion of the free pigment and is probably only loosely bound to the outside of LHCII, where it probably plays a role in controlling the organization of the complex within the lipid membrane. Also, in this position violaxanthin would be able to enter into the reactions of the xanthophyll (violaxanthin) cycle.

Excited O_2 is destroyed in an epoxidation reaction involving the ring structure of the carotenoids and an enzyme. Singlet O_2 is also removed, via the carotenoid triplet state, to give ground-state oxygen. There are differences between carotenoids in their efficiency: e.g. *al trans*-lycopene quenches (destroys) singlet O_2 faster than β-carotene. Carotenoids also absorb 380–520 nm radiation, which is rather energetic and may damage biological systems; they offer protection against a wide range of damaging cellular products not only in photosynthesis but in general metabolism in plants and animals. Carotenoids themselves are destroyed only with excessive energy load. The structure with nine or more double bonds, is essential for efficient energy dissipation, seven double bonds or less being ineffective.

3.4.4 The xanthophyll cycle

Xanthophylls are involved in a number of energy and reductant regulatory processes in plants and are closely associated with photosynthesis. It has long been known that when algae and higher plant leaves are illuminated, violaxanthin, a di-epoxide xanthophyll (5,6,5',6'-diepoxizeaxanthin), is converted (in an enzyme-catalyzed de-epoxidation reaction) via antheraxanthin (5,6-monoepoxizeaxanthin) to zeaxanthin (no epoxide group). In dim light or darkness the zeaxanthin is converted back (by an enzymatic epoxidation) to violaxanthin; ATP or an energy-rich state is required for this. There are a number of equivalent mechanisms in algae, for example a cycle based on diadinoxanthin and diatoxanthin operates in *Euglena*. The cycle does not occur in photosynthetic bacteria, suggesting that it is related to protection from photochemical damage. Considerable attention has been directed to understanding the cycle and its role in plants, since it has practical importance. Manipulation of the amounts of the cycle components and the activity may provide a way of protecting the photosynthetic system from damage in adverse environments, for example under very bright light, particularly during water or cold stress, which prevent the normal use of the energy captured by chlorophyll (see CO_2 assimilation, Section 12.4). The cycle is summarized in *Figure 3.7d*. In darkness the violaxanthin is present in the thylakoid membrane, with very little or no antheraxanthin or zeaxanthin. There is no change in pH across the thylakoid membrane, and the electron transport chain and electron acceptors (ferredoxin, pyridine nucleotides) are oxidized. When the thylakoids are illuminated, electron transport occurs, intermediates of the electron transport chain and its acceptors become reduced, and the change in pH increases. The lumen becomes acidic and the stroma alkaline. In weak light the energy absorbed by chlorophyll is used efficiently to make ATP and NADPH, which are consumed in

Figure 3.7. (Opposite) Mechanisms of energy regulation in photosynthesis by carotenoids. (a) The direct reaction with excited chlorophyll; (b) the reaction with excited states of oxygen by epoxidation; (c) reaction of carotenoids with excited states of oxygen; (d) reaction with NADPH and O_2 in the xanthophyll cycle

biochemical reactions. Any excess energy and triplet states or active oxygen species are removed by the carotenoids associated with the RC and photosystems. However, when the number of photons captured rises and the electron transport rate increases, the capacity of the reactions for making ATP and NADPH or for using them is exceeded: the rate of photosynthetic CO_2 assimilation, for example, becomes saturated. Under these conditions the change in pH becomes very large and the transfer of electrons to oxygen increases greatly. Thus, the potential for damage is considerable. Under these circumstances the violaxanthin cycle operates to dissipate energy and reluctant. As *Figure* 3.7 shows, the epoxide groups of violaxanthin are removed by the de-epoxidase located in the thylakoid lumen, converting it to zeaxanthin, via antheraxanthin. The enzyme has a molecular mass of 54 kDa and an optimum pH of 5.2; it is inactive at pH above 7. Thus, it is active in the light when the lumen pH falls (see Chapter 5). The de-epoxidation requires reductant.

Reductant is provided by ascorbate as well as reduced glutathione (see Section 5.4.3) and both will increase concentration in the lumen. The epoxidation occurs at the stromal side of the membrane; the reaction is catalyzed by a mixed-function oxygenase which works optimally at pH 7.5, the condition of the illuminated stroma. The reaction consumes NADPH, producing water and making $NADP^+$ available to accept more electrons. As the two enzymes are located in the two compartments and have different pH optima, maximum activity of the xanthophyll cycle depends on the light-induced proton gradient across the thylakoid. Under these conditions ATP is synthesized. It is likely that a major function of the xanthophyll cycle is to regulate the ratio of NADPH to ATP, reducing the possibility of photo-damage and optimizing conditions for membrane function and chloroplast biochemistry. Importantly, chloroplast envelopes are rich in violaxanthin in darkness and zeaxanthin in the light and may regulate chloroplast stromal conditions relative to the cell cytosol. Probably zeaxanthin also regulates the dissipation of excess excitation in the antenna chl of the photosystems, as indicated by the blockage of de-epoxidation by thiol compounds such as dithiothreitol, which inhibit a large part of the 'high energy state quenching' of chlorophyll fluorescence by radiationless dissipation in the antenna. Probably zeaxanthin interacts with other mechanisms of regulation. Carotenoids are thus an important 'safety valve' dissipating the excess energy of the excited pigments and the products of the reaction with O_2. This limits the danger of photodestruction of tissues in intense light, when O_2 concentration is high and the normal acceptor of electrons is deficient.

3.5 Action spectra and two light reactions

The absorption spectrum is the number of photons captured by photosynthetic pigments as a function of wavelength and the action spectrum is the rate of photosynthesis (e.g. O_2 evolution or CO_2 absorption) or other response resulting from the capture. An action spectrum of O_2 release by algal cells is given in *Figure 3.8*. Action spectra show the efficiency of energy use. From the Stark–Einstein law of equivalence the rate of a reaction is proportional to the number of photons absorbed, so that efficiency may be determined from an action spectrum. However, if the action spectrum is the product of several reactions, as in photosynthetic CO_2 assimilation or O_2 release, the action spectrum of the

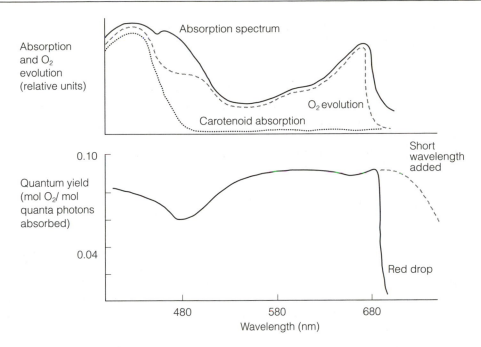

Figure 3.8. Absorption and action spectra for O_2 evolution by the alga *Chlorella*. Red light is ineffective in O_2 evolution ('red drop'). Supplementation of red light (>690 nm) with shorter wavelength light (<690 nm) increases the O_2 evolution (−−−). In the blue region, where carotenoids absorb, the efficiency of O_2 production is poor

basic reactions may be obscured. Ideally a process directly linked to energy consumption is measured, independent of other processes. For example, measurement of photosynthetic action spectra by O_2 or CO_2 exchange of whole leaves, algae, *et cetera*, is complicated by respiration and photorespiration, which must be assessed in deriving the true action spectrum. Also the response at different wavelengths can only be compared when light intensity is limiting (i.e. the process is linearly dependent upon the amount of light captured) not at saturating light intensity where other factors limit. If all pigments captured energy and delivered it to RC chlorophyll with equal efficiency, and each wavelength was equally effective in promoting photosynthesis, then the chemical process would be the same at all wavelengths, and the action and absorption spectra would have the same shape. However, several pigments may capture energy with different efficiencies at different wavelengths, so that absorption and action spectra do not match. Thus, in *Figure 3.8* there is close agreement between the action and absorption spectra for chloroplasts between 570 and 680 nm, but in the blue region where carotenoids absorb, less O_2 is produced per unit of light absorbed, as energy transfer from carotenoids to chlorophyll is inefficient. Algae evolve O_2 at different wavelengths of light (*Figure 3.8*) with an almost constant quantum yield (mol O_2 per mol photons absorbed) up to about 650–690 nm, at which point the rate decreases precipitately despite absorption of photons by chl *a*. This phenomenon is known as the 'red drop'. Adding short wavelengths together with red light greatly enhances O_2 evolution compared with the two wavelengths given separately (*Figure 3.8*); this is the enhancement or Emerson effect. This effect shows that there is cooperation between two pigment systems, called photosystems (abbreviated PS), one absorbing short, the other long wavelengths. The two photosystems are arranged in series, the shorter-wave absorbing photosystem (PSII) preceding the longer (PSI). The order of wavelengths is important. Where the 'red drop' occurs only PSI operates and the PSII does not. The long-wavelength absorbing PSI 'primes' the system and the shorter-wavelength absorbing PSII delivers energy to it, with the photosystems acting in series. Identification of the two photosystems was a major achievement and forms the basis of much of our understanding of the biophysics of photosynthesis. Before discussing the characteristics of PSI and II in Chapter 5, evidence for groups of chlorophylls functioning as semi-discrete, energy-gathering units, called oxygenic photosynthetic units, will be considered.

The concept of photosynthetic units developed from studies of oxygen production. Algae (*Chlorella*) were illuminated with brief (10^{-5} s) flashes of intense light to saturate photosynthesis. This permitted only a single photon capture per chlorophyll molecule ('single turnover flash') and prevented the energy from being used chemically during the flash. A dark period of about 40 ms at room temperature was required for maximum O_2 production per flash. Thus, photosynthesis was divisible into light and dark reactions which together gave a maximum efficiency of light utilization. In the cold a longer (0.4 s) dark period was needed for maximum O_2 evolution due to slower chemical processes. However, the maximum O_2 yield per flash was the same despite the different rates. When O_2 evolution was measured following flashes of different energy, separated by a dark period greater than 40 ms (at 25°C), O_2 per flash increased to a maximum of one O_2 molecule for about 2400 molecules of chlorophyll activated. The ratio of O_2 evolved to photons absorbed (the quantum yield) in dim light was about 1 O_2 molecule per 8 or 10 photons absorbed, but at saturating light 2400 chlorophylls were excited to produce one O_2. Therefore, to evolve one O_2, 2400 chlorophyll molecules cooperated to collect eight photons, *ca* 300 chlorophylls per photon, and supply the energy to the chemical processes. The group of chlorophylls, reaction center and other pigments have been called a 'photosynthetic unit', although since the complexity of the system has been realized and the fact that the different photosystems occur in variable ratios, the term 'unit' is inappropriate. However, the linkage between photosystems implied by the term must still be considered. The eight quanta are captured by the two photosystems, so the energy of four photons per photosystem must be transferred to each reaction center for 1 O_2 molecule, 4 e^- and 4 H^+ to be evolved and liberated from 2 H_2O molecules. A series coupling of two photosystems is suggested; each is energized twice, first by a short wave photon (PSII) and again by a longer wave photon (PSI). Some antenna chlorophylls gather photons and transfer the excitation to the reaction center, increasing the effective area per chlorophyll and allowing quanta to be absorbed in very dim light. This explains the gush of O_2 in experiments where the chance of capturing a photon was small; the eight quanta must first be 'accumulated' before electrons can be removed from water. However, energy gathered in one photosynthetic unit can be passed to other units or photosystems ('spillover') under normal conditions, so that units are not functionally distinct. The photosynthetic unit corresponds loosely to the chlorophyll–protein complexes in thylakoid membranes (Section 4.5) associated with an individual photosystem. A thylakoid disk 500 nm in diameter may contain 10^5 chlorophylls associated with some 200 electron transport chains, each with one PSI and one PSII unit – 250 chlorophylls per photosystem. Chlorophyll *b* passes excitation energy on to chl *a* and then to P680 (PSII) or P700 (PSI) RCs. However, present concepts of photosynthetic organization are more dynamic; the size of the antenna and number of reaction centers and the proportion of PSI and II per chloroplast thylakoid are not fixed, but vary with species and conditions. Also, the photosynthetic unit is not the functional unit for all thylakoid processes, for example, ATP synthesis is related to the whole thylakoid. Quantum yield (also called quantum efficiency) measures the ability of photons to produce chemical change and is the number of O_2 molecules evolved (or CO_2 fixed) per quantum of light absorbed. Its reciprocal is the quantum requirement, that is, the number of quanta needed per O_2 produced or CO_2 consumed. The quantum yield of photosynthesis has been controversial: Warburg and associates claimed 1 O_2 evolved per 4 quanta absorbed but from 8 to 10 quanta is now accepted. The value of quantum yield is very important as it provides a basis for understanding the energetics of photosynthesis and interpreting the mechanism. To reduce one CO_2 requires 460 kJ mol^{-1} of quantum energy. Three quanta of red light (680 nm, 174 kJ mol^{-1}) would suffice if efficiency of capture and conversion approached 90%, which is thermodynamically impossible. Even four quanta of red light would require 66% efficiency at minimum. Blue light, although more energetic, is no more effective than red light. Four quanta per CO_2 fixed or O_2 evolved conflicts with experiment and the concept of two light reactions. It is now accepted that eight photons are needed per O_2 as a theoretical minimum and more may be required, depending on conditions. The energy available from eight red photons is $8 \times 2.9 \times 10^{-10}$ J or 1.4 MJ mol^{-1}. Energy fixed in carbohydrates is 470 kJ mol^{-1}, so the efficiency is $(0.47 \times 10^6)/(1.4 \times 10^6) \times 100$ or 34%; ATP formulation increases efficiency to 38%, a good efficiency for many energy conversions. However, under intense natural illumination with only about 50% of radiation as PAR (400–700 nm), overall efficiency is less than 5%; algae in dim light achieve 10% efficiency.

References and further reading

Bryant, D.A. (ed.) (1994) *The Molecular Biology of Cyanobacteria*. Kluwer, Dordrecht.

Demmig-Adams, B. and Adams, W.W. III (1996) The role of xanthophyll cycle carotenoids in the protection of photosynthesis. *Trends Plant Sci.* **1**: 21–26.

Döring, G., Renger, G., Vater, J. and Witt, H.I. (1969) Properties of the photoactive chlorophyll in photosynthesis. *Zeutschrift für Naturforschung B24b*: 1139–1143.

Dörr, F. (1983) Photophysics and photochemistry. General principles. In *Biophysics* (eds W. Hoppe, W. Lehmann, H. Markl and H. Ziegler). Springer, Berlin, pp. 265–288.

Eskling, M., Arvidsson, P.-O. and Akerlund, H.-E. (1997) The xanthophyll cycle, its regulation and components. *Physiol. Plant.* **100**: 806–816.

Frank, H.A. and Cogdell, R.J. (1996) Carotenoids in photosynthesis. *Photochem. Photobiol.* **63**: 257–264.

Gilmore, A.M. (1997) Mechanistic aspects of xanthophyll cycle-dependent photoprotection in higher plant chloroplasts and leaves. *Physiol. Plant.* **99**: 197–209.

Green, B.R. and Durnford, D.G. (1996) The chlorophyll–carotenoid proteins of oxygenic photosynthesis. *Annu. Rev. Plant Physiol. Mol. Biol.* **47**: 685–714.

Havaux, M. (1998) Carotenoids as membrane stabilizers in chloroplasts. *Trends Plant Sci.* **3**: 147–151.

Holzwatth, A.R. (1991) Excited state kinetics in chlorophyll systems and its relationship to the functional organisation of the photosystem. In *The Chlorophylls* (ed. H. Scheer). CRC Handbooks, Boca Raton, FL.

Horton, P. , Ruban, AN. and Walters, R.G. (1996) Regulation of light harvesting in green plants. *Annu. Rev. Plant Physiol. Plant Mol. Biol.* **47**: 655–684.

Karrasch, S., Bullough, P. A. and Ghosh, R. (1995) 8.5 Å projection map of the light harvesting complex I from *Rhosospirillum rubrum* reveals a ring composed of 16 subunits. *EMBO J.* **14**: 631–638.

Kok, B. (1961) Partial purification and determination of oxidation–reduction potential of the photosynthetic chlorophyll complex absorbing at 700 nm. *Biochim Biophys Acta* **48**: 527.

Malkin, R. (1996) Photosystem I electron transfer reactions – components and kinetics. In *Oxygenic photosynthesis: the Light Reactions* (eds D.R. Ort and C.F. Yocum). Kluwer Academic, Dordrecht, pp. 313–332.

Michel-Beyerle, M.-E. (ed.) (1990) *Reaction Centers of Photosynthetic Bacteria*. Springer Series in Biophysics, Vol. 6. Springer, Berlin.

Nechushtai, R., Eden, A., Cohen, Y. and Klein, J. (1996) Introduction to Photosystem I: reaction centre function, composition and structure. In *Oxygenic Photosynthesis: the Light Reactions* (eds D.R. Ort and C.F. Yocum). Kluwer Academic, Dordrecht, pp. 289–311.

Noguchi, T., Inoue, Y. and Satoh, K. (1993) FTIR studies on the triplet state of ~P680 in the Photosystem II reaction center: triplet equilibrium within a chlorophyll dimer. *Biochemistry* **32**: 7168–7195.

Ort, D.R. and Yocum, C.F. (1996) Electron transfer and energy transduction in photosynthesis: an overview. In *Oxygenic Photosynthesis: the Light Reactions* (eds D.R. Ort and C.F. Yocum). Kluwer Academic, Dordrecht, pp. 1–9.

Satoh, K. (1996) Introduction to the photosystem II reaction center – isolation and biochemical characterization. In *Oxygenic Photosynthesis: the Light Reactions* (eds D.R. Ort and C.F. Yocum). Kluwer Academic, Dordrecht, pp. 193–211.

Scheer, H. (1991) *The Chlorophylls*. CRC Handbooks, Boca Raton, FL.

Truscott, T.G. (1990) The photophysics and photochemistry of the carotenoids, Y. *Photochem. Photobiol. B: Biol.* **6**: 359–371.

Vermaas, W.F.J. (1994) Evolution of heliobacteria: implications for photosynthetic reaction center complexes. *Photosynthesis Res.* **41**: 285–294.

Architecture of the photosynthetic apparatus

4.1 Introduction

Structural components of the photosynthetic mechanism form a hierarchy of organization in which components of different size and complexity cooperate. The structures developed in photosynthetic organisms, from bacteria to higher plants, for capturing and converting light energy to chemical form are similar at the molecular and supramolecular levels of organization (Chapter 1). In all photosynthetic systems, a membrane containing the light-capturing and energy-conversion components separates the inside of the cell from what is, in terms of the developmental morphology of the photosynthetic membrane, effectively the outside of the cell. This reflects the common early evolutionary history of plants, upon which later evolution has built. In eukaryotes, and particularly vascular plants, the hierarchy may be simplified into:

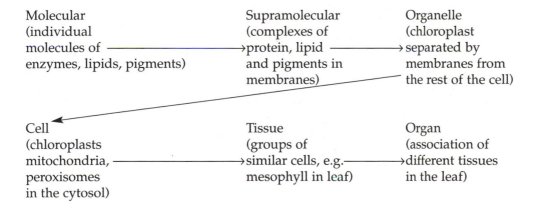

Molecular (individual molecules of enzymes, lipids, pigments) → Supramolecular (complexes of protein, lipid and pigments in membranes) → Organelle (chloroplast separated by membranes from the rest of the cell)

Cell (chloroplasts mitochondria, peroxisomes in the cytosol) → Tissue (groups of similar cells, e.g. mesophyll in leaf) → Organ (association of different tissues in the leaf)

4.2 Photosynthetic prokaryotes

This diverse group differs in structure, as well as pigment composition (Chapter 3), and in the source of reductant for CO_2 assimilation.

4.2.1 Basic photosynthetic structure

The basic structure may be regarded as that of *Halobacterium* (see *Figure 1.2*), where a closed lipid membrane separates an internal space from the environment. In many photosynthetic bacteria, extensive invagination of the external cell membrane into the cell's interior has produced, during evolution, a large surface area for light capture, H^+ diffusion and ATP synthesis. These invaginations have probably given rise to the closed thylakoid vesicles of current plant chloroplasts by sealing of the membranes, thus cutting off the internal space from the rest of the cell and the external environment. Compartmentation of the chloroplast and nucleus from the rest of the cell is a major distinction between eukaryotes and prokaryotes. Further differences at tissue and organ level are very apparent in algae, which span an enormous range of size and complexity. Some are single cells, micrometers in size, others multicellular fronds, meters long, and differentiated into organs. Vascular plants are all multicellular, of varying complexity, and varying greatly in size from millimeters to many meters. Progression of organization from simple membranes to organelles and cells is marked by great variation in structure, but with underlying similarities, for example between the light harvesting and RC components of the photosynthetic complexes, thylakoids, *et cetera*.

4.2.2 Photosynthetic bacteria

The photosynthetic bacteria, as a group, have photosynthetic membranes of greatly different extent and form. In the purple sulphur bacteria (e.g. species of *Rhodopseudomonas* and *Rhodospirillum*) they range from simple, flat membranes to complex branched and lobed structures but all are derived from, and continuous with, the cytoplasmic membrane. Green sulfur bacteria (e.g. *Chlorobium* species) have extensive flattened sacks or vesicles (called chlorosomes) which, in *Chlorobium limicola*, contain up to 10 000 bchl *c* molecules each. The chlorosome's membranes are 5 nm thick, and contain up to 50% protein in the form of granular patches and 5% pigments with most pigment molecules densely packed in arrays within the chlorosome. The chlorosome is attached to, and also transfers excitation energy to, the reaction center complex via an antenna of bchl *a* and protein associated with the reaction center in the photosynthetic membrane. The cyanobacteria, although evolving oxygen, lack membrane-bound organelles but have extensive double thylakoids arranged in parallel on which the pigment–protein antenna complexes, the phycobilisomes, are regularly distributed. These are described in Section 3.4.2. The RCs of the phycobilisomes are very similar to those of the purple bacteria and the higher plant. Clearly the great similarities between different photosynthetic organisms reflect evolution from a single type of basic mechanism with parallel evolution in different groups.

4.3 Photosynthetic eukaryotes

Algae and higher plants are a large, diverse group differing in structure and pigment composition (Chapter 3), but all have thylakoid membranes enclosed in an envelope forming the chloroplast and thus effectively separated from other parts of the cell. Structure of thylakoids and chloroplasts in green algae resembles that of vascular plants although the range of forms is greater, with discoid chloroplasts in some diatoms (*Melosira*) and brown algae (*Zonaria*). Such chloroplasts may be very thin, extremely numerous and closely packed (e.g. in the siphonous algae *Codium* and *Caulerpa*). The chloroplasts may be stellate, e.g. in *Ectocarpus* and *Zygnema*; the latter has extremely branched chloroplasts with a very large surface to volume ratio. In contrast, there is a single cup-shaped chloroplasts in *Chlamydomonas*. The algae often form very complex thalli with extensive laminae for light capture, so resembling vascular plants in size and complexity but adapted to very different environmental conditions. The reasons for such diversity in structure at all scales are not further considered here (see Anderson, 1999).

4.3.1   Leaf structure

Leaves are the site of most higher plant photosynthesis. Other organs (e.g. stems, petioles and cladodes) contain or may form functional photosynthetic systems, probably very similar to those in leaves, which may be of considerable significance to the plants assimilate supply under particular ecological conditions. The structure of leaves provides the necessary conditions to maintain the metabolic functions, including photosynthesis; for example, epidermis and cuticle minimize water loss. Leaf structure, shape and cell distribution are genetically determined but change, within limits, with growth conditions, allowing adjustment to the environment. A semi-quantitative estimate of the number and size of component parts of an 'average' higher plant leaf (*Table 4.1*) is based on data for spinach, tobacco and wheat. The mesophyll tissue is differentiated in many species (*Figure 4.1a*) into palisade and spongy mesophyll cells, both with extensive air passages contiguous with the substomatal cavities. Rate of exchange of gases between cells and the atmosphere is related to their surface area and to the geometry of the intercellular spaces of the mesophyll. Capture of light also depends on the mesophyll structure, for example, the number and density of cell layers. In some plants, domed epidermal cell surfaces act as lenses, focussing light into the mesophyll. Photosynthetic processes appear to be the same in palisade and spongy mesophyll cells of C3 plants, but in C4 plants (Chapter 9) the mesophyll is clearly differentiated into morphologically distinct tissues with different photosynthetic functions. Of course, in leaves there is extensive nonphotosynthetic tissue, such as the xylem and phloem transport pathways for water and parenchyma which may store assimilates. These have functions, and metabolism, which are very different to the photosynthetic cells. There is clear evidence that the light conditions within a leaf are not uniform, for example gradients in intensity and spectrum can be detected with very fine fiber-optic probes inserted to different positions. So it is important to bear in mind that leaves are heterogeneous when considering their behavior.

4.3.2   Cell structure

Individual photosynthetic cells in algae and most higher plants perform all the primary, and many secondary, reactions of photosynthesis, requiring only water and nutrients. Mesophyll cells may be isolated from leaves by enzymes, which break the bonds holding together the walls of the adjacent cells. Protoplasts are freed from the cell wall by cellulases – enzymes which lyse bonds in the cellulose matrix of the wall. Cells and protoplasts will photosynthesize actively when given CO_2, light and nutrients, if the

Table 4.1. A semi-quantitative analysis of the photosynthetic system in an 'average' C3 plant leaf

Leaf characteristics

Area of leaf (extremely variable: consider a unit area)	1 m^2
Thickness of leaf	3 × 10^{-4} m
Volume of (1 m^2) leaf	3 × 10^{-4} m^3
Fresh mass of (1 m^2) leaf	0.17 kg
Dry mass of (1 m^2) leaf	0.04 kg
Volume of cells (assuming density of dry matter = 1.2 × 10^3 kg m^{-3})	2.2 × 10^{-4} m^3
Air space in leaf = total − cell volume	0.8 × 10^{-4} m^3
As percentage of total volume	27%

Leaf cell number per m^2 leaf

Leaf has: one layer of palisade mesophyll cells in transverse section and in parallel section contains	4 × 10^9 cells
five layers of spongy mesophyll cells in transverse section and in parallel section each layer contains	0.6 × 10^9 cells
Total spongy mesophyll cell number	3 × 10^9 cells
Total mesophyll cell number	7 × 10^9 cells

Cell dimensions

Palisade cell	
'average' diameter	2 × 10^{-5} m
'average' length	8 × 10^{-5} m
volume (for cylinder with hemispherical ends)	2.9 × 10^{-14} m^3
total volume of all palisade cells	1.2 × 10^{-4} m^3
Spongy mesophyll cell	
'average' diameter of cell	3 × 10^{-5} m^3
volume of cell	1.4 × 10^{-14} m^3
total volume of all spongy mesophyll cells	4.2 × 10^{-5} m^3
Total volume of mesophyll cells	1.6 × 10^{-4} m^3

Chloroplasts in cells

Assume that each palisade cell contains 100 chloroplasts:	
total chloroplasts m^{-2}	4.0 × 10^{11}
Assume each spongy mesophyll cell contains 50 chloroplasts:	
total chloroplasts m^{-2}	1.5 × 10^{11}
Total number of chloroplasts m^{-2} leaf	5.5 × 10^{11}

Chloroplast

A chloroplast approximates a hemisphere of diameter	5 × 10^{-6} m
Volume of one chloroplast	3.3 × 10^{-17} m^3
Total chloroplast volume in 1 m^2 of leaf	1.8 × 10^{-5} m^3
Of cell volume:	
chloroplasts occupy	8%
vacuoles occupy	80%
cytoplasmic volume	4.8 × 10^{-5} m^3
Of cytoplasmic volume:	
chloroplasts occupy	38%
Chloroplast envelope area chloroplast^{-1}	5.9 × 10^{-11} m^2
Chloroplast envelope area m^{-2} leaf	32 m^2
Chloroplast envelope area mg chlorophyll^{-1}	0.064 m^2
Total volume of chloroplast stroma	1.2 × 10^{-5} m^3

Table 4.1. *Continued*

Thylakoid system

A typical chloroplast contains 60 grana stacks and each has 15 thylakoids/granum, i.e. 30 membranes. Diameter of 1 'end' of granum is	4.5×10^{-7} m
Area of vesicle (not allowing for ends)	1.6×10^{-13} m²
Total area of membranes in one stack	5.0×10^{-12} m²
Total area of grana membranes in one chloroplast	3×10^{-10} m²
If four stromal thylakoids the width of the granum pass across the diameter of chloroplast the stromal thylakoid area is	2.4×10^{-10} m²
Total thylakoid area, grana + stroma per chloroplast	5.4×10^{-10} m²
Total thylakoid area m⁻² leaf	300 m² (other estimates 835 m²)
Thickness of thylakoid membranes (approx.)	7.5×10^{-9} m
Volume of thylakoid membranes	4.1×10^{-18} m³
Thylakoid dimensions:	
area of end surface of a granum	1.6×10^{-13} m²
space between membranes is approx.	8.0×10^{-9} m
volume of individual vesicle lumen	1.3×10^{-21} m³
volume of lumen in one stack	1.9×10^{-21} m³
volume of grana lumen in a chloroplast	1.2×10^{-18} m³
volume of stromal thylakoids in a chloroplast	9.0×10^{-19} m³
total volume of thylakoid lumen in a chloroplast	2.1×10^{-18} m³
total volume of thylakoid lumen m⁻² leaf	1.2×10^{-6} m³
total volume of membrane + lumen in a chloroplast	6.2×10^{-18} m³
of chloroplast volume lumen occupies	20%

Chlorophyll content of leaves

Average content of chlorophyll m⁻²	0.5 g
Molecular mass of chlorophyll *a*	894
Chlorophyll content	5.6×10^{-4} mol m⁻²
or	1.9 mol m⁻³ leaf
Chlorophyll/chloroplast	9×10^{-13} g
or	1.0×10^{-15} mol
Volume of thylakoid lumen (g chlorophyll)⁻¹	2.3×10^{-6} m³ g⁻¹
Volume of chloroplast (g chlorophyll)⁻¹	3.6×10^{-5} m³ g⁻¹
or volume (mol chlorophyll)⁻¹	0.033 m³ mol⁻¹
Volume of thylakoid membrane (g chlorophyll)⁻¹	4.6×10^{-6} m³ g⁻¹
Area thylakoid membrane (g chlorophyll)⁻¹	600 m² g⁻¹ (other estimates 1670 m² g⁻¹)
Concentration of chlorophyll in thylakoid	2.2×10^{5} g m⁻³
or	2.4×10^{2} mol m⁻³ (0.24 M)
No. chlorophyll molecules chloroplast⁻¹	6.7×10^{8}
Area membrane (chlorophyll molecule)⁻¹	1.2×10^{-18} m²
Area per 'head' of chlorophyll molecule	2.2×10^{-18} m²

Miscellaneous values

Chlorophyll molecules per photosystem I and II	300
Concentration of NADP reductase in stroma	8×10^{-5} M
Coupling factor (CF$_1$)	0.45 g (g chlorophyll)⁻¹
Molecular mass of CF$_1$	325 kDa

Table 4.1. *Continued*

CF$_1$ granum^{-1}	200
ATP content illuminated chloroplasts (nmol ATP mg chlorophyll^{-1})	40
ADP content (nmol ATP mg chlorophyll^{-1})	12
ATP concentration	1.5 mM
ADP concentration	0.5 mM
ATP/ADP ratio	3
NADPH concentration	0.1 mM
RuBP carboxylase/oxygenase concentration of enzyme sites	4 mM
Electron transport chains per thylakoid 'disk'	200
Dry mass per chloroplast	20×10^{-12} g
Volume of water to total volume	75%
Protein content (percentage of dry mass of chloroplast)	60%
Lipid content (percentage of dry mass of chloroplast)	20%
Chlorophyll (percentage of dry mass of chloroplast)	4%
Carotenoids (percentage of dry mass of chloroplast)	0.9%
Nucleic acids (percentage of dry mass of chloroplast)	2.5%
Soluble products of photosynthesis (percentage of dry mass chloroplast)	7.5%
Mg concentration in chloroplast stroma (light)	30 mM
Mg concentration in chloroplast stroma (dark)	15 mM
Inorganic phosphate in chloroplasts (mol P$_i$ mg chlorophyll^{-1})	3
Inorganic phosphate concentration	100 mM
RuBP concentration	0.1–2 mM

osmotic potential of the medium is correct. Cells and protoplasts from C3 plants function relatively independently, although metabolites may be exchanged between cells and tissues. In C4 plants with considerable morphological and functional differentiation between photosynthetic tissues within a leaf cooperation has evolved between the different types of photosynthetic cells and their autonomy has been largely lost. However, by providing metabolites (e.g. organic acids) thought to be transferred between tissue types (see Chapter 9) to isolated cells of different types, they may continue to function. It is also possible to reconstitute their activities, providing an important test of the concepts of how C4 photosynthesis operates.

The ultrastructure of photosynthetic cells is shown in *Figure 4.1b*. In the thin layer of cytoplasm around the large central vacuole are chloroplasts, mitochondria and peroxisomes, which are all involved in aspects of photosynthetic metabolism, together with inclusions such as pigment granules and starch grains. The number of chloroplasts per cell is very variable, approximately 20–100 in higher plants. Chloroplast distribution in cells differs with species and changes with conditions, such as illumination. C3 plants have typically flattened spherical or lens-shaped chloroplasts some 3–10 μm in greatest dimension. Mitochondria are prominent, with characteristic membranes, the inner often folded into cristae. Peroxisomes have a single limiting membrane and dense granular contents (largely protein), without distinctive features in electron micrographs, and are often closely associated with chloroplasts. The number of mitochondria and peroxisomes varies and may depend on the function of the cell and on the environment during growth. Mitochondria provide ATP and organic acids to the cytosol and chloroplasts,

(a)

(b)

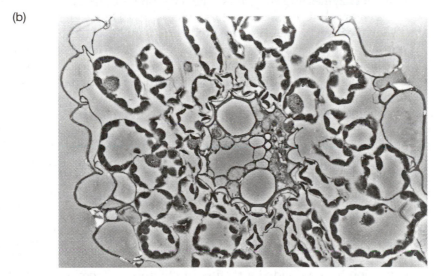

Figure 4.1. (a) Scanning electron micrograph of a transverse section of a bean (*Vicia faba* L.) leaf ($\times$ 216). Electron micrographs courtesy of IACR-Rothamsted, Harpenden, Herts, UK (b) Transverse section ($\times$ 400) of wheat (*Triticum aestivum* L.) leaf showing the loosely packed mesophyll cells arranged around a central vascular bundle. The cells contain chloroplasts lining the cell walls.

depending on conditions within the cell. Peroxisomes metabolize products from the chloroplast (e.g. glycollate) in cooperation with mitochondria, for example in photo-respiration (Chapter 8).

4.4 Chloroplasts

Chloroplasts perform all the primary (e.g. light capture and electron transport leading to NADPH and ATP synthesis) and most of the secondary processes (e.g. synthesis of 3-carbon phosphorylated compounds from CO_2) of photosynthesis. They also synthesize many proteins and other components. Because of the importance, their structure will be considered in detail. Chloroplasts are bounded by a continuous envelope and contain all the membrane-bound light-harvesting chlorophyll and other pigments, proteins and redox compounds involved in transport of electrons and synthesis of ATP. They also con-tain the soluble enzymes and substrates required for CO_2, NO_3^- and SO_4^{2-} assimilation by photosynthesis together with its products. Chloroplasts may be isolated by breaking the plasmalemma of protoplasts, released by breaking the cell wall as mentioned earlier, or by breaking the cell wall mechanically, in a buffer solution (pH about 7). Non-permeating osmotic substances (e.g. sorbital or sucrose, 0.3 M) are used to maintain the osmotic potential, and ionic and pH requirements are met by using buffers. Chloroplasts are sep-arated from other organelles by centrifugation (90 s at 3000g) and then suspended in buffered solution. However, mechanical damage and unphysiological conditions may impair the envelope membranes. Intact chloroplasts assimilate CO_2 faster than damaged ones because they retain enzymes and co-factors. The envelope is detected by the bright appearance of isolated chloroplasts in phase-contrast microscopy. However, the envelope may appear intact yet have been broken and resealed. Ferricyanide, an electron acceptor in electron transport, cannot penetrate intact membranes, so measuring its reduction $(Fe(CN)_6)^{3-} + e^- \rightarrow (Fe(CN)_6)^{4-}$ in a spectrophotometer at 420 nm indicates the state of the envelope. Manipulation of isolated chloroplasts has been very important in analyz-ing their functions. Separated thylakoids and envelope may be disrupted by detergents or ultrasonically; the components are separated by column chromatography, density gra-dient fractionation, gel electrophoresis, *et cetera*, and chemically analyzed to determine the components of light-harvesting systems and electron-transport chains. The position of proteins or other components within the membranes is found using specific anti-bodies. If an antibody only affects, for example, electron transport in disrupted thylakoids, then the target protein (antigen) is inside the membranes. Similarly, chemical probes of known function are used to penetrate the membranes, changing processes and indicating the position of components. X-ray analysis, electron microscopic and scanning electron and confocal microscopy may all be used on particles, membrane pieces or chloroplasts and cells to determine the size, shape and orientation of components. These techniques in combination have provided a picture of chloroplast membrane and thy-lakoid vesicle structure and function.

4.4.1 Chloroplast internal structure

Electron microscopy shows the chloroplast to consist of an envelope made up of two separate membranes, enclosing a complex of membranes, the thylakoid system. Indeed, the thylakoids form a very extensive, highly branched and folded, and probably continuous, membrane system. Parts of the membranes are often joined or stacked into grana which appear as discrete piles of plates (or layers of membranes in thin transverse sections of the chloroplast) in electron micrographs, or when fluorescing (*Figures 4.2 and 4.3*). Thylakoid membranes are constructed of lipid with many protein complexes embedded; the lipids contrast with the background in electron micrographs, when stained with lipophilic electron-dense osmium. The space between the double membrane envelope and the thylakoid membranes is the chloroplast stroma.

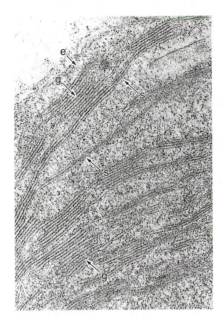

Figure 4.2. Structure of the thylakoid membranes in higher plant chloroplasts, illustrated by an electron micrograph (×34 000) of sugar beet (*Beta vulgaris* L.). The transverse section shows the double chloroplast envelope (e, indistinct) at the top-left enclosing the stroma and the thylakoid membranes forming granal stacks (g) and unstacked stromal thylakoids (s). Courtesy of Dr M. Kutík, Charles University, Prague, Czech Republic

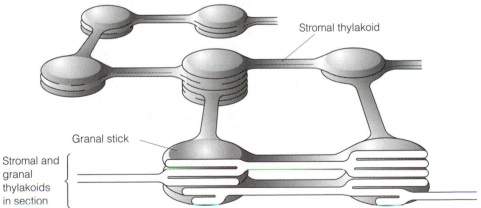

Figure 4.3. Schematic of the 3-dimensional structure of thylakoids in C3 chloroplasts, combined with a section through them, illustrating current views of how granal stacks are formed. Several different lobes of the extensive membrane become appressed into grana, and are joined by the single stomal membranes. (Reproduced from Arvidsson and Sundby, 1999, with permission from CSIRO Publishing)

The envelope. Two membranes form the envelope; each is about 5.6 nm thick and they are separated by the intra-envelope space (*ca* 10 nm), with areas of high electron density between the membranes which are possibly contact points involved in transport, for example of proteins between cytosol and stroma (see Chapter 10). The membranes are lipid bilayers of galactolipids (some 75%) with very unsaturated fatty acids (galactosyl glycerides and phosphatidyl choline), containing carotenoids but no chlorophyll. Monogalactosyl-diglyceride (MGDG) makes up 50% of the membranes, but has a very different function in the two envelope membranes. There are glycolipids giving rise to galactolipids and sulpholipids. Up to 95% of these are polyunsaturated fatty acids, such as linolenic acid (18: 3), and others have 16: 3 fatty acids. Phosphocholine is a major envelope phospholipid making up 30–35% of the outer membrane glycerolipid. The significance of the differences in lipid and fatty acid composition is not understood, but they are probably critical for lipid-protein interactions, membrane fluidity and responses to temperature.

The membranes are not identical in structure or function. The outer cytoplasmic membrane allows many substrates to pass freely, whereas the inner (stromal) membrane is highly selective, allowing passage of only some solutes by special enzyme systems called translocators. Protein particles in both membranes are complexes associated with the transporters and also with transport of other proteins (see Chapter 10) across the envelope, synthesis of lipids and other components, etc. The cytoplasmic membrane has a density of 1.08 g cm^{-3}, a lipid/protein ratio of 2.5–3 and 9 nm diameter occur at 1500–3300 particles µm^{-2} in both halves of the lipid bilayer. The stromal membrane has 7500–10 000 particles µm^{-2}, a correspondingly greater density (1.13 g cm^{-3}), and a smaller lipid/protein ratio of (1: 1.2) than the outer membrane. Protein particles of 7 and 9 nm diameter occur at a density of 1×10^3 and 1.8×10^3 µm^{-2} in the halves of the lipid bilayer next to the stroma and intermembrane space, respectively. There are many polypeptides (more than 75 and possibly several hundred) of 10–120 kDa, with major components of 54, 37, 14 and 12 kDa, possibly involved in metabolite transport (translocators or transporters, see Section 7.6). A 30 kDa polypeptide is probably the triosphosphate-P$_i$ translocator and a 37 kDa component is protochlorophyllide oxidoreductase (Section 8.12).

The stroma. The chloroplast stroma is not an homogeneous aqueous solution of small molecules and dilute proteins. In electron micrographs it contains indistinct granules and particles, which are mainly proteins, since the stroma is a dense protein gel, with about 0.4 g protein cm^{-3}. The most abundant protein is ribulose bisphosphate carboxylase/oxygenase (Rubisco, see Section 7.2.2), which forms over half of the protein; in some conditions, such as water stress or air pollution, it may form crystals. Through this densely packed matrix small molecules of metabolites must diffuse and small and large proteins must pass, either in the normal course of metabolism or during construction and repair of the chloroplast structures when proteins move from the cytosol to the stroma and thylakoid membranes and lumen. The movement of metabolites is achieved either by diffusion or via transport systems, or by associations of related enzymes into reaction complexes, which places products of one reaction in the correct position as substrates for transport proteins. Evidence for the association of components into cooperating complexes and for groupings of enzymes with the thylakoid membranes is growing (see Süss and Sainis, 1997). In addition to Rubisco, there is a large concentration of Rubisco activase. Also, all the other enzymes of the photosynthetic carbon reduction cycle are in the stroma, together with the enzymes and terminal redox carriers of the electron transport chain; ATP synthase also protrudes from the thylakoids into the stroma. Other inclusions are products of the photosynthetic processes; for example, starch granules up to 2 µm long accumulate in the stroma and displace the thylakoid membranes, and globules of lipids and plastoquinone accumulate, often markedly so under stress conditions or in old leaves. The DNA and RNA of the chloroplast genome and protein-synthesizing system and associated transport proteins occur in chloroplasts which synthesize many of their constituent proteins.

The thylakoids.   The most noticeable features of chloroplasts in electron micrographs (*Figure 4.2*) are the thylakoids (from the Greek θυλακοειδζς for 'sack-like'), an extensive membrane vesicle system. In transverse section the thylakoids appear as parallel pairs of continuous membranes separated by a space, the thylakoid lumen, which is 5–10 nm wide and contains few identifiable features. Thylakoid membranes frequently associate into granal stacks, interconnected by pairs of membranes, called stromal thylakoids (or alternatively intergranal connections or frets), which are in contact with the stroma on both sides. The interface between the appressed membranes is the partition region. In C3 plants over 60% of the thylakoid surface is typically in the grana. The outer and end membranes of granal stacks and the stromal membranes, but not the partition regions, have direct contact with the stroma.

Several models of the three-dimensional structure of the thylakoid system have been suggested from analysis of serial sections of chloroplasts. Thylakoid membrane vesicles in grana are stacked and flattened, but not closed, sacs (*Figures 4.3–4.5*) interconnected with the other membranes. The vesicles join the stromal lamellae at different points around the periphery of a granum. The structure derives from the folding and joining of separate sheets of lamellae which are interconnected and probably originate from a single point (*Figure 4.4*), the prolamella body, in the developing chloroplast. The thylakoid system appears to be a single interconnecting giant closed vesicle with continuous lumen, a feature of great importance in electron transport and ATP generation. It is dynamic, changing form and relative position within the chloroplast. This may be related to the movement of materials within the chloroplast. Perhaps the formation of grana is a mechanism that delimits parts of the system into separate volumes: this may be of value in concentrating H$^+$ ions, for example, which are of importance in the synthesis of ATP. It also improves PSII function and provides protection against photoinhibition under different light intensity, and is thus regulated by the redox state of the plastoquinone pool and the cytochrome *b/f* complex (Chapter 5). Probably, the flexible organization also increases efficiency of many other aspects of chloroplast metabolism. Aspects of these processes are considered later.

The composition of the thylakoid lumen is not known, but proteins of the water-splitting complex and the light-harvesting complex for example, protrude from the membranes into the lumen and occupy part of the volume. They probably form hydrogen bonds with protons and other components of the stroma, and buffer the lumen, for example against changes in proton concentration. Grana differ in extent, complexity and size between species, and with conditions during growth, for example with bright illumination there is less granal stacking. Even isolated thylakoids *in vitro* stack and unstack, according to the ionic concentration and light quality. A semi-quantitative summary of the size of an average thylakoid system is given in *Table 4.1*.

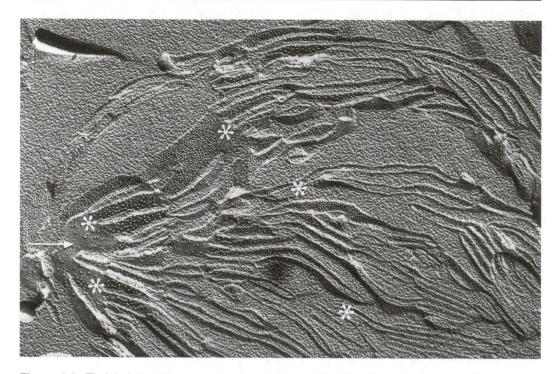

Figure 4.4. Thylakoid membranes of an isolated spinach chloroplast after freeze-fracture, showing progressive branching of membranes (*) from a common point (→). Electron micrograph (×50 000) courtesy of Professor L.A. Staehelin, University of Colorado, Boulder, CO, USA

Internal composition and structure of thylakoid membranes. A single bilayer membrane is 5–7 nm thick and consists of lipid (50% of the mass) together with proteins, pigments and other major components which are vital for photosynthesis. Thylakoid lipids are a complex mixture; some 80% is glycolipid containing galactose, such as MGDG diglyceride (50% of total lipid on a molar basis) and digalactosyl-diglyceride (DGDG, 25%), which have neutral hydrophobic 'heads'. DGDG is only in photosynthetic membranes, and changing the proportion of DGDG in thylakoids decreases PSII efficiency and energy transfer (probably by altering the pigment–protein interactions in the membranes) but does not alter oxygen evolution. The remaining lipid is mainly phospholipid (10%); phosphatydylglycerol influences membrane fluidity especially at low temperatures. Sulfolipids (5%, such as sulfoquinovosyl diacylglyceride) which are charged at pH 7, also affect membrane characteristics. The synthesis and structure of the lipids are considered in Chapter 8. The fatty acids of lipids are highly unsaturated. Linolenic acid (C18:3) is the predominant fatty acid and *trans*-3-hexadecanoic acid (C16:1) acylated to phosphatidyl glycerol is specific to thylakoid membranes; its function may be structural. The fatty acid 'tails' form a nonaqueous, hydrophobic central core to the membrane, whilst the hydrophilic heads are at the surface. The outermost half of the membrane, next to the stroma, is some 3–5 nm thick; the inner, next to the lumen, is 2–3 nm thick. As the two most abundant lipids are highly unsaturated, the membrane is very fluid at physiological temperatures with no cholesterol or other sterol to cause rigidity. Vitamin E (α-tocopherol) is a lipophilic constituent which may provide structure to the

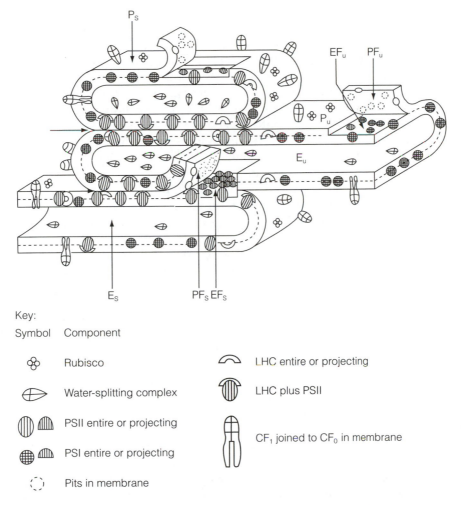

Figure 4.5. Diagram of the thylakoids of a higher (C3) plant chloroplast. Part of a granal stack and stromal thylakoid is shown. Coupling factor (CF_1) and ribulose bisphosphate carboxylase (Rubisco) are attached but not in the partition region ($\rightarrow$). In the lumen of the granal thylakoids are particles of the photosystem II, water-splitting complex. The partition region contains particles on the endoplasmic fracture face (EF_s) identified with PSII and the light-harvesting protein complex. Smaller particles on other membranes may be PSI and the base of coupling factor. See the text for explanation of membrane surfaces and particles

membrane and the protein complexes in it. However, α-tocopherol is only present in small concentrations, probably because it interacts with membrane lipids and decreases the fluidity, but it also has significant photoprotective function. Thylakoid membranes are subjected to intense radiation in an environment in which oxygen is produced, as well as highly energetic pigments, very reduced intermediates of electron transport and so on. They are therefore very liable to damage from, for example, active oxygen species, which may cause lipid peroxidation, thus destroying the membrane, or damaging protein complexes such as PSII by photoinhibition. Carotenoids and particularly xanthophylls are lipophilic, and probably associated with the periphery of the light-harvesting complexes. Carotenoids may have a role in preserving the structure of the complexes and regulating their position within the membrane, acting as stabilizers, optimizing energy exchange and increasing the quenching of chlorophyll fluorescence. The violaxanthin cycle (Section 3.4.4), a mechanism for the dissipation of excess excitation energy in thylakoids, requires movement of the xanthophyll molecules between the enzymes of the cycle which are located on the stromal and luminal sides of the membrane. Violaxanthin is equally distributed in granal and stromal thylakoids, and suggests that the enzyme reactions of the violaxanthin cycle take place in the lipid environment of the membrane, not at the water–lipid membrane interface. Dipolar carotenoids, such as zeaxanthin are of the correct size (3 nm long) to span across both halves of the lipid bilayer (3 nm thick) with the hydrophilic, polar, head groups in the head-group region of the membrane lipids and the lipophilic, long-carotene portion of the molecule perpendicular in the membrane. Perhaps this provides stability with increasing rigidity, regulating the membrane lipid environment in relation to energy regulation. Electron spin resonance of membranes with spin-label probes shows that thylakoids containing zeaxanthin are less fluid than those treated with dithiothreitol to block zeaxanthin formation. Also, increased zeaxanthin content protects thylakoids from disruption and leakiness to ions at high temperatures and decreases lipid peroxidation. In contrast, β-carotene in membranes seems to increase fluidity: the increased β-carotene/chlorophyll ratio in cold-hardened plants may reflect the need for greater movement of complexes in the cold, although there is no indication of this in thylakoids.

Thylakoid membranes are particularly fluid compared with other membranes in plants; this is probably essential for the photosynthetic mechanism, with the abundant pigment–protein complexes moving within the lipid layers laterally and vertically, and also rotating. The lateral diffusion coefficient of lipids is 10^{-10} m^2 s^{-1} and that of proteins 5×10^{-11} m^2 s^{-1}. Distances over which pigment–protein complexes move are small (1–1000 nm) so displacements of the order of 10–100 nm occur rapidly, particularly if the proteins are charged, enabling the thylakoid to change its structure and function to optimize processes, for example ion and water fluxes across the membrane and energy distribution between light-transducing complexes.

Particles in thylakoid membranes.

The surface structure of thylakoid membranes is observed by electron microscopy of isolated membranes and internal structure after freeze-fracturing (by cutting) frozen membranes. During cutting, the membrane bilayer separates along the line of weakness caused by the hydrophobic tails of the membrane lipids, exposing particles within the membrane (Figure 4.5). The surfaces of membranes in electron micrographs of freeze-fractured thylakoids are denoted by their contact with the stroma, that is protoplasmic surface (PS), or the lumen, that is endoplasmic surface (ES), and their fracture surfaces are PF and EF, respectively. Membranes from stacked (granal) or unstacked (agranal) regions are shown by subscript s and u, respectively (Branton *et al.*, 1975). On the outer surface of stromal (PS$_u$) and of grana thylakoids (PS$_s$) in contact with the stroma, are particles of Rubisco, loosely attached and easily removed. Most prominent are flattened, club-shaped particles, 10 nm in diameter, of ATP synthase (CF1). Rubisco and CF1 are extrinsic (i.e. external) proteins that occur only in contact with the stroma, not in the partition region. Smaller (9 nm diameter) particles on PS are exposed parts of intrinsic proteins (i.e. occurring within the membrane), for example the base part of CF, called CF_0. Particles from within the membrane may also project rather indistinctly out of the surface, and form a lattice with rows 8–9 nm apart and interparticle distances of 10 nm. On the very smooth inner lumen surface, ES_s, of granal thylakoids are rectangular (10 × 15 nm) particles of four (sometimes two or six) subunits each *ca* 5 nm in diameter (*Figure 4.6*). In artificially unstacked thylakoids these large particles may be arranged in a lattice, when they always have four units of 18 × 20 nm spacing on the ES_s face. The stromal thylakoid inner surface (ES_u) is much more textured than ES_s and has only a few smaller (4–6 nm diameter) particles just projecting above the surface. More than 80% of the large particles of EF faces occur in granal regions and 20% or less in stromal areas.

Particle distribution on fractured membranes has been analyzed mainly on spinach (Staehelin *et al.*, 1996). *Figure 4.6* shows an electron micrograph of the freeze-fracture faces. On EF_s are many (*ca* 1500 particles μm^{-2}) large particles in two populations of 15 and 11 nm diameter, 60–70% and 30–40%, respectively, of the total population. Depending on the conditions (salt concentrations, *et cetera*) these particles, which appear lobed,

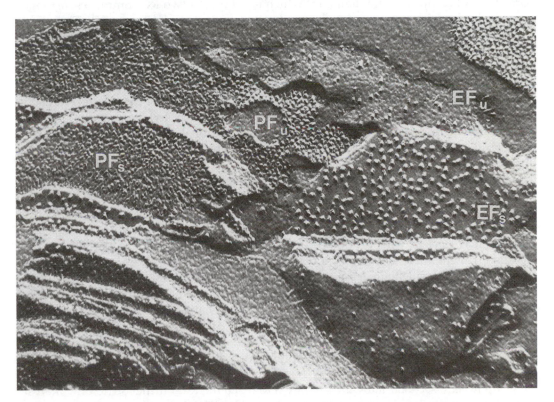

Figure 4.6. Freeze-fractured granal stacks of isolated thylakoids of a spinach chloroplast, showing the EF and PF faces of stacked and unstacked areas. Note the characteristic large (15 nm) particles on EF_s. The stacks are linked by a sheet of unstacked membrane. Electron micrograph (×185 000) courtesy of Professor L.A. Staehelin, University of Colorado, Boulder, CO, USA

may be arranged in very regular arrays or lattices forming a uniform population about 16 nm diameter. There are also about 700 smaller particles μm^{-2}, 10 nm diameter, of two subunits. On the stromal face, EF_u, there are fewer (450–570 μm^{-2}) smaller (10 nm diameter) particles which are not arranged in a lattice, and are more scattered than on the EF_s face.

The PF faces in both stromal and granal thylakoids contain particles: PF_s has 3500–4500 particles μm^{-2} of 8 nm average diameter (range 5–12 nm). PF_u has 3600 particles μm^{-2}, in two groups of 8 and 11 nm average diameter and less deeply embedded than those in the PF_s faces. Deep pits in the PF_s are left by the large particles on the EF_s face tearing out of the PF_s lipid bilayer during fracture (*Figures 4.5 and 4.6*). Four (sometimes two or three) smaller PF_s particles are regularly arranged around pits and would be closely associated with large EF particles. The groups, separated by about 2.5 nm, form the lattices in granal areas when they penetrate the ES and PS faces.

Particles in membranes have been identified with PSI and II, and the light-harvesting chlorophyll–protein and cytochrome *b–f* complexes. The large EF particle probably has a core of a PSII complex (8 nm diameter) which is a dimer of two RC complexes, arranged symmetrically, with antenna complexes attached to each end (*Figure 4.7*). This dimer (Section 4.5) is associated with two, four or six units of light-harvesting complex in granal thylakoids, but only one in stromal membranes. Smaller particles in the PF_u and PF_s surfaces may be chlorophyll–protein complexes of PSI, light-harvesting complexes and CF_0. Variations in size may be due to the association of light-harvesting complexes with different numbers of other complexes or with components of the electron transport chain. Stromal thylakoids contain most of the PSI and granal thylakoids most PSII; both are heterogeneous in size. PSIα in the margins of the granal stacks has a larger antenna than PSIβ in the stromal thylakoids. Granal PSII are called PSIIα, with those in the stroma, called PSIIβ, having smaller antenna than the PSIIα and some partially inactive. The protein components of the chl *a/b* binding proteins are very similar. LHCII contains three types of polypeptides: I is encoded by *Lhcb1*, a multigene family, giving very similar proteins, and II (*Lhcb2*) and III (*Lhcb3*) are products of smaller families (Chapter 10). The ratios of I, II and III vary from 10 to 20:3:1. LCHII contains two proteins. LHCII forms trimers (it is unclear whether homo- or heterotrimers) in thylakoids, perhaps as they mature. Trimers are associated with the dimeric PSII particles, at the periphery, where they act as the main antennae delivering excitation energy to the attached CCIIa/CCIIb inner antennae which supply the two RC cores. PSI is monomeric and two copies of each of the four LHCI chl protein complexes are permanently attached around the core *in vivo*. The detailed structural knowledge is contributing to understanding how photons are captured and excitation energy used by plants. The structures allow the excitation energy to be transmitted throughout PSI and II antenna complexes in 10–50 ps, so that the losses of energy are small (but measurable from fluorescence), but the excitation is held for times up to nanoseconds, allowing excitation to reach the RC. Further structural analyses will help to clarify how the functions are achieved. In PSII there is no gradient of energy between parts of the antenna, and differences in absorption between different complexes may show that the chlorophylls are in different combinations with protein and allow the complexes to form energetically linked groupings. The system provides for great efficiency in energy capture and use, particularly in dim light.

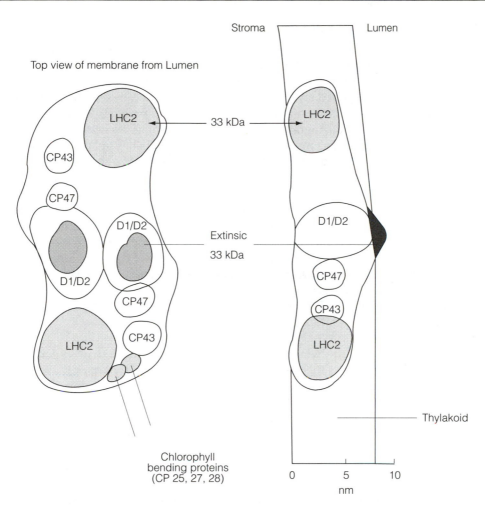

Figure 4.7. Schematic of the structure of the PSII dimer complex with a luminal surface view, and laterally, from within the plane of the thylakoid membrane. The 33 kDa protein which binds to the water-splitting complex is exposed to the lumen, together with other structural proteins

Large particles span the lipid bilayer from inside to outside surfaces in stacked membranes. Unstacked membrane particles are not arranged in regular lattice arrangements and do not span the membrane. Perhaps forces on or in the membrane cause the large particles to stand upright, make the lipid layer thinner, and produce arrays. However, the smaller PF particles do not project into the thylakoid lumen. Thylakoid membranes are 'sided' in construction: the water-splitting complex is in the lumen; a PSII chlorophyll–protein complex, a cytochrome *b*–*f* complex and a light-harvesting complex span the membrane; PSI chlorophyll–protein complex is on the outer side of the membrane; finally enzymes of carbon metabolism and ATP synthesis occur on the outer (stromal) surface. This asymmetry allows thylakoids to transport electrons from water in the lumen to redox components in the stroma and to accumulate protons in the lumen.

4.4.2   Membrane structure of C4 plants

Considerable differences occur between C4 species in chloroplast structure; here maize is taken as an example. The photosynthetic tissue in a leaf is divided into bundle sheath and spongy mesophyll (*Figure 4.8a and b*). The bundle sheaths consist of large cells with thick walls (a dark-staining suberized layer within the wall is a noticeable feature). Chloroplasts are few, large (15–20 μm diameter) and prominent with many parallel, densely packed thylakoids with no obvious grana (agranal), although some contact occurs around the vascular bundles. The mesophyll has more relatively small granal chloroplasts (*Figure 4.8b, c*). Mesophyll cell chloroplasts are comparable to those of C3 plants already described, with many discoid chloroplasts of about 8 μm diameter, double-membrane envelope, thylakoids arranged in parallel and many obvious grana. A dense layer of tubules, the chloroplast reticulum, lines the envelope, which is closely appressed to a dense layer of cytoplasm next to the cell wall; groups of plasmodesmata connect the bundle sheath and mesophyll cells. Thylakoid membranes of the two types of chloroplast differ in macromolecular construction (*Figure 4.9*). The granal and stromal areas of the C4 chloroplast (*Figure 4.8b and d*) are similar to the equivalent areas of C3 plants. The EF_s face is extensive with large particles (15 nm diameter) on a smooth background. On the PF_s face the particles are smaller and more numerous. EF_u faces lack 15 nm particles and have 10 nm diameter particles, similar to the PF_u particles, but at much lower density.

Although the bundle sheath thylakoids lack obvious grana, the restricted contact areas have 15 nm EF_s particles and PF particles of 8.5 nm, only slightly larger than the mesophyll particles, and appear to be rudimentary grana, of comparable subunit structure to C4 mesophyll and C3 chloroplasts. Particle density is similar in stacked mesophyll and in contact areas in the bundle sheath, but as there is much less stacked area in bundle sheath compared to mesophyll chloroplasts, the number of larger particles per thylakoid is only one-tenth. However, the PF particles are similar in number. Stromal lamellae of bundle sheath and mesophyll chloroplasts have 10 nm particles on EF_u and PF_u surfaces. The small PSII activity and few large EF particles in bundle sheath thylakoids suggest that the EF particle is a PSII unit (dimer of the core PSII particle) together with the light-harvesting chlorophyll–protein complex which is responsible for chloroplast stacking. Thus, the basic structure of thylakoids is similar in different plants and tissues but with quantitative differences in composition related to function.

4.5   Chlorophyll and protein complexes

The macromolecular structure of thylakoids has been determined from studies in which chlorophyll–protein complexes are removed from membranes with detergent, the anionic sodium dodecyl sulfate (SDS), for example, and separated by electrophoresis on polyacrylamide gels (PAGE) containing SDS. As separation of complexes depends on extraction and conditions during electrophoresis, and monomers of complexes may combine into larger units (oligomers) which give distinct bands in gel separation, a variable number of bands results, and the size and composition of membrane components may be difficult to resolve. Current models of particle organization are reviewed by Thornber *et al.* (1994), Staehelin and van der Staay (1996) and Hankamer *et al.* (1997). There is now acceptance that there are three main chlorophyll–protein complexes in the membrane,

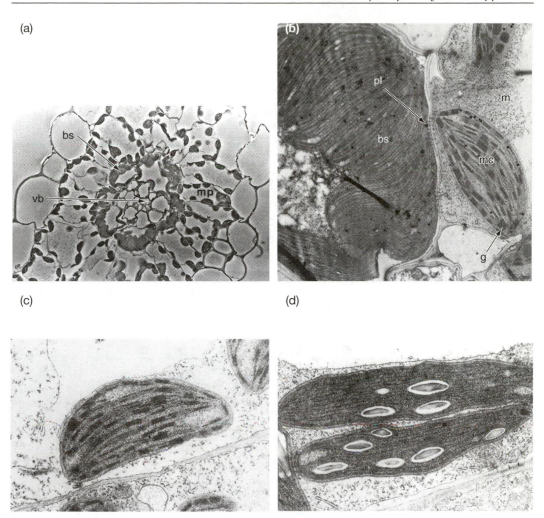

Figure 4.8. (a) Transverse section of a maize (*Zea mays* L) leaf (×440) with the vascular bundle (vb) surrounded by a bundle sheath (bs) of large cells with prominent chloroplasts. Cells of the mesophyll parenchyma (mp) have smaller, less densely packed chloroplasts. (b) Section of maize bundle sheath cell with bundle sheath chloroplast (bs) and mesophyll cell (m) with chloroplast (mc) showing chloroplast dimorphism. Plasmodesmata (pl) traverse the wall between the two cell types (g = grana). Electron micrograph (×3560) by courtesy of IACR-Rothamsted, Harpenden, Herts, UK. (c) Transverse section of a mesophyll chloroplast from maize, showing distinct grana, and stromal thylakoids. (d) Transverse section of a bundle sheath chloroplasts of maize, lacking grana; there are several prominent starch granules (st) within each chloroplast and dark plastoglobuli (pg). Magnification of (c) and (d) ×27 200. (Micrographs (b) courtesy of IACR-Rothamsted, Harpenden, Herts, UK and (c) and (d) courtesy of Dr. M. Kutík, Charles University, Prague, Czech Republic)

71

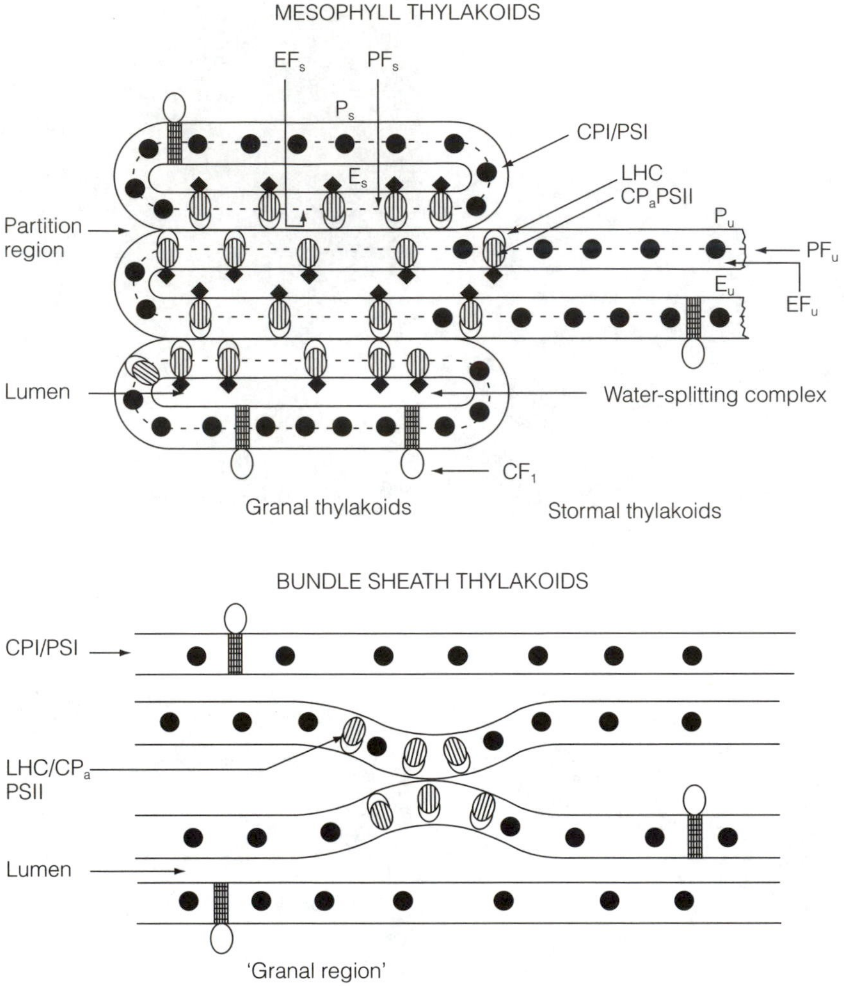

MESOPHYLL THYLAKOIDS

Granal thylakoids   Stormal thylakoids

BUNDLE SHEATH THYLAKOIDS

'Granal region'

Figure 4.9. Diagram of maize mesophyll and bundle sheath thylakoids. Only small areas of contact (grana) occur between bundle sheath lamellae compared with extensive grana in the mesophyll. Particles of PSII and associated light-harvesting complex are in grana of both types (compare to *Figure 4.5*)

which contain 90% of the chlorophyll (*Table 4.2* and *Figure 4.10*). They correspond to membrane particles and are PSI, PSII (which are photochemically active) and light-harvesting chlorophyll complex (LHC) which has only antennal (i.e. light energy capture) function.

4.5.1 PSI complex

The PSI core is now called CCI (also called P700 chlorophyll *a* complex or chlorophyll–protein complex I, CPI for short) and is associated with a light-harvesting chlorophyll *a/b*–protein complex, now called light-harvesting complex or LHCI, composed of four

Table 4.2. Chlorophyll–protein complexes of the photosystems (PSI and PSII) and the light-harvesting complexes (LHC) associated with them in thylakoids of higher plants (after Thornber et al., 1994)

Name: Current	Name: Alternative	Molecular mass (kDa)	Chlorophyll a/b ratio	Percentage total chlorophyll	Number of: Chlorophylls	Number of: Polypeptides (mass kDa)
PSI	CP1	230	6.0	38		Many
CCI core	CP1, CHLa-P1	120	chl a only	20	60–110	Many
Light-harvesting complex						
LHC 1a	LHCPIa	65–25	1.4	18		Two (24 and 22) in trimers, xanthophyll, no neoxanthin or carotenes.
LHC 1b	LHCPIb	65–25	2.3	18		Two (21) trimers, lutein, one (17) violoaxanthin only
LHC 1c		17	chl a/b?			one (11) carotenoids
LHC 1d		—				binds co-factors
PSII						
CCII-RC	CPIII	80	chl a and pheo only	1		Seven (47, 43, 34, 32 or D2, 30 or D1, 9, 4) dimers
CCIIa	CPIV, CP47	55	chl a only	5		Seven (47, 43, 34, 32, 33, 9, 4) dimers
CCIIb	CHL a/b, CP43	50	chl a only	5		Seven (47, 43, 34, 32, 33, 9, 4) dimers
Light-harvesting complex						
LHCII a	CHL a/b, CP29	35	2.3	4		One (31) lutein, violoaxanthin and neox 1:1:1
LHCII b	CHL a/b, CP11	72	1.3	40		Three (28, 27, 25) lutein, violoaxanthin and neoxanthin
LHCII c	CP27	30	1.8	4		One (26) lutein, some violoaxanthin 1:1 and neox
LHCII d	CP24	24	0.9	4		One (21) lutein
LHCII e		13				One (26) highly enriched xanthophylls

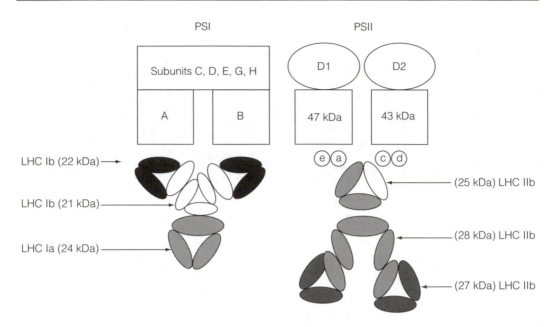

Figure 4.10. Schematic of the arrangement of PSI and PSII core dimer particles in the thylakoids and the associated light harvesting complexes (see *Table 4.2*) formed of trimers of different subunit composition

different types of complexes (a–d; see *Table 4.2*), which has only antenna function and no photochemical activity. The LHC particles and the arrangement and proportions have been resolved (see Thornber *et al.*, 1994) and are summarized in *Table 4.2* and *Figure 4.10*. LHC of PSI is made of LHCIb subunits with different but similar masses: 21 and 22 kDa, and LHC1a of 24 kDa, arranged in trimers (homo- or heterotrimers) linked to the PSI core complex. The PSII antenna is composed of homo- or heterotrimers of LHCIIb (subunits of 25, 27 and 28 kDa), linked to the 43 and 47 kDa polypeptides. Also, other LHC subunits (a, c, d and e) are associated: they are involved in structure and energy transfer. Light-harvesting complex, mainly found in granal thylakoids of plants with chl *b*, contains 50% of chl *a* and all chl *b* (ratio *a:b*, 1:3) and approximately one carotenoid per four to six chlorophylls. Within the LHC complex, groups of chl *b* molecules, probably arranged within excitation transfer distance, deliver energy with high efficiency to groups of chl *a* molecules more loosely arranged and transferring energy by the Förster mechanism. LHC passes energy to the reaction centers but as the energy and conditions within the thylakoid change so the energy transfer between the complexes changes allowing energy transfer between PSI and PSII ('spill-over' but not now regarded as an important energy dissipation mechanism). The chl *a:b* ratio of the complete units reflects the proportion of different subunits which have rather fixed ratios. All the proteins bind carotenoids, (e.g. lutein) and the xanthophylls neoxanthin and violaxanthin which are involved in the regulation of energy dissipation in the thylakoid by the violaxanthin cycle in the antenna.

Thus, a large LHC antenna of chl *a* and *b* is linked closely to the antenna chl *a* of the reaction center complexes of both photosystems. PSI is almost restricted to membranes exposed to the stroma and absent from the interior of stacked membranes. Probably 80%

of PSII is in the granal regions, sheltered from the stroma. Illumination causes PSI to associate with PSII and LHC, from which it receives more energy, at the edge of the granum. Groups of LHC linked to the core dimers of PSI and PSII, may distribute energy within a single group of particles (a 'puddle' model) or co-operatively (a 'lake' model) with excitation energy passing between groups, but the efficient energy distribution achieved under different conditions is a consequence of the interaction of the complexes. This allows greater flexibility in photosynthetic response to environment and to physiological conditions in the photosynthetic apparatus.

4.5.2 PSII complex

The PSII complex has a core of CCII (often called CPIII), with associated subunits serving as the internal chl *a* antenna of PSII, and the associated chlorophyll–protein complexes LHCII a–e. CCII is probably a dimer of 230 kDa mass, each composed of two polypeptides of 50 or 60 and 70 kDa, and contains about 30–40% of the total chlorophyll *a* (no chl *b*), 10 mol chl *a* per mole of protein and one P700 molecule to every 100 chl *a* molecules. P700 is probably surrounded by a tightly bound inner and loosely bound outer chl *a* antenna, with about five β-carotenes and two xanthophylls. A quinone (naphthoquinone, phylloquinone or tocopheryl quinone but no plastoquinone) and perhaps iron–sulfur centers are bound in the complex; these transport electrons. Some CPI may be composed of P700 and 5–10 monomers of protein–chlorophyll, a functional unit. CPI occurs throughout the plant kingdom in all organisms; photosynthetic bacteria have an equivalent and very similar photosystem complex. CPI predominates in membranes in contact with the stroma.

The PSII complex occurs in all oxygen-evolving plants (and is thus essential for O_2 evolution) and contains 10% of the total chlorophyll, mainly or only chlorophyll *a* for the inner or close antenna. The (*Figure 4.7*) structure is a dimer made up of two parts, the D1/D2 reaction center core with attached 33 kDa proteins of the water-splitting complex (see Chapter 5) and a CP47 and CP43 subunit antenna, plus associated cytochrome b_{559}. A trimer of LHCII forms the main antenna, at each end of the elongated dimer particle. Other polypeptides containing chlorophyll *a* and *b*, are also linked, and minor polypeptides like the 23 kDa protein have a role in the organization and energy transfer. It has been suggested that there may be two types of PSII particles with different amounts of antenna, forming $PSII_\alpha$ and $PSII_\beta$. $PSII_\alpha$ contains more chl *a* in the antenna than $PSII_\beta$ and several $PSII_\alpha$ complexes may join together, but the $PSII_\beta$ complexes are not combined together. The more $PSII_\alpha$, the less grana stacking occurs. Possibly $PSII_\beta$ complexes are in the stromal membrane and $PSII_\alpha$ in the stacked regions; maybe these different forms are complexes in stages of synthesis or repair with the D1 protein being removed and inserted into the complex in stromal thylakoids before the repaired, fully functional unit moves into the granal region.

4.5.3 Light-harvesting chlorophyll proteins

The structure of LHCII from pea has been described at 0.34 nm resolution by X-ray crystallography, by Kühlbrandt *et al.* (1994), and provides a general model for the pigment–protein complexes (*Figure 4.11*). There are three membrane-spanning helices,

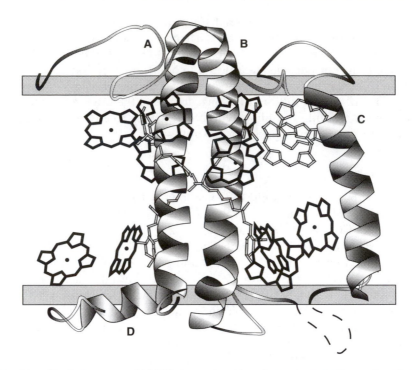

Figure 4.11. Simplified structure of LHCII determined by electron crystallography, showing the three protein chains spanning the thylakoid membrane, and the locations of chlorophyll molecules (Reprinted with permission from *Nature*, Kühbrandt *et al*., Atomic model of plant light-harvesting complex by electron crystallography. 1994; **367**: 614–621. Copyright 1994, Macmillan Magazines Limited)

with the first and third crossing at 30° to the membrane normal and linked by ion pairs of arginine on one helix and glutamate on the other. The two helices and their N-terminal loops have a two-fold symmetry with two carotenoids linked to the loops at each end of the helix and eight chl molecules. Twelve chl were identified, seven are probably chl *a*, close to carotenoids allowing efficient energy transfer, five are chl *b*, three of them being attached to the central helix. The magnesium atoms of three chl are ligated by carbonyl groups of glutamate residues which cross-link the first and third helices. Other specific linkages of chlorophyll to the protein structure, and the shielding of the carotenoid head structures by the loops connecting the helices and exposed at the membrane surface, shows the precision with which the structure is organized for efficient energy capture and transfer. Excitation energy can be delocalized over the ring and passed to any one of a number of photosytems and reaction centers.

Most of the LHC is hydrophobic and buried in the membrane. However, a segment (2 kDa) of 20 amino acids on the major polypeptides is exposed to the stroma, and a segment of each is exposed to the lumen. The exposed portion in the stroma may be removed by adding protease enzyme (trypsin). The amino acid sequence at the C-terminal of the peptide segment is (lysine$^+$ or arginine$^+$) lysine$^+$, arginine$^+$, serine, alanine, threonine[+], threonine[+], lysine[+], lysine[+], with positive charges as indicated. Despite these positive charges, the thylakoid surface has a net negative charge (one charge per 6 nm^2) so that the membranes are mutually repelling and dissociate unless cations (e.g. Mg^{2+}) are present to shield the negative charges. This allows the positive-charged portion of LHC to join onto some negative charges and cause thylakoid stacking. The threonine residues on the 2 kDa peptides can be phosphorylated by a specific kinase, abolishing the positive charge and so altering the balance of charges on the membrane surface, changing stacking and distribution of particles in the granal and stromal thylakoids and also energy distribution.

References and further reading

Anderson, J.M. (1999) Insights into the consequences of grana stacking of thylakoid membranes in vascular plants: a personal perspective. *Australian J. Plant Physiol.* **26**: 625–639.

Arvidsson, P-A. and Sundby, C. (1999) A model for the topology of the chloroplast thylakoid membrane. *Australian J. Plant Physiol.* **26**: 687–694.

Bennett, J. (1991) Protein phosphorylation in green plant chloroplasts. *Annu. Rev. Plant Physiol. Plant Mol. Biol.* **42**: 281–311.

Blankenship, R.E., Madigan, M.T. and Bauer, C.E. (1995) *Anoxygenic Photosynthetic Bacteria.* Kluwer Academic, Dordrecht.

Boekema, E.J. and Rögner, M. (1995) Electron microscopy. In *Advances in Photosynthesis,* Vol. 3, *Biophysical Techniques in Photosynthesis* (eds J. Amesz and A.J. Hoff). Kluwer Academic, Dordrecht, pp. 325–336.

Branton, D. *et al.* (1975) Freeze-etching nomenclature. *Science* **190**: 54–56.

Hankamer, B., Barber, J. and Boekema, E.J. (1997) Structure and membrane organization of photosystem II in green plants. *Annu. Rev. Plant Physiol. Plant Mol. Biol.* **48**: 641–671.

Havaux, M. (1998) Carotenoids as membrane stabilizers in chloroplasts. *Trends Plant Sci.* **3**: 148–151.

Hinshaw, J.E. and Miller, K.R. (1989) A novel method for the visualization of outer surfaces from stacked regions of thylakoid membranes. In *Techniques and New Developments in Photosynthesis Research* (eds J. Barber and R. Malkin). Plenum Press, New York, pp. 111–114.

Kühlbrandt, W., Wang, D.N. and Fujioshi, Y. (1994) Atomic model of plant light-harvesting complex by electron crystallography. *Nature* **367**: 614–621.

Leegood, R.C. and Walker, D.A. (1983) Chloroplasts (including protoplasts of high carbon dioxide fixation ability). In *Isolation of Membranes and Organelles from Plant Cells* (eds J.L. Hall and A.L. Moore). Academic Press, London, pp. 185–210.

Ort, D.R. and Yocum, C.F. (eds) (1996) *Oxygenic Photosynthesis: the Light Reactions.* Kluwer Academic, Dordrecht.

Pierson, B.K. and Olson, J.M. (1987) Photosynthetic bacteria. In *Photosynthesis, New Comprehensive Biochemistry,* Vol. 15 (ed. J. Amesz). Elsevier, Amsterdam, pp. 21–42.

Rögner, M., Boekema, E.J. and Barber, J. (1996) How does photosystem 2 split water? The structural basis of efficient energy conversion. *Trends Biochem. Sci.* **21**: 44–49.

Staehelin, L.A. and van der Staay, G.W.M. (1996) Structure, composition, functional organization and dynamic properties of thylakoid membranes. In *Oxygenic Photosynthesis: the Light Reactions* (eds D.R. Ort and C.F. Yocum). Kluwer Academic, Dordrecht, pp. 11–30.

Süss, K-H. and Sainis, J.K. (1997) Supramolecular organization of water-soluble photosynthetic enzymes in chloroplasts. In *Handbook of photosynthesis* (ed. M. Pessarakli). Marcel Dekker, New York, pp. 305–314.

Thornber, J.P. , Cogdell, R.J., Chitnis, P. , Morishige, D.T., Peter, G.F., Gomez, S.M., Ananadan, S., Preiss, S., Dreyfuss, B.W., Lee, A., Takeuchi, T. and Kerfeld, C. (1994) Antenna pigment–protein complexes of higher plants and purple bacteria. In *Molecular Processes of Photosynthesis,* Vol. 10, *Advances in Molecular and Cell Biology* (ed. J. Barber). JAI Press, Greenwich, pp. 55–118.

Tiede, D.M. and Thiyagarajan, P. (1995) Characterization of photosynthetic supra-molecular assemblies using small angle neutron scattering. In *Advances in Photosynthesis,* Vol. 3, *Biophysical Techniques in Photosynthesis* (eds J. Amesz and A.J. Hoff). Kluwer Academic, Dordrecht, pp. 375–390.

Trebst, A. (1994) Dynamics in photosystem II structure and function. In *Ecophysiology of Photosynthesis* (eds E.-D. Schulze and M.M. Caldwell). Ecological Studies 100, Springer, Berlin.

Wild, A. and Ball, R. (1997) *Photosynthetic Unit and Photosystems. History of Research and Current View. (Relationship of Structure and Function.)* Backuys Publishers, Leiden.

Electron and proton transport

5.1 Introduction

Excitation of the 'special pair' chlorophylls in the RCs of anoxygenic and oxygenic photosynthetic organisms, and transfer of electrons to the primary acceptors, produces a strongly reduced intermediate which is used to reduce carbon dioxide (CO_2), and nitrate (NO_3^-) and sulfate (SO_4^{2-}) ions. Although the light reactions appear very similar, the source of electrons to replace those ejected from the special pair differs between organisms which do or do not produce O_2 (oxygenic and anoxygenic, respectively). Further, not only are there differences in the source of electrons supplied to the reaction center within the anoxygenic types, but mechanisms of electron movement differ also. The anoxygenic mechanism reflects the evolutionary diversity of prokaryotic organisms; oxygenic systems are more uniform. This chapter considers the processes in some detail.

5.1.1 Bacterial electron and proton transport

One of the simplest light capture and energy transduction systems known, and a model for how early photosynthetic organisms may have worked, is a light-driven proton pump involving protein conformation changes but without e$^-$ movement, linked to ATP synthesis. This occurs in the bacterium *Halobacterium halobium* (Kushner, 1985). Light energy 'drives' protons out of the cytosol into the external medium (*Figure 5.1*). Diffusion of the H$^+$ back into the cell through ATP synthase provides a part of *H. halobium*'s energy needs, but it is not fully autotrophic. The organism grows in brackish water in light; with deficient oxygen (but not if O_2 is completely absent) purple patches 1 µm long form over more than half of the cytoplasmic membrane. The patches are of bacteriorhodopsin, pigment–protein molecules (similar to visual rhodopsin in the retina of the vertebrate eye), arranged in a crystalline lattice. Retinal, which absorbs light, is an aldehyde of vitamin A bound by a lysine group to a protein (an opsin-like polypeptide of 26 kDa) by a protonated Schiff's base. The protein has seven α-helical segments forming a crescent and spanning the membrane with the N-terminal of three to six amino acids on the external side of the membrane and 17–24 forming the C-terminal at the cytoplasmic surface. Retinal is positioned about one-third of the way to the outer membrane surface. When the protonated chromophore in the 13-*cis*-all-*trans*-retinol form absorbs light

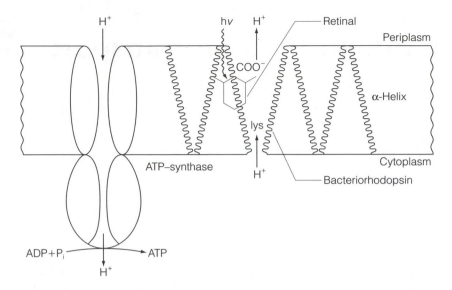

Figure 5.1. The light-driven proton pump of the bacterium *Halobacterium halobium*, based on light capture by retinal combined with a protein, which spans the membrane. Protons are expelled from the cytoplasm into the external medium and diffuse back into the cell via ATP synthase (CF)

at 570 nm it loses a proton, forming 13-*cis*-all-*trans*-retinal and absorption changes to 412 nm. The H^+ is transferred by the protein (which alters conformation and has been called a 'proton wire' because it carries protons), from the inside of the cell to the external medium. The deprotonated pigment then absorbs a proton from the cytoplasm. Thus, the simple mechanism acts as a pump of H^+ analogous to other photosynthetic H^+ transport, but without electron transport. The H^+ from the medium diffuses back into the cell through ATP synthase. The ATP supports metabolism, particularly when the respiratory activity in the bacterium is limited. This mechanism illustrates how photosynthetic systems may have evolved. Its presence in a 'primitive' bacterium suggests an early form of exploitation of radiation, although it may be a later evolutionary addition.

5.1.2 Photosynthetic bacterial reaction center

The photosynthetic RC of the photosynthetic purple bacterium *Rhodopseudomonas viridis* was the first membrane protein complex to be crystallized (Deisenhofer *et al.*, 1985). This was achieved by solubilizing membranes in solutions of particular detergents with long alkyl chains and small polar heads together with careful addition of small ampiphilic molecules such as benzamidine to prevent denaturation of these very hydrophobic proteins. The crystals were analyzed by X-ray crystallography to 0.3 nm resolution. The detailed structure (Michel and Deisenhofer, 1986) shows (*Figure 5.2*) that the RC forms a 7 nm diameter protein complex with the three proteins H, M and L and 14 co-factors. In detail it contains four bchl, two bphaeo *b* molecules and two quinones, arranged symmetrically around a two-fold rotation axis between the two bchl molecules and through

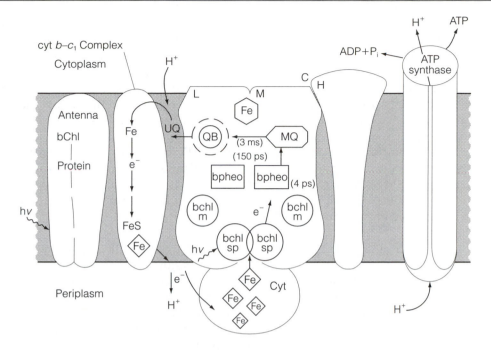

Figure 5.2. Schematic illustration of the arrangements of electron transport components in the bacterial reaction center of *Rhodopseudomonas viridis*. Two proteins, L and M, carry the bacteriochlorophyll special pair (bchl sp), which absorbs energy and ejects electrons. The electrons pass to a bacteriopheophytin (bpheo), with another acting as a 'chaperone', and on to menaquinone (MQ) and to a bound ubiquinone (QB) before transport to the cyclic electron transport path in the cyt b–c_1 complex. Other components of the system are: antenna (low-molecular-mass polypeptides, <10 kDa); reaction center polypeptide subunits L (light, 28 amino acid residues), M (medium, 30 amino acids), and H (heavy, 46 amino acids); mass of subunits 20–30 kDa; ATP synthase (CF_0–CF_1). cyt, cytochrome; UB, ubiquinone; ⟨Fe⟩, heme iron; [Fe] ferrous iron atom magnetically coupled to MQ which is linked to a histidine residue; bchlm, bacteriochlorophyll monomer; bpheo, bacteriopheophytin, one of which is linked to the chain of MQ. Transfer times for electrons are given in brackets.

a non-heme iron atom. The bchl molecules form the 'special pair' on the periplasmic side of the membrane. The Fe atom is between the two quinones on the cytoplasmic side of the membrane. There are two electron paths from the special pair, each with bchl, a bphaeo and a quinone. However, the quinones function in series, not parallel, and only one path is used in photosynthesis; the other may have carried electrons early in evolution but its present role is perhaps to 'fine tune' the RC. The accessory (i.e. not the RC) bchl is close to the carotenoid and may provide a path for quenching excited states of the bchl. The oxidized RC, after e^- ejection, is reduced again by electrons from two low and two high redox potential heme groups which are in a row. All these components are linked to the L and M subunits of the RC proteins, which are homologous to D1 and D2 proteins of PSII. The subunits possess five helices of 19 or more amino acids, without charge and very hydrophobic, containing many S amino acids and histidines which bind the pigments. These helices cross the membrane and are arranged with two-fold symmetry. Because of the different distribution of amino acids in the segments of the polypeptides the periplasmic and cytoplasmic membrane surfaces have four negative and four positive charges, respectively. This may be important for organization and orientation of the complex in the membrane, and also for e-transport and development of electrical potential across the membrane. The peptide segments on the membrane surface form flat planes orienting the complex and perhaps holding it steady. There is an H subunit with one helix crossing the membrane and its carboxy-end is bound to the L–M complex on the cytoplasmic side of the membrane. Pigments are bound into hydrophobic regions within the L and M subunits. The histidine residues join in five-fold coordination to the Mg of the special pair bchl. A few water molecules are found inside the membrane and may be important for hydrogen bonding in the protein. This very symmetrical and organized structure is important for e^- transport.

The RC special pair of chlorophylls (Section 3.2) is arranged parallel to the membrane normal. This complex is contained within the inner antenna, formed by a symmetrical ring, 12 nm external and 7 nm internal diameter, and composed of 16 subunits. Each subunit is formed by a U-shaped protein structure of α and β polypeptides with two molecules of bchla absorbing at 875 nm situated in the U. This highly symmetrical arrangement provides a very large energy capture system with, of great importance, a large π-electron system which rapidly distributes excitation energy to the RC from wherever it is captured. This inner antenna is probably linked to a main antenna of eight light harvesting (LH_2) complexes. Each is formed of a ring of nine units made from an α- and β-polypeptide, enclosing a lipid-filled space, with two bchl 800–850 molecules positioned between the polypeptides. The very close arrangement of the pigments and the complexes is essential for ensuring the optimum proximity between pigments for energy transfer, probably by the Förster mechanism. Each LH_2 contains 18 bchl a 800–850 and the whole antenna some 162 bchl a; the ratio of bchl 800–850 to bchl 875 is about 10:1. This provides rapid access of the short-wavelength-absorbing pigment to the long-wavelength-absorbing π-electron system and then to the RC, so ensuring rapid and efficient energy transfer, with losses less than 10%.

5.1.3 Electron transport in the anoxygenic photosynthetic bacteria

In Chapter 3 the structure of the purple bacterial light-capturing system was described and the mechanisms of e^- transport are considered further here. Ejection of an electron from P870 to one of the bchl molecules occurs within a few picoseconds of light capture (*Figure 5.3*). Then two e^- are transferred via bacteriochlorophyll to ubiquinone, thus decreasing the back reactions. Ubiquinone picks up two protons from the inside of the cell and transfers them to the outer side of the membrane. These protons diffuse back into the cytoplasm through the ATP synthase enzyme complex or coupling factor, where ATP is synthesized from ADP and inorganic phosphate (see Chapter 6). The electrons pass to cytochrome (type c_2) catalyzed by an oxidoreductase of three subunits, a cyt b apoprotein, a cyt c_1 and an FeS protein. The apoprotein has eight hydrophobic sections spanning the membrane with four histidine residues which bind two heme molecules, as transmembrane e^- carriers. Other hydrophilic subunits are found on the membrane surface (Dutton, 1986).

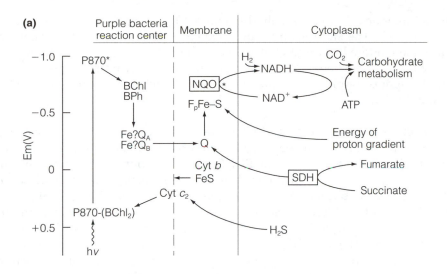

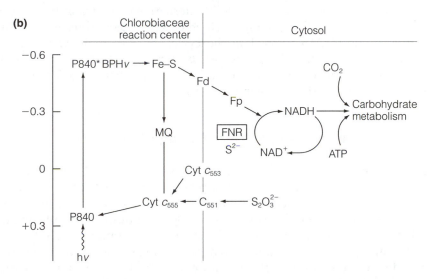

Figure 5.3. Electron transport in the photosynthetic bacteria. (a) Purple bacteria, showing the flow of e⁻ in the reaction center, reduction of quinones in the membrane and passage of e⁻ to CO_2 catalyzed by NAD quinone oxidoreductase (shown at *). As the available energy (Em) is insufficient, energy from the proton gradient is coupled to the enzyme. Electrons from substrates are fed into cyt c_2 and via a b–c_1 complex (quinol cytoreductase). (b) Electron flow in Chlorobiaceae. Reduced Fe–S centers pass e⁻ to NAD^+ for reduction of CO_2 via the enzyme ferredoxin nucleotide reductase (FNR). Electrons from substrates reduce the oxidized reaction center via cytochrome 551 (c_{551}). SDH, succinate dehydrogenase; NQO, NAD^- quinone oxidoreductase

Electron flow in the photosynthetic bacteria differs between bacterial types. The purple bacterial and chlorobial forms are shown in *Figure 5.3a and b*, respectively. In the purple bacteria, electrons from substrates reduce cytochrome c_2 and are passed back to the oxidized RCs. Alternatively, electrons from metabolism of organic acids (e.g. conversion of succinate to fumarate by succinate dehydrogenase (SDH)), may contribute to the reduction of quinone (Q). The electrons from quinone reduce NAD^+, a reaction catalyzed by NAD^- quinone oxidoreductase (NQO). Because the redox potential for the reaction is unfavorable for the reduction of NAD^+ by ubiquinone, energy from the proton gradient formed in the light is coupled to the enzyme in order to achieve the reduction. NADH is the reductant for CO_2.

The green bacteria (Chlorobiaceae and Chloroflexaceae) contain bchl *c*, *d* or *a* as main photosynthetic pigments organized in chlorosomes (see Chapter 4), forming a very large antenna, 1000 molecules to one RC. One of the four polypeptides in chlorosomes is a 3.7 kDa protein which carries bchl *c*; it is probably a dimer and has 10–16 chlorophyll molecules attached. A water soluble protein, which binds bchl *a*, forms a basal plate between the chlorosomes and the RC. The RCs of *Chloroflexus* are almost identical to those of purple bacteria, but other green bacteria have different structures. Electrons from the RC (*Figure 5.3b*) pass to ferredoxin (Fd) and these reduce NAD^+ via a flavin-NAD reductase enzyme on the cytoplasmic side of the membrane. Electrons from substrate, e.g. SO_2, O_3^{2-}, pass to a cytochrome on the opposite side of the membrane and reduce the oxidized RC. Alternatively, electrons from ferredoxin pass to the cytochrome via the quinone—cytochrome complex, thus forming a cyclic electron pathway. The NADH is used in the cycle, together with ATP generated by the ATP synthase, as described for purple bacteria, but reduction of NAD^+ is achieved without the need for energy from the proton gradient. In *Chlorobium*, glycogen and hydroxybutyric acid are major products.

5.2 Electron transport in oxygen-producing photosynthesis

Photosynthesis of the prokaryotic cyanobacteria, as well as that of eukaryotic algae and higher plants, produces oxygen and the basic process is similar in them all; indeed the eukaryotic chloroplast is of prokaryotic origin. A form of the Hill and Bendall 'Z' scheme of the sequence of processes and electron transport leading from water-splitting through $NADP^+$ reduction is given in *Figure 5.4* and the structural components related to it in *Figure 5.5*. Photon capture by the photosystem antennae and excitation transfer to PSII and PSI provide the energy for oxidation of water and electron movement to acceptors, which donate e^- to biochemical processes, and for passage of protons into the thylakoid lumen, for synthesis of ATP. The electron transport system may be considered in five parts: (1) a water-splitting complex; (2) a photosystem II complex; (3) an electron carrier chain; (4) a PSI complex; and (5) a group of e^- carriers which reduce electron acceptors ($NADP^+$, O_2). Electron transport starts with the capture of photons by chlorophylls and accessory pigments. Transfer of the energy to RCs of PSI and PSII excites the dimer chlorophylls and causes ejection of electrons to acceptors, starting e^- transport along the chain of redox components. Excitation of P680 of PSII results in an oxidized RC P680$^+$. PSII is defined as that part of oxygenic photosynthesis catalyzing transfer of e^- from water to plastoquinone (PQ):

$$4\,H^+_{stroma} + 2\,PQ + 2\,H_2O \xrightarrow[\text{PSII}]{4h\nu} 2\,PQH_2 + O_2 + 4\,H^+_{lumen} \qquad (5.1)$$

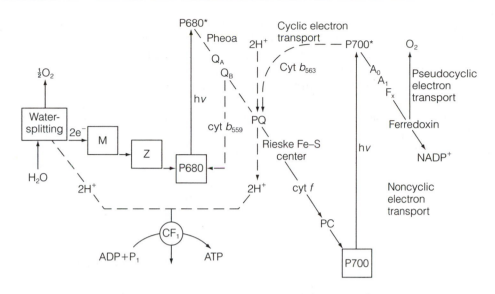

Figure 5.4. Scheme of electron transport in the thylakoid membrane from water to NADP+ with two light reactions, PSII and I in series; M and Z are intermediates in water-splitting; pheophytin (pheo), quinone acceptors Q_A, Q_B and PQ (plastoquinone), cytochrome f (cyt f) and plastocyanin (PC) pass e−. A_0, A_1 and F_x are intermediates in e− movement to ferredoxin. Proton transport leads to synthesis of ATP. Electrons also cycle from PSI back to PQ in the chain or pass to oxygen

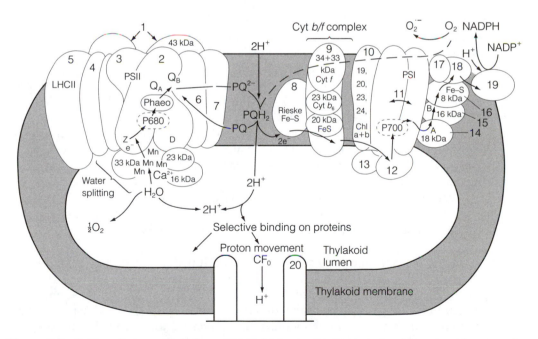

Figure 5.5. Schematic representation of the light-harvesting and photosystem complexes and electron transport chain in the thylakoid membrane. The number of protein complexes and their relation is shown. The mass of components is indicated. The components are listed in *Table 5.1*

85

with transfer of H^+ from the stromal to lumen side of the thylakoid membrane. This oxidized PSII is reduced by e^- from a water-splitting complex via intermediate states M and Z (*Figure 5.4*), components of the water-splitting complex and the electron carrier system between it and the RC. The energized e^- passes, from more to less negative potential, to the primary acceptor pheophytin and then in sequence to the quinone acceptors Q_A, Q_B and PQ. Quinones are important carriers of e^- and H^+ in many biological processes. From PQ the electron passes to cytochrome b/f and plastocyanin before reducing an oxidized PSI RC. Here it is energized again by excitation energy derived from photon energy trapped in the chlorophyll matrix, and passed via intermediate states A_0, A and B to oxidized ferredoxin (Fd) and $NADPH^+$, which are reduced and are able to enter into biochemical reactions in the chloroplast stroma. Electron transport chains bridge the thylakoid membranes, allowing electrons removed from water held in the water-splitting complex inside the thylakoid lumen to pass across the membrane to ferredoxin on the stromal side. Plastoquinone in the membrane is reduced by the electrons; the H^+ from the stroma attaches to reduced plastoquinone and is carried to the lumen, where it is released and the plastoquinone oxidized. Thus, electron transport is coupled to a plastoquinone cycle which carries ('pumps') H^+ from the stroma to the thylakoid lumen in the reverse direction to electron transport, increasing H^+ concentration in the thylakoid lumen and forming the proton concentration gradient, the energy of which drives ATP synthesis.

5.2.1 Photosystem II complex

The structure and function of PSII at the molecular and atomic levels of organization are now well understood, particularly since description of the purple photosynthetic bacterial RC, which is a valuable analog. The H, M and L protein subunits of the bacterial RC are analogous to the proteins of higher plant and cyanobacterial PSII, but, of course, the bacterial structure lacks the oxygen-evolving complex. PSII from higher plants consists of a chlorophyll pigment antenna, an RC complex with a molecule of P680 joined to structural proteins and components linking P680 to the water-splitting enzyme complex and to the electron acceptors of P680. Purified RC complexes have been analyzed by methods such as separation of proteins by SDS polyacrylamide gel electrophoresis with urea to denature and dissociate the subunit structure, coupled with Coomassie blue staining to identify the bands. However, early preparations were complex, with antenna chlorophyll interfering with the optical, EPR and X-ray spectroscopic signals used to analyze the structure. Recently it has been possible to isolate an RC core complex (the D1/D2/cyt b_{559} complex; see below), which contains very little chlorophyll and pheophytin but can still perform photochemistry; this has greatly facilitated analysis of RC processes.

Table 5.1 lists the main components of PSII so far identified. Some 22 polypeptides are associated with PSII. Firstly the reaction core: this consists of one of each of the two polypeptides, D1 and D2 (D for 'diffuse' appearance on gels) of different mass and with five transmembrane segments (but their functions are not equal), and is thus a heterodimer in construction. D1 is homologous to the L and D2 to the M subunit polypeptides of the bacterial RC complex. D1 contains a tyrosine (Y = Tyr161), often referred to as Y_z, which takes the electrons from the water-splitting complex, and binds pheophytin and Q_B. Q_B accepts electrons from reduced Q_A. D2 has the Y_D tyrosine, and binds the

Table 5.1. Components of the light-harvesting and electron transport complexes of the oxygenic plant thylakoid membranes as depicted in *Figure 5.5*

1.   Antenna protein–pigment complex
2.   32 kDa, D1 herbicide binding protein of the reaction center
3.   32 kDa, D2 reaction center protein
4.   Cytochrome b_{559}, 9 kDa b_{559} type 1 and 4 kDa b_{559} type 2 proteins
5.   Light-harvesting antenna
6.   10 kDa docking protein
7.   22 kDa stabilizing protein (intrinsic membrane protein)
8.   20 kDa Rieske Fe–S center
9.   Cytochrome b_6–f complex with polypeptides
10. Light-harvesting protein–pigment (chlorophyll *a* and *b*) complex of PSI and polypeptides
11. PSI reaction center, with two 70 (?) kDa polypeptides
12. Plastocyanin, 10.5 kDa
13. Plastocyanin binding protein (10 kDa)
14. Fe–S protein A, 18 kDa
15. Fe–S proteins B, 16 kDa
16. Fe–S protein 8 kDa
17. Ferredoxin binding protein
18. Ferredoxin
19. Ferredoxin, NADP oxidoreductase
20. Coupling factor, CF_0, membrane subunits

plastoquinone molecule, Q_A. Both bind P680, a pair of chl *a* molecules bound to D1 and D2, forming the 'special pair' but with weaker exciton coupling than the bacterial RC special pair. D1 is a product of the *psb*A gene and D2 of *psb*D. The RC also contains α-cytochrome b_{559} and β-cytochrome b_{559}, with hemes involved in photoprotection: these can exist in a number of different redox states and act as secondary electron donors and acceptors to several components of PSII. Another protein of unknown function, of 4.8 kDa, has one membrane-spanning helix with a charged C-terminus, which suggests a role in regulation. I protein, encoded by gene *psb*I is in the RC. The PSII core contains some 16 proteins: CP47 and CP43 (see *Table 4.2*) transfer excitation energy to the RC and are very hydrophobic with six transmembrane helices with amino- and carboxy-terminals on the stromal surface. Both also have large loops between the fifth and sixth helices exposed to the lumen, which bring CP47 into contact with the 33 kDa protein of the water-splitting complex and probably form the binding site of the manganese complex. CP47, CP43 and the 33 kDa protein are encoded by *psb*B, *psb*C and *psb*O, respectively, and are required for oxygen evolution. CP47 contains 20–22 chl *a* and two to four β-carotenes, and CP43 20 chl *a* and five β-carotenes. The 33 kDa protein may position chloride atoms close to the oxygen-evolving site. The α- and β-subunits of cytochrome b_{559} and several proteins of small mass are also required, for example the 23 and 16 kDa extrinsic proteins (genes *psb*P and *psb*Q) are involved in binding calcium (an obligatory co-factor for water oxidation) and chloride (also essential). These components function together with the water-splitting complex. Proteins H (photoprotection), K (PSII stability and assembly) L and X (Q_A function), M, N, R (donor and acceptor side functions), S (chaperonin in the antenna), T and W (both of unknown function), have rather unclear functions. They are encoded by *psb* genes with the same letter. Excitation energy is passed to the CP47 and

CP43 in the PSII core by Lhcb4 (also called CP29), Lhcb5 (CP26) and Lhcb6 (CP24), encoded by *lhcb*4, 5 and 6, respectively. Light harvesting is by the antenna complex, of Lhcb1, 2 and 3 (encoded by *lhcb*1, 2 and 3), and associated chl *a* and *b*, about 250 per RC, β-carotene, lutein, neoxanthin and violaxanthin.

5.2.2 Electron-accepting side of PSII

Under physiological conditions P680$^+$ oxidizes the donor, Z, a tyrosine residue Yz at amino acid position 161 on the D1 protein, as the kinetics of Z$^+$ formation are in phase with the reduction of P680$^+$. In turn, Z$^+$ is reduced by electrons from the water-splitting complex M. The Yz gives rise to a special EPR signal having transients with different decay rates called signal II$_{very\ fast}$ (20 μs), and II$_{slow}$ (1 s); II $_{very\ fast}$ results from transfer of e$^-$ to P680$^+$. The signal is not from a quinone but is a radical located on the e$^-$ donor side of PSII. The II$_{very\ fast}$ signal is stopped by removal of Mn^{2+} (and possibly polypeptides of the water-splitting complex) and also by DCMU (see *Table 5.2*). Thus, the tyrosine is bound close (by 1–1.5 nm) to the RC, allowing e$^-$ to transfer very rapidly within the membrane. The signal II$_{slow}$, which is stable in the dark, is probably from another tyrosine, Y$_D$, at position 161 on the D2 protein.

The P680 donates e$^-$ to a pheophytin molecule (generally called H$_A$ as it is the primary acceptor) located on the D1/D2 heterodimer. There is a second pheophytin (H$_B$) not directly involved in e$^-$ transport and on a part of the protein which is functionally distinct. H$_B$ may have a role in modifying the electronic structure of the RC, as in the bacterial RC. Probably D1 and D2 can change their conformation according to whether the quinone binding site (Q$_B$) is filled and also both can be phosphorylated, which is a way of modifying the efficiency of the electron transport system. Purple bacterial RCs (*Figure 5.2*) contain a third polypeptide called the H subunit, which does not function in e$^-$ transfer but probably conforms the L/M heterodimer. PSII may have a similar structure, called the quinone-shielding protein, which has homology in gene sequence and similar molecular mass. It has one transmembrane sequence and a very large N-terminus, probably on the stromal side of the membrane. Another 22 kDa protein may also serve to regulate PSII.

The core of PSII has been isolated as elliptical particles from cyanobacteria, green algae and higher plants of similar composition and structure. A combination of protein analysis by sophisticated electrophoretic techniques and electron microscopy (e.g. single particle averaging and electron crystallography) has shown the structure in detail. The monomeric PSII core complex has a molecular mass of *ca* 250 kDa, is about 12 × 10 nm and 8 nm thick, and contains one copy of each of the D1, D2 (together with the RC and electron transport components), cyt *b*$_{559}$, CP43, CP47 and the 33 kDa extrinsic proteins plus some low-molecular-mass polypeptides and *ca* 40 chl *a* molecules. The D1/D2 heterodimer is at one end of the ellipse and the CP43 and CP47 at the other. Oxygen is evolved from the particles. The PSII core monomers link anti-parallel, perhaps by CP47, forming PSII dimers, 20 × 12 × 8 nm, molecular mass 450 kDa, with the same components in the same proportion as the monomer. The two D1/D2 units in contact and a CP43 subunit form the antenna at each pole. During the synthesis of PSII, CP43 is incorporated last. This dimer resembles the structure of particles in thylakoid membranes shown by freeze-fracture (*Figure 4.6*). Further agreggation of the dimer with one set of light-harvesting chlorophyll-binding proteins (Lhcb1, 2, 4, 5) gives a PSII–LHCII complex 22 × 14 × 8 nm (*ca* 560 kDa mass). Yet further addition of another Lhcb forms the PSII–LHCII complex with two sets of RCs and two antenna complexes with about 200 chl *a* and perhaps four chl *b*, about 30 × 13 × 8 nm in dimensions and 700 kDa mass. The extrinsic 33 kDa subunit probably protrudes 3 nm from the luminal surface of the thylakoid membrane (*Figure 4.7*), indeed much of the hydrophilic protein of PSII is exposed to the lumen, but not to the stroma. The extrinsic 33 kDa protein shields the water-splitting Mn complex and projects into the lumen, along with the 23 kDa proteins which bind calcium and chloride ions involved in water splitting. Light energy is probably captured by the CP43, and the excitation energy passed to CP47 and then to the D1/D2 heterodimer. Energy can then be passed to either of the RCs of the dimer, so increasing efficiency if the rate of excitation transfer is greater than the rate of utilization by an individual RC. Fluorescence analysis shows cooperation between PSII monomers. The PSII complexes have been identified with the particles in the surfaces, both freeze-etched and freeze-fractured, of thylakoids (Chapter 4): the ES is the dimeric PSII–LHCII complex, which in the fractured membrane exposes the two 23 kDa proteins of oxygen evolving complex of each, giving the tetrameric appearance (see *Figure 4.6*).

Even larger scale interaction between PSII core complexes and antennae components occurs in cyanobacteria. Phycobilisomes on the membrane surface are linked to the PSII subunits in the membrane (Chapter 3, *Figure 3.6*), in the ratio 1:2, respectively, and match in size. Dense packing of the phycobilisome-PSII units increases the efficiency of energy capture and utilization. The system is dynamic, changing with light (see 'state-transitions', Section 5.6), perhaps allowing control over energy capture and exchange in relation to the cellular conditions. Higher plant thylakoids are stacked, enriched in PSII or unstacked, with PSI, and antenna complexes with chl *a* and *b*. PSII in the stroma is probably dimeric with proteins (e.g. CP29, CP26, CP24, see *Table 4.2*) that bind chlorophyll and connect the dimer to a trimer of LHC2. Probably the dimer is symmetrical, as mentioned, with an LHC2 subunit linked at both ends to the CP43 and 47 proteins. The CP29, 26 and 24 monomeric poypeptides bind only about 15% of the PSII chlorophyll, but may regulate excitation transfer from the LC2 complexes, which form the major, outer antenna with over 60% of the chlorophyll, to the RC. With one CP29, CP26 and CP24, and two to four trimers of LHC2 in the 'close' antenna there would be 230–250 chlorophylls per RC. Probably there is extensive interconnection between the antenna components, thus increasing the efficiency of energy capture and of excitation transfer. The system is also dynamic, changing with light and energy status of the thylakoid via protein phosphorylation, which alters the association (for details see Rögner *et al.*, 1996).

Reaction centers of PSII contain at least one tightly bound heme, cyt b_{559}, which has two subunits a and b (encoded by *psb*E_L and *psb*F_L) of 9 and 4 kDa, each with a histidine group which probably coordinates heme binding into the protein subunits. This cyt b_{559} can exist in two forms of different potential, high and low, but these do not correlate with PSII activity and probably have a protective function in decreasing the chance of photoinhibition, which damages the D1 protein particularly. It may also be involved in proton pumping during cyclic electron flow. Within the core complex, P680 is crucial to electron transfer capacity. Yet the structure of P680 is poorly understood (e.g. whether it is a monomer or dimer of chl *a* (Chapter 3). Probably P680 is a pair of chl *a* molecules which interact strongly, especially in the singlet state, and weakly in the triplet state. The chlorophyll may be oriented with its macrocycle parallel to the membrane plane. Separation between components of the RC is probably crucial to the efficiency of energy transfer. The distribution of energy between P680, H_A and Q_A is approximately in equilibrium. Q_A is, on evidence of the electrochemical shift in the blue and green spectral region of pheophytin, closer to pheophytin than Q_B; the distance between the Fe and the Q_A quinone ring is 0.7 nm. Y_D, although not involved in electron transport, interacts with manganese in the M complex within distances of 3–4 nm and sits almost midway between the thylakoid surfaces.

A major characteristic part of PSII is the 'regulatory cap' which is involved in water splitting. The cap is composed of three extrinsic proteins (hence called EP) of 33, 23 and 16 kDa and a minor protein of 5 kDa mass, which are hydrophilic and bind in equimolar amounts, probably two copies of each, to the luminal surface of the thylakoid. The cap may regulate passage of ions (Ca^{2+} and Cl^-) and water into the Mn complex. EP33 links to the RC and to the antenna, and EP23 and EP16 may shield it from the lumen and link to the distal antenna, judged by antibody studies. The binding sites may be of different charge; when removed the tetrameric complexes on the ES surfaces of thylakoids, are lost, leaving the PSII dimer.

Electron transport in a PSII reaction center.

A reaction center will accept or 'quench' excitation if it is reduced (P680), that is it contains an electron which can be ejected by excitation, but will not use excitation when oxidized (P680$^+$). Then excitation migrates to reduced RCs. Ejection of an electron occurs within nanoseconds; the fewer reduced RCs the longer the time that excitation dwells in the antenna and the greater the chlorophyll fluorescence. If Q is reduced, fluorescence yield is about 3%, but it increases to 12% if Q is oxidized. Events at the RC are summarized by the following sequence with Z the donor, P680 the reaction center, pheophytin (pheo) the primary acceptor and Q_A and Q_B secondary acceptors:

$$\text{Z.P680.pheo.Q}_A \xrightarrow{\;h\nu\;} \text{Z.P680}^+\text{.pheo.}^-\text{Q}_A \longrightarrow \text{Z}^+\text{.P680.pheo.Q}_A^- \longrightarrow \text{Q}_A \longrightarrow \text{Q}_B^{2-} \quad (5.2a)$$

$$\text{Z.P680.pheo.Q}_A \xrightarrow{\;h\nu\;} \text{Z.P680}^+\text{.pheo.}^-\text{Q}_A \longrightarrow \text{Z}^+\text{.P680.pheo.Q}_A^- \longrightarrow \text{Q}_A \longrightarrow \text{Q}_B^{2-} \quad (5.2b)$$

P680$^+$ is a very powerful oxidant (+1.2 V or greater), able to remove electrons from water (+0.8 V) and produce a relatively weak reductant (−0.6 V). The state of Z.P680.pheo.Q$^-$ is nonquenching, as electrons cannot be transferred even if P680 contains an electron. Recombination of pheo- with P680$^+$ possibly gives rise to the high variable fluorescence. Charge separation forms the radical pair P680$^+$.pheo$^-$, which can be detected very efficiently by ESR (quantum yield greater than 90%). For efficient separation, back-reactions of the electron with P680$^+$ must be limited; this is achieved by loss of energy as heat, delocalization of the electron and formation of a triplet-like state, and most importantly by spatial separation by rapid transfer (in a few hundred picoseconds) to the secondary quinone acceptor, Q_A. Q_A passes electrons to another acceptor, Q_B, which is bound to a

32 kDa polypeptide when oxidized but when reduced it diffuses into the PQ pool. Q_B is tightly bound to the protein and is reduced (in 200 μs), twice in sequence, then protonated. $Q_B H_2$ is not bound and diffuses into the pool of plastoquinone and is replaced by an oxidized Q. Inhibitors, such as the herbicide DCMU, may bind next to or at the site of Q_B attachment and thereby alter the energy charge at the site and prevent e^- transport. Different herbicides attach to different amino acids at the herbicide binding site, for example atrazine (urea/triazine type of inhibitor) targets a serine (264 on the D1 protein), and the phenol-type inhibitors attack histidine. Herbicides such as atrazine and DCMU bind to the same site or closely linked sites. Herbicide resistance frequently arises by mutation in populations of plants (weeds) subjected to frequent treatment with herbicides, and often a single amino acid change is all that is required for resistance, for example serine 264 changes to a glycine, conferring atrazine resistance, so herbicide resistance develops rapidly and is a major economic concern. Other inhibitors of electron movement in PSII include silicomolybdate and related ions, which may accept electrons from pheophytin and are thus insensitive to DCMU. The peptide probably regulates H^+ movement to Q_B, its reduction by added reductants, and also controls herbicide activity. Trypsin destroys the protein and influences granal stacking.

The M complex and water splitting.

The water-splitting enzyme complex is on the lumen side of the thylakoid membrane exposed to the aqueous medium, and is designated M. The type, structure and number of the proteins and associated electron donors and acceptors, particularly of manganese, the most important metal in the complex, and their binding to structural components is an active area of research. The complex forms a four-electron gate and governs the S states (see *Figure 5.7*) of the water-oxidizing process, passing electrons to Y_Z to reduce $P680^+$, producing four H^+ and four e^-, but holding the charge until four positive charges are 'accumulated' when water is split and O_2 is released. Manganese is an essential component of the complex with four atoms per PSII essential for active O_2 production, but their location on the PSII proteins or regulatory cap and how they function is speculative. In thylakoids two Mn fractions related to O_2 production have been identified: one fraction of two Mn atoms is tightly bound to PSII and probably involved in charge accumulation, and the other is more loosely bound and removed by Tris buffer, chelating agents (EDTA), or by heating, which stops O_2 liberation but not electron transport. Mn^{2+} added back to thylakoids stimulates O_2 evolution only after chemical reduction or illumination. This 'photoreactivation' requires phosphorylation and uses only PSII reactions; there is no interaction between PSII centers. Photoreactivation has two dark reactions: one changes Mn^{2+} to Mn^{3+}, the next Mn^{3+} to Mn^{4+}, and it is possible that these are required so that the metal can bind to a site in the configuration needed for water-splitting. Manganese is probably in clusters at two binding sites and may be in groups of one and three or in binucleate structures with one pair bound closely together (0.27 nm) the other less so, as judged from X-ray absorption and near infrared optical spectroscopy and EPR. It is possible that four Mn ions may form a distorted cubane structure, bound in part to a histidine on D1. Manganese ions are also bound to the O and N of the protein matrix; changes in S states are probably not related to physical rearrangement of their position. The regulatory cap has been discussed; the EP33 shields the Mn ('Mn stabilizing protein') but its regulation is complex. Removal of EP33 prevents the function of Mn but if high concentrations of ions are present changes of S states may still occur, although O_2 evolution may be slowed or stopped. Two cysteines in EP33 form a disulfide bridge, which, if reduced, changes the protein con-formation, and this prevents active O_2 evolution. Oxidation of the sulfide bridge can restart the process. Calcium and chloride ions are essential co-factors for PSII and their presence in the M complex is probably controlled by the regulatory cap. For example, if EP16 and EP23 are removed then O_2 evolution may decrease by up to 50%, but adding Ca^{2+} and Cl^- stabilizes the processes, probably by affecting the binding sites. Calcium is essential for O_2 evolution and there is a calcium-binding calmodulin-type protein which influences behav-ior. How chloride ions function is unknown; there may be several binding sites and not only is Cl^- effective but also bromide and nitrate ions, suggesting that electrostatic inter-actions determine the function. The regulatory cap possibly controls water entry and access of other components of the lumen to the active site of water splitting. Water mole-cules bind directly to Mn, judging from the use of isotopically labeled water and its analogs (NH_2OH, NH_3). Ammonia, which is isoelectronic with water, inhibits O_2 evolu-tion, the S states and EPR signals. It is not known if water is continuously bound at all stages of the oxidative cycle to Mn ions or to the protein matrix close to them or if inter-mediate redox states involve Mn ions and proteins; perhaps the histidines on the protein regulate the redox changes and electron transfer. Valency changes in the Mn clusters (and possibly associated proteins) are detected as EPR signals which are related to Mn in the S_2 state. The EPR $g = 2.0$ multiline signal is due to a multinuclear Mn complex, either two different conformations of a four-Mn cluster or a variable number of different valency Mn ions in the S_2 states. Changes in valency result in loss of the EPR signal. X-ray absorption-edge energies measurement suggests that the Mn oxidation state increases in the $S_0 \rightarrow S_1$ and $S_1 \rightarrow S_2$ but not in the $S_2 \rightarrow S_3$ states, or it is hidden. Currently the sequence of valency changes and associated electron and proton transfers are not understood.

Manganese is a period IV transition metal with oxidation states of $+2$ (most common) up to $+7$. Two electrons lost from each of two linked Mn atoms would give four oxidizing equivalents ($4+$). Mn^{3+} is stable in complexes and could act as intermediary oxidant; Mn^{4+} is also stable and water splitting could take place by single-electron-transfer steps. The reactions and stable state should not be too long lived. The Mn^{2+} complex with O_2 is unstable, so that O_2 is rapidly released, as required for fast water splitting. Oxygen must bind without releasing singlet or other forms of O_2 which could damage the protein. Manganese fulfils these requirements; ESR studies at low temperature on previously heated thylakoids, which release Mn^{2+}, show that manganese is involved in the S-states changes, and the physical entity corresponding to intermediate S. The molecular mechanism of H_2O binding to the complex and the role of ions such as chloride and nitrate, which act around PSII and are required for O_2 evolution, are unknown. It is important to know how water splitting is carried out *in vivo* because it may show how to produce H_2 for fuel from an unlimited source – H_2O.

Water splitting and cycle of S states.

When dark-adapted algae are illuminated with short flashes of bright light, separated by darkness, a characteristic pattern of O_2 evolution results (*Figure 5.6*). The first two flashes evolve little or no O_2, the third a large 'gush' and the fourth a smaller amount of O_2 than the third but more than the first, that is, a periodicity of four. Oscillations are damped and after some 20 flashes the yield per flash is constant. This pattern is characteristic of algae and chloroplasts. Oxygen production is slower just after illumination than with longer illumination and

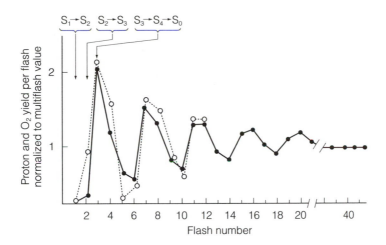

Figure 5.6. Oxygen evolution and proton release by chloroplasts given short (2 µs) intense flashes of light separated by darkness. The number per flash is expressed relative to the production after many flashes. (●—●) O_2 evolution; (○--○) H^+ evolution

this 'lag' period is inversely proportional to light intensity × time, showing that activation of an intermediate of water splitting is needed for O_2 production. The states induced by light flashes are not stable and light captured by one photosystem cannot contribute to another photosystem. Four photons captured by PSII co-operate to dissociate water; they cause the RC of PSII to eject four electrons. Four oxidation equivalents accumulate on an intermediate, S, before they accept four electrons from two molecules of water, releasing O_2. If S^{4+} is the oxidized component which reacts with water:

$$2\,H_2O + S^{4+} \longrightarrow S + 4\,H^+ + O_2 \tag{5.3}$$

and S is a 'charge-accumulating' chemical device. As photon capture is infrequent in dim light, the intermediate oxidized states must remain stable for sufficient time for four positive charges to accumulate and water oxidation to occur. A model by Kok *et al.* (1970) explains the periodicity of O_2 as a cycle of S states. It proposes that S accumulates four oxidizing equivalents solely from P680 and that only O_2 is liberated in a single process. If the water splitting and PSII complex, written as S.Z.P680.Q, is equivalent to S_0, then S_1 is S^+.Z.P680.Q^- with the RC refilled with an electron and one oxidizing equivalent accumulated on S; further flashes cause the sequence shown in *Figure 5.7*.

Events involving PSII reaction center and water splitting are therefore:

(1) activation of the RC chlorophyll, P680 and charge separation;
(2) reduction of acceptor and rapid donation of an electron from the water-splitting complex, S, via Z to $P680^+$;
(3) repetition until S_4 is formed, releasing 4 H^+; and
(4) removal of four electrons from two water molecules by S_4, and liberation of O_2.

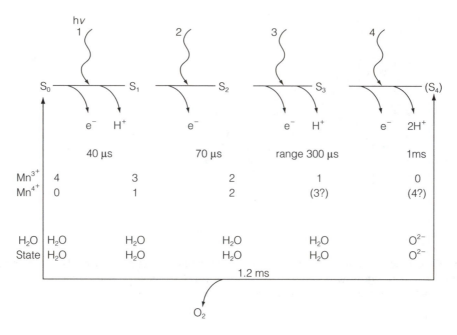

Figure 5.7. Sequence of events in water splitting and O_2 evolution in PSII. The S states are related to the electron and proton release, manganese changes and states of the water molecules which are split (speculative, see text). The cycle is driven by light (hν) capture

To explain why most O_2 is released on the third flash (rather than the fourth as expected), it is assumed that in darkness S is 75% in the state S_1 and 25% in state S_0. The damped oscillations observed experimentally are thought to result from 5% of the light flashes causing double hits and two oxidations and 10% of the photons not hitting a target, so averaging the states. The rate of conversion of each step $S_0 \rightarrow S_1$ and $S_1 \rightarrow S_2$ is probably 40–70 μs, $S_2 \rightarrow S_3$ about 300 μs, but $S_3 \rightarrow S_4 \rightarrow S_0$ requires 1 ms so the O_2 evolving step is rate limiting. It is not clear if water oxidation limits the process or the rate of O_2 release. In step (4) all oxidation occurs at the $S_3 \rightarrow S_4 \rightarrow S_0$ step. The S states decay in darkness; S_2 and S_3 are unstable and are deactivated to S_1 by addition of electrons, probably from reduced compounds via the electron transport chain and cyclic electron transport. Electrons flow from PQ via Q to P680 then to S_2, but not to S_3. States S and S_1 equilibrate in darkness and S_0 goes to S_1 if artificially reduced by ferricyanide or DCPIP and ascorbate. Hydroxylamine (NH_2OH) and ammonia bind tightly to the O_2^- evolving complex and donate electrons, removing two oxidizing equivalents and increasing the lag in O_2 evolution which cannot be reversed by washing. The S_2 and S_3 states are rapidly destabilized by a diverse group of very lipid-soluble, weak-acid anions with -NH and -OH groups, called ADRY agents from acceleration of the deactivation reactions of Y (alternative sign for the water-splitting system), an example being Ant 2p [2(3-chloro-4trifluoreomethyl)anilino-3,5-dinitrothiophene]. ADRY reagents transport electrons from Q over the energy barrier to S.

The mechanism of O_2 evolution involves more than accumulating four oxidizing equivalents before O_2 release. Measurements of pH changes in thylakoids treated with uncoupling agents (Section 6.6) to prevent accumulation of H^+ caused by the electron flow through the thylakoid membrane show a complex release of protons (H^+) in relation to the light flashes (*Figures 5.6 and 5.7*). Single protons are released at the S_0 to S_1 and S_2 to S_3 steps, whilst 2 H^+ are released from S_3 to S_0 via S_4. Release of H^+ neutralizes the charge accumulation on S, so it is not justified to call it a 'charge-accumulating' device.

5.2.3 Electron transport beyond the primary acceptor of PSII

Energized electrons from P680 reduce Q_B to the semiquinone and anionic plastoquinone which accepts protons to give hydroquinone. Reduced Q_B diffuses from the binding site and enters the PQ pool, which (as PQ is very lipid soluble) is almost exclusively in thylakoid membranes. Quinones are important in many biological electron transport processes, carrying e^- and H. Several forms exist but PQ B and C are usually less than 20% of the amount of PQ A and their structures are poorly known. PQ is a substituted benzoquinone with two methyl ($-CH_3$) groups attached to the ring and a long chain of nine isoprene groups; other types of PQ have three and four isoprene units. PQ A is colorless, absorbs light in the ultraviolet (260 nm) and is reduced at +0.1 V via plastosemiquinone to plastohydroquinone, which absorbs at 290 nm. The pool of PQ engaged in electron transport in the light is several times larger than that of other carriers with a molar ratio to chlorophyll of 1:10, with about 40 molecules of PQ per PSII RC; this large pool accepts e^- from several PSII units, acting as a 'buffer' with PSI. PQ is oxidized by cyt *f*, which itself is mainly oxidized under bright light because electrons are more rapidly removed by PSI than they are supplied from PQ, that is, the rate limiting step lies between PQ and cyt *f*. The large pool of PQ and rapid oxidation by cyt *f* ensure that PQ is not completely reduced in bright light; this enables PSII to function and donate e^-, even though the transport rate from Q to cyt *f* is slow.

PQ also has the important ability to transport protons across the thylakoid membrane (*Figure 5.4*). Reduced (anionic) PQ on the stromal side picks up protons and carries them to the lumen, thus increasing the H^+ concentration inside as a consequence of electron transport. The mechanism is, probably, that PQH_2 diffuses laterally between PSII and cyt *f* (distances of the order of 100 nm) within the membrane, losing H^+ due to the electrical charge on the membrane. If PSII and PSI are unequally distributed between the granal and stromal thylakoids, transport of electrons between them over some distance may be by diffusion of PQ. The cycle of PQ reduction and oxidation, which transfers H^+ across the thylakoid, is central to the generation of the pH gradient component of the proton motive force for ATP synthesis (Chapter 6) and for coupling to e^- transport. Mitchell (1975) suggested that PQH_2 loses e^- to Rieske Fe–S centers and H^+ to the lumen of the thylakoid. The Rieske Fe–S centers are part of the cyt b–c_1 or b_6–f complex in the photosynthetic membrane. The PQH then 'dismutates', giving e^- to cyt b_{563} and to cyt b_{559} low-potential form, and losing H^+ to the lumen. The cytochrome cycles e^- back to reduce PQH on the stromal side of the thylakoid and picks up more H^+. This process is called the proton motive 'Q cycle' or quinone cycle. It describes electron and proton transfer in the thylakoid and involves the cyt b_6–f complex. The sequence is that reduced plastoquinone, in the thylakoid membrane, loses an electron to the Rieske Fe–S center on the lumen side of the membrane, and two protons diffuse into the lumen (possibly fewer than 2 H$^+$ may be transported per electron). The semiquinone formed transfers its electron to the heme in the lumen side of the membrane, followed by reduction of a heme on the stromal side, thus moving the negative charge across the membrane. The heme on the stromal side reduces a quinone on the stromal side, which attracts a proton, forming a semiquinone. Further oxidation of reduced PQ at the lumen side results in another reduction of the lumen-side heme and transfer of the electron to the stromal heme, which then reduces quinone on the stromal side. In this state, a second proton is attached and the PQH$_2$ diffuses back into the pool, cycling to the lumen side in a continuation of the cycle. This is supported by evidence from the spectroscopic changes in hemes, that the transfer of electrons from the lumen-side heme to the stromal is spontaneous due to differences in their mid-point redox potentials, about 100–140 mV. The stoichiometry of the Q-cycle is important for assessing the relationship between electron transport and ATP synthesis. It provides a role for the cytochromes and could also give 2 H$^+$ per e$^-$ transported. Presently, however, the experimental evidence is equivocal: one and two H$^+$ per e$^-$ have been measured. Some 90% of the ATP required for CO$_2$ assimilation can be provided by linear electron transport to NADP$^+$ (1 H$^+$/e$^-$), assuming 3 H$^+$/ATP, but with a ratio of 4 extra ATP is required. In C4 plants the ATP production is also inadequate with a ratio of 3 H$^+$/ATP. A Q-cycle would overcome the limitation of protons. Increased ATP production via cyclic electron transport could involve electrons from reduced ferredoxin transferring back to the heme on the stromal side of the membrane and entering the cycle.

Electrons pass from PQ to cyt f via a Rieske Fe–S center, with characteristic ESR signal and midpoint potential of +0.2 V, which functions in a one-electron transfer,

$$PQH_2 + Fe\text{–}S_{oxidized} \longrightarrow PQH + H^+ + Fe\text{–}S_{reduced} \tag{5.4}$$

The two e$^-$ from PQH$_2$ go to separate Fe–S molecules as the Rieske centers are single e$^-$ carriers; it has been suggested that they carry H$^+$ also. Mutants of *Lemna* without the Fe–S center lack noncyclic electron transport. The cytochrome b_6–f complex is inhibited by a number of compounds, such as stigmatellin, antimycin and DBMIB; the latter binds to the plastoquinol oxidation site.

The cytochrome b_6–f complex, a plastoquinol:plastocyanin oxidoreductase, is composed of redox polypeptides cyt b_{563} (called b_6) and cyt f and the Rieske iron–sulfur proteins, in ratio 2:1:1. The complex is a dimer of 210 kDa mass with 22–24 transmembrane helices, contributed in the monomer by seven polypeptides, four from *pet*B, three from *pet*D, and one each from *pet*A, C, G, X and L. Cytochrome f is an elongated (7.5 × 3.5 × 2.5 nm) protein of molecular mass about 65 kDa, with two domains made of β-strands with heme bound by four histidines (very strongly conserved in all sequences) to the N-terminus, on the lumen side of the thylakoid. The two hemes are orientated perpendicular to the membrane surface, separated by about 1.2 nm. Cytochrome f has an absorption band at 554 nm when reduced and a potential of +0.37 V. The redox characteristics of cyt f are not altered by incorporation in the complex. There is a 1.1 nm long chain of five water molecules buried in the molecule. They are conserved in the protein from different sources, and are hydrogen-bonded with a conserved group of amino acid residues, suggesting a major role for the structure in the transport of H^+, indeed the five water molecules are called a 'proton wire' as they may provide a long-distance pathway for H^+ to enter the membrane through the parts of the protein complexes exposed at the stromal surface, to move through the Rieske protein without affecting its redox activity, and to leave the molecule at the lumen side. Cytochromes are organelle-encoded and the Rieske iron–sulfur protein is nuclear encoded. Three other small (5 kDa) polypeptides are present, and there is currently much interest in explaining how the components of the complex are moved to the site of assembly, inserted into the membrane and folded, and the heme inserted (see Chapter 10). Cytochrome may move towards the outside of the thylakoid on illumination as the membrane changes shape and becomes thinner, increasing its ability to accept electrons from the stromal side during cyclic electron flow.

Cytochrome *f* donates single electrons to plastocyanin (PC), a 40 kDa protein of four polypeptides each with one Cu atom coordinated by two N atoms from histidine and two S atoms from cysteine in an eight-stranded β-barrel. This site is characterized by hydrophobic surfaces and two regions of negative charge, giving two routes of electron transfer to the Cu; both are close to a tyrosine amino acid (Tyr83) on PC. PC has a redox potential of 0.37 V and occurs in the thylakoid lamella of oxygenic tissue, in the ratio of one molecule to 400 chlorophylls. PC is blue when oxidized, with a major absorption band at 597 nm, and is inhibited by mercury ($HgCl_2$) which blocks the -SH groups, stopping all electron flow to PSI. PC accepts one electron from cyt *f* and may form a pool of electrons which can be passed to PSI. Electron transfer from cyt *f* to PC is facilitated by close contact and electrostatic interaction between them, related to the structures of both molecules. Cytochrome *f* exhibits a molecular structure which is called a 'docking-site' for PC, a positively charged group of lysine amino acids which allows the two molecules to link in a particular conformation, possibly in a sequence, and provides a fixed spatial structure to facilitate the electron transfer. The PC may be released from the cyt *f* by protonation after it is reduced.

The important role of metals as co-factors in electron transport should be emphasized. Many of the components of the electron transport chain are metalloproteins (e.g. cytochromes, Fe–S proteins, plastocyanin, ferredoxin), but this type of metal–protein interaction is essential for the function of enzymes in many biochemical processes, for example MGDG synthesis in thylakoids, action of carbonic anhydrase, and ferredoxin–thioredoxin reductase activation of chloroplast enzymes. Indeed, the list of metalloproteins is long and there is great variety in the metals (e.g. copper, iron, zinc, manganese, magnesium) and how they are held in the protein, coordinated directly with the apoprotein, as inorganic clusters, or with inorganic co-factors. The mechanisms of action are many, but the essential point is that the metal–protein complex provides an electronic structure which is energetically favorable for electrons to move from substrate to product, or between components in electron transport chains. It is not yet clear how metalloprotein complexes are made and assembled. Clearly the metal must be available, as well as the protein; often this is made in the cytosol and transferred into, for example, the thylakoid membrane by means of transit peptides (Section 10.6). Then the holoenzyme is assembled. Distinct transport and assembly mechanisms are probably required for the task. Construction of iron–sulfur centers in PSI and ferredoxin is an example: they occur in the chloroplast, the rates of synthesis of components must match plant development, specific enzymes are required coded in the chloroplast genome or nucleus and iron, stored as ferritin in etioplasts, is used as the chloroplast develops. To understand the complexity of the processes which occur just to ensure the accurate formation of photosynthetic components provides a fascinating challenge.

5.3 Photosystem I

The structure of PSI (*Figure 5.5*) is formed by 13 different protein subunits, seven extrinsic and hydrophilic and six hydrophobic and intrinsic to the membrane, with a total of 28 helices spanning the thylakoid §membrane. Five polypeptides are chloroplast encoded, the others nuclear. This photosystem has a heterodimeric reaction center of two polypeptides, PSI-A and PSI-B of 82 and 83 kDa mass, each with 11 helices traversing the membrane. PSI-A and PSI-B are homologous (45% identity) and encoded by the genes *psa*A and *psa*B, respectively. The polypeptides bind the reaction center chl *a* dimer (P700), the initial electron acceptors A_0 (chl *a*), A_1 (phylloquinone) and A_2 (a 4Fe–4S center), plus the other chlorophylls and β-carotenes. The antenna has about 100 chl *a* per P700, and an associated chl *a/b* antenna, LHC1 (*Table 4.2*), made from four different subunits. The structure of PSI is not like the photosynthetic purple bacterial RC, but is more like that of green-sulfur bacteria. The chlorophyll dimer and quinones are paired and arranged along a two-fold axis with the single 4Fe–4S center above, in a cage formed from the transmembrane helices for both of the polypeptides. The approximately 100 chl *a* molecules are attached to the outside of the protein cage at different points and orientated perpendicular to the membrane axis. The β-carotenes are orientated with respect to the chlorophyll iron–sulfur proteins (nonheme iron) and proteins which bind and structure the units are found in the complex. Other intrinsic polypeptides include products of genes *psa*I, J, K and L (of 4, 5, 8 and 18 kDa mass, respectively); their functions are unclear. Extrinsic proteins include *psa*C, which binds the Fe–S centers, *psd*D, E (responsible for the position of ferredoxin) and F (plastocyanin docking). The polypeptides have masses of 9, 16, 9 and 17 kDa, respectively. The 10 kDa *psa*G subunit is of unknown function, and the 10 kDa *psa*H may link PSI and LHC1. All theses proteins are on the stromal side of the membrane. On the lumen side of the thylakoid, *psa*I, J, K and L (masses 4, 5, 8 and 18 kDa) and a 9 kDa protein have no clear function. On the lumenal side plastocyanin (*pet*E) is associated with PSI and, in the stroma, ferredoxin (*pet*F) is a major electron donor and FNR (*pet*H) is responsible for $NADP^+$ reduction.

Excitation from the antenna leads to oxidation of the P700 RC chlorophyll. Electron movement around PSI is so fast that there is little fluorescence from the antenna at room temperature, although at the temperature of liquid nitrogen ($-196°C$) fluorescence at 730 nm is observed from PSI antenna chlorophyll. The primary acceptor, X, is possibly a pheophytin anion, or a chlorophyll monomer in a special environment within a polypeptide. A secondary acceptor is found in the form of ESR signals characteristic of 2Fe–2S or, more probably, 4Fe–4S centers of an 18 kDa protein; it is called Fe–S center B (Fe-S.B). Another Fe–S center (Fe-S.A) may take electrons from Fe–S.B. With oxidation–reduction of P700, an optical spectroscopic signal occurs at 430 nm (hence called P430), possibly from the 4Fe–4S centers. The Fe–S centers delocalize electrons, stabilizing them for long enough to enable chemical reactions to take place. The different Fe–S units associated with PSI may allow e^- to be transported to different processes depending on conditions; if $NADP^+$ is limiting then e^- may pass to the electron chain via another center.

5.3.1 Donation of electrons to ferredoxin

The very negative mid-point potential of the reduced carriers enables Fd to be reduced. Ferredoxins are electron carrying iron–sulfur redox proteins, found in animals and plants, with Fe at the active center (but not as heme), usually as 2Fe–2S (or sometimes 4Fe–4S). Plant ferredoxins are small polypeptides of approximately 10 kDa mass and reddish brown in color, and transfer one electron at each step. They have a redox potential of -0.43 V (soluble ferredoxin), characteristic absorption spectra below 500 nm and ESR spectra which are important for their identification and quantitative estimation. Ferredoxins in the thylakoid membrane attached to PSI accept electrons from the secondary acceptors of PSI, but soluble ferredoxin is situated on the stromal surface of the thylakoid and may receive electrons from a ferredoxin-reducing substance (FRS), but there is some doubt about the role of this protein. Ferredoxins from chloroplasts pass electrons to many biological processes, back to PQ in the electron chain, for example, giving cyclic electron transport, or reducing $NADP^+$, the normal route in photosynthesis:

$$2 \text{ ferredoxin}^- + NADP^+ + H^+ \longrightarrow NADPH + \text{ferredoxin} \qquad (5.5)$$

Ferredoxin reduces $NADP^+$ via the flavoprotein enzyme ferredoxin $NADP^+$ oxido-reductase located in the stromal side of the thylakoid. The higher plant enzyme is of 34 kDa mass, contains one molecule of FAD, an -S-S- bridge and four -SH groups, one of which binds in the molecule and is essential for catalytic activity. It forms a complex with ferredoxin and $NADP^+$ bound to a lysine amino acid. The reductase also catalyzes other reactions with ferredoxin or NADPH; for example, NADPH reduces cyt f or NAD^+, and NADPH may be oxidized by transferring an electron to ferricyanide, dyes, *et cetera* (diaphorase activity). It is therefore an important control point in the electron chain and may redistribute electrons to other substrates if the normal acceptor, $NADP^+$, is in short supply. Ferredoxin also provides electrons to reduce sulfate and nitrate (Chapter 7) and activates some enzymes of the photosynthetic carbon-reduction cycle. Reduced ferredoxin reacts with molecular oxygen forming the superoxide radical O_2 and H_2O_2 in the Mehler reaction. Ferredoxin also links electron transport with the chemical reactions and provides flexibility in metabolism.

$NADP^+$ and NADPH, the oxidized and reduced forms of the photosynthetic reductant with characteristic spectra, are water-soluble molecules. The structure (which resembles that of ATP) is complex and maintains a stable redox state under cellular conditions, allowing specific binding to enzymes, *et cetera*. Reduction involves a 2 e^- transfer and one proton 'adds on'. The simplest model of electron flow from PSI to $NADP^+$ is:

$$P700 \rightarrow A_1 \rightarrow (A_2) \rightarrow (A + B) \rightarrow Fd_{sol} \rightarrow NADP^+ \tag{5.6}$$

but models with parallel branches have been suggested, allowing more flexible dispersal of electrons if $NADP^+$ limits electron flow.

5.4 Energetics of electron transport

To remove an electron from water (+0.8 V potential) to ferredoxin (−0.43) requires two photons acting in series (Chapter 3). *Figure 5.8* shows the Hill and Bendall 'Z' scheme (so-called from its appearance) of electron transport. P680 ejects e^- to Q at about 0 V potential and P680$^+$ (potential +1.2 V) removes e^- from H_2O via S and Z. Electrons from Q pass to P700 at +0.4 V; the coupling of electron transport to ATP synthesis is considered in Chapter 6. P700 ejects an electron to ferredoxin at −0.43 V, an accumulated energy of 1.2 V. However, the total energy in the two photoacts is +0.8 to −0.2 V in PSII and +0.4 to −0.8 V in PSI, a total of 2.2 V. Of the 1 V lost, part is recovered in ATP synthesis. The efficiency of photoactivation and electron transport is $1.4/2.2 \times 100 = 64\%$ or greater. With a minimum of two photons per e^- transported, eight quanta of red light are needed per oxygen released, an energy of 23.4×10^{-19} J. The energy accumulated by four electrons passing to $NADP^+$ and synthesizing 2.7 ATP is around 8.8×10^{-19} J so maximum efficiency is 38%. However, if the efficiency with respect to the total solar spectrum, not

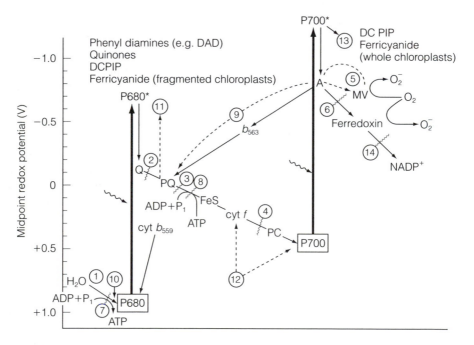

Figure 5.8. The 'Z' scheme of photosynthetic electron transport and redox potential diagram for the photosystems and components of the electron transport chain. Numbers show the places at which inhibitors act, or at which e^- can be added or removed from the chain. Some compounds acting at these points are listed in *Table 5.2*

just red light, is calculated then the maximum efficiency of the photochemical processes at the RCs is about 20%.

5.4.1 Rates of processes in electron transport

Physical reactions within the pigment bed are much faster than electron transport processes. Photon capture and excitation migration to the RC are fast (10^{-15} and 5×10^{-12} s, respectively). Fluorescence from chlorophyll occurs in 10^{-9} s when PSII is reduced, but is slower when it is oxidized. Fluorescence from PSI is very limited as e^- transport from the RCs occurs in picoseconds. Electron transfer from RCs to acceptors requires about 20×10^{-9} s. What happens between 10^{-9} and 10^{-12} s is unknown. An electron is transferred from water to $NADP^+$ in 20 ms. Water splitting (0.2 ms) and electron movement from the PSII acceptor to plastoquinone (400 µs) are fast in comparison with e^- movement from plastoquinone to cyt f, which requires about 20 ms and is the rate-limiting step in the process; the large pool of PQ minimizes this limiting step in e^- transport. The pool of plastoquinone is of variable size, increasing under conditions such as bright light and low temperature when the donor is highly reduced and the pool turns over relatively slowly. Electron transfer from plastocyanin to PSI takes 0.2 ms and from PSI to $NADP^+$ 10 µs. The plastoquinone 'pump' of H^+ across the thylakoid controls the rate of H^+ gradient development, a six-fold difference in rate (16–90 ms) occurring when changing from conditions of high to low back-pressure of H^+; proton transport is probably slower (60 ms) than PQ to cyt f transfer and the development of ΔpH requires 30 s or so. Protons diffuse across the thylakoid in 5 s if CF_0–CF_1 (see Chapter 6) is not functioning. Proton transport rates across the membrane may be controlled by the herbicide-binding protein of PSII.

5.4.2 Artificial electron donors and acceptors

Many artificial redox compounds donate electrons to or accept them from the electron transport chain. They have been important in analysis of photosynthetic processes since Hill's discovery in 1937 that ferricyanide accepts electrons from water with oxygen evolution (now called the Hill reaction, and the substances Hill reagents):

$$4\ Fe(CN)_6^{3-} + 2\ H_2O \xrightarrow[\text{Chloroplasts}]{\text{Light}} 4\ Fe(CN)_6^{4-} + 4\ H^+ + O_2 \qquad (5.7)$$

Changes in the absorbance spectra of ferricyanide or a dye like DCPIP indicates the oxidation–reduction state of the electron acceptors and donors and electron flow. As their redox potentials are known, the potentials of the electron chain components may be determined. Differences in lipid solubility, molecular size, *et cetera*, can also be exploited to indicate where a reaction is taking place in the thylakoid or chloroplast. Electron transport is blocked by compounds (see *Table 5.2*) which remove electrons from different parts of the chain or are nonfunctional analogs of compounds in the chain. Hydroxylamine, as mentioned in Section 5.2.2), may occupy the position of H_2O in the water-splitting complex and inhibit water splitting. The viologens, for example, methyl-viologen transfer

Table 5.2. Sites of inhibitor activity and electron donation or acceptance identified by numbers on the electron transport scheme (*Figure 5.7*), the processes affected and examples of compounds affecting them

Site	Process	Compound
Inhibitors		
1	Water splitting	NH_2OH; TRIS
2	e^- transport to PQ	DCMU; atrazine
3	e^- and H^+ transport by PQ	DBMIB; DAD
4	e^- transport in Fe–S centers	HCN; mercurichloride; amphotericin B
5	e^- removal from PSI and autocatalytic production of oxygen radicals	Bipyridyl herbicides (Paraquat)
6	Reduction of $NADP^+$	DSPD
7 and 8	Uncoupling ATP synthesis from e^- transport	NH_4^+; FCCP; CCCP; DCCD; Dio-9; DNP; phlorizin
14	$NADP^+$ reduction	2-Phosphoadenosine diphosphate ribose
Artificial electron donors and acceptors		
9	e^- carrier from PSI to PQ giving cyclic flow	PMS
10	e^- donor to PSII	Catechol, DPC, ascorbate; phenylenediamine
11	Acceptor from PSII	DCPIP at PQ; KFeCN (fragmented chloroplasts); DAD
12	PSI donors	Ascorbate; DPC; PMS; DAD reduced; DCPIP
13	Acceptors from PSI	DCPIP; KFeCN (whole chloroplasts); Paraquat; MV (acceptor from F_A)

Abbreviations

Atrazine	2-Chloro-4-(2-propylamino)-6-ethylamine-5-triazine
NH_2OH	Hydroxylamine
DCMU	3(3,4-Dichlorophenyl)-1,1-dimethylurea (Diuron)
DBMIB	2,5-Dibromo-3-methyl-6-isopropyl-*p*-benzoquinone (Dibromothymoquinone)
DCPIP	2,6-Dichlorophenolindophenol
HCN	Hydrogen cyanide
DAD	2,3,5,6-Tetramethyl-*p*-phenylenediamine (Diaminodurol)
CCCP	Carbonylcyanide-*m*-chlorophenylhydrazone
Dio-9	Antibiotic
DSPD	Disalicylidenepropanediamine
DPC	Diphenylcarbazide
PMS	Phenazinemethosulfate (5-methylphenazonium-methylsulfate)
FCCP	Carbonylcyanide-*p*-trifluoromethoxyphenylhydrazone
DCCP	Dicyclohexylcarbodiimide
DNP	Dinitrophenol
TRIS	Tris (hydroxymethyl) aminomethane
MV	Methylviologen

electrons to O_2, forming singlet oxygen, which destroys lipids (Section 5.4.3). Viologen recycles, so the process is autocatalytic and destroys tissues rapidly in the light. DCMU is a quinone analog and blocks the acceptor of PSII or between Q and PQ, thus separating PSII from PSI. One molecule of DCMU per 100 chlorophylls completely stops electron transport. Atrazine stops e^- flow at PQ by inhibiting the binding of Q to the protein complex. DBMIB is a structural analog of PQ and prevents electron transfer to cyt f. An analog of $NADP^+$, phosphoadenosine diphosphate ribose, blocks ferredoxin $NADP^+$ reductase. The antibiotic DIO-9 inhibits ATP synthesis at CF1.

Inhibitors of PSI include antimycin A, atrazine and *o*-pheanthroline, and quinone-related compounds which block the electron acceptor site A1 (which is vitamin K1) in the core complex. The Fe–S centers are blocked by mercury compounds, such as mercuric chloride. Electrons are donated to PSI by phenyldiamine, and to cyt f by DCPIP plus ascorbate. Using DBMIB to block electron flow, DCPIP (plus ascorbate for reduction) donates electrons to PSI and methyl viologen is the acceptor; these reagents provide a test system for PSI and the effects of conditions on it. PSII is measured by electron flow to DCPIP with transport blocked by DBMIB. Open-chain electron transport is measured with water as donor and viologen or $NADP^+$ as acceptor. Cyclic electron transport is measured in broken chloroplasts which have lost soluble ferredoxin, by adding PMS which carries electrons from PSI back to PC or cyt f. The mechanisms of interaction between many inhibitors and herbicides and the electron transport chain are now understood at the molecular level since their structures have been determined. Some inhibitors are commercial herbicides such as DCMU, the viologens (methyl viologen is called 'Paraquat' commerically) and atrazine.

5.4.3 Formation of reactive forms of oxygen by photosynthesis

Oxygen in its diatomic form (O_2), which contains two unpaired electrons with parallel spins ('triplet' oxygen), is not only a product of photosynthesis but is 'assimilated' into different forms as a consequence of it. Light energy excites chlorophyll, principally of PSI, resulting in high energy electron excitation in long-lived molecular form which can react with oxygen with the formation of reactive O_2 species. An overview is given in *Figure 5.9* (see Asada, 1999; Noctor and Foyer 1998):

$$O_2 \xrightarrow{e^-} \underset{\text{superoxide}}{O_2^{\cdot-}} \xrightarrow{H^+} \underset{\substack{\text{perhydroxyl} \\ \text{radical}}}{HO_2^{\cdot-}} \xrightarrow{e^- \ H^+} \underset{\substack{\text{hydrogen} \\ \text{peroxide}}}{H_2O_2} \xrightarrow{e^- \ H^+} \underset{\substack{\text{hydroxyl} \\ \text{radical}}}{OH^{\cdot}} \xrightarrow{e^- \ H^+} 2H_2O \qquad (5.8)$$

In photosynthesis, excited pigments (e.g. triplet chlorophyll, ^{3}chl) donate energy to O_2 giving $^1O_2$ (singlet oxygen):

$$\text{chl} \xrightarrow{h\nu} {}^1\text{chl}^* \qquad {}^3\text{chl}^* + O_2 \longrightarrow \text{chl} + {}^1O_2^* \qquad (5.9)$$

or by electron transfer, producing superoxide:

$$\text{chl} \xrightarrow{h\nu} {}^3\text{chl}^* + O_2 \longrightarrow \text{chl}^+ + O_2^{\cdot-} \qquad (5.10)$$

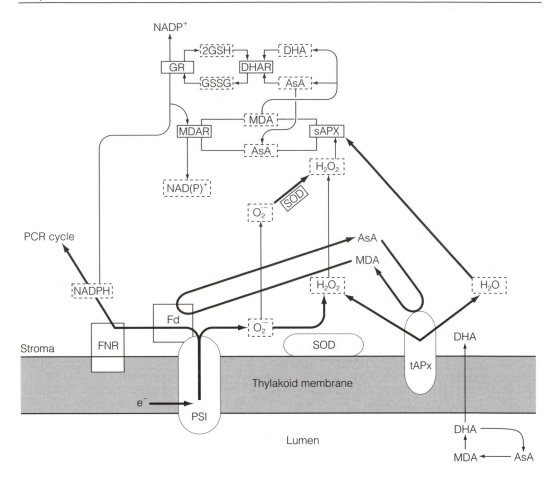

Figure 5.9. Schematic of production and dissipation of active oxygen (singlet, O_2^-) by the Asada ascorbate pathway associated with the thylakoid membranes and the stroma of chloroplasts. Electron transport in PSI leads to reduction of O_2. Superoxide dismutase (SOD) breaks the superoxide down to hydrogen peroxide (H_2O_2), then thylakoid and stromal ascorbate peroxidase (sAPX and tAPX, respectively) are engaged in the removal of singlet oxygen. Also, the ascorbate and glutathione cycles are required to dissipate the peroxide formed by SOD breakdown. MDA, monodehydroascorbate; MDAR, monodehydroascorbate reductase; DHA, dehydroascorbate; DHAR, dehydroascorbate reductase; GR, glutathione reductase; GSH and GSSG, reduced and oxidized glutathione; AsA, ascorbate

Photosystem I may pass e^- to O_2 directly or via intermediates, paraquat for example, which reduces O_2 to superoxide and recycles to carry more e^- (the basis for its herbicidal activity), or from physiological intermediates such as ferredoxin which produces superoxide, although probably the 4Fe–4S clusters in the PSI complex are the donors.

The rate of superoxide synthesis is not large, 20 μmol mg chl^{-1} h^{-1} for a PSI complex with 600 molecules of chlorophyll per RC, and the K_m for oxygen is micromolar. In leaves, where oxygen reduction is shown by uptake of $^{18}O_2$ in the light when photorespiration is prevented by very large CO_2 concentration, the K_m for oxygen is about 80 µM (compared with the concentration in water in equilibrium with air of 250 µM), so photoreduction is saturated in air containing some 15% O_2 and in brighter light than is the case for thylakoids. Photoreduction of O_2 increases greatly when normal routes (e.g. CO_2 assimilation) for use of electrons from PSI are inhibited. In some conditions rates of the active oxygen cycle can be large, 20% in tropical trees, and estimates of 30% of total electron flux in leaves in bright light or water-stressed wheat, corresponding to 25 e^- per P700 s^{-1} for superoxide synthesis and 50 e^- per P700 s^{-1} for its reduction to water.

Production of superoxide may be linked to ferredoxin and peroxide (H_2O_2) from the Mehler reaction:

$$Fd_{red} + O_2 \rightarrow Fd_{ox} + O_2^{\cdot -}$$
$$Fd_{red} + O_2^- + 2H^+ \rightarrow H_2O_2 + Fd_{ox} \tag{5.11}$$

Hydrogen peroxide and O_2^- react to give the hydroxy radical ($^\cdot OH$) in a reaction catalyzed by metal salts:

$$H_2O_2 + O_2^{\cdot -} \xrightarrow{Fe_{salt}} O_2 + {^\cdot OH} + OH^- \tag{5.12}$$

The hydroxyl radical ($^\cdot OH$) is more reactive than the other oxygen radicals, because it reacts very quickly (at a rate determined by its diffusion) with nearby proteins, hence there is need for very rapid destruction. Active oxygen species are removed by enzymes and by redox-coupled reactions involving ascorbic acid and glutathione. Superoxide reacts with H^+ in the presence of the enzyme SOD, in alkaline conditions, to give H_2O_2 which is destroyed by catalase:

$$O_2^{\cdot -} + O_2^{\cdot -} + 2\,H^+ \xrightarrow{SOD} H_2O_2 + O_2$$
$$2\,H_2O_2 \xrightarrow{Catalase} 2\,H_2O + O_2 \tag{5.13}$$

Both enzymes act very rapidly to destroy these potentially damaging forms of active oxygen, resulting in a cycle in which oxygen released ultimately from water is metabolized to water, consuming reductant generated by the photosynthetic system in the process. One molecule of O_2 is reduced to two molecules of H_2O by 4 e^-, which come from H_2O, so the balance is no net O_2 use but electrons consumed. Thus it acts as a mechanism which can balance the primary processes of energy capture and reductant generation with the demands for them by metabolism (e.g. CO_2 assimilation). For this to function effectively, the enzyme system must be able to use active oxygen species at small concentrations (to prevent accumulation to damaging concentrations) and fast enough to use them as they are produced. Photoreduction of oxygen has a half-life of 4×10^{-2} s and the conversion of superoxide to water via hydrogen peroxide a half-life of less than 2×10^{-3} s, so the rate of breakdown is much faster than that of synthesis, allowing the concentration of active oxygen species to remain very small. The redox potentials, E'_0, of the different components of the oxygen reduction processes are, in volts at pH 7.0 and a molar concentration of oxygen, O_2, -0.16; $O_2^{\cdot -}$, $+0.39$; H_2O_2, $+0.85$; $^\cdot OH$, $+1.30$; and H_2O, $+2.30$.

SOD is found throughout the plant cell, including chloroplasts, and disproportionates superoxide very rapidly, so superoxide does not react with ascorbate or glutathione directly. SOD in chloroplasts differs from that in other compartments, and is predominantly the copper–zinc-containing isoform, but Fe–SOD is also common, restricted to the stroma. SOD is a homodimer, each 16 kDa subunit (a β-barrel structure) containing one Cu and one Zn atom. In the viscous, high-protein-content stroma, the rate of reaction is probably one-tenth slower than that in water, limited by diffusion of superoxide to the enzyme. There is about one molecule of SOD per PSI, with most clustered around the stromal thylakoids, with magnesium ions aiding attachment. This close proximity allows the reaction to occur within 5 nm of the sites of superoxide production. Antisense decrease in SOD content leads to photobleaching. In addition, some catalase is in the chloroplast, probably bound to PSI, as it is not soluble. These two enzymes (together with the carotenoid quenching of reactive forms of oxygen) are an important defense against highly reactive oxygen states. The potential for damage to components of the chloroplast should not be underestimated. In the lumen where oxygen is evolved, the concentration of O_2 is very large in close proximity to the D1 protein, which is rapidly damaged under conditions where the intermediates of PSII are highly reduced and the lifetime of the energetic states is long. Superoxide reacts with unsaturated fatty acids, causing lipid peroxidation and thereby destroying membranes and, with chlorophyll, causing photobleaching. Accumulation of malondialdehyde indicates destruction of lipids. Superoxide also oxidizes sulfur compounds, NADPH and ascorbic acid, or reduces cyt c and metal ions; such reactive O_2 species must be rapidly removed. SOD is a fast enzyme which, together with catalase, protects the thylakoid.

The metabolism of forms of active oxygen is coupled to the oxidation–reduction of ascorbic acid and the tripeptide glutathione, using reduced NAD(P)H, so the redox state of the electron transport pathway, pyridine nucleotides, and the H_2O_2 and superoxide content are regulated:

$$H_2O_2 + \text{ascorbate} \xrightarrow[\text{peroxidase}]{\text{Ascorbate}} 2 \text{ monodehydroascorbate} + 2H_2O \qquad (5.14)$$

Ascorbate is of major importance in the regulation of redox processes in plants, including the synthesis of NAD(P)H and the balance with ATP. It also has a particular role in removing active oxygen in the cell and chloroplast. Ascorbate is at 20–50 mM concentration in cells, but the site and mode of synthesis are unclear. Suggested sites are the mitochondria and cytosol. Two routes have been suggested for synthesis from glucose: one, the 'inversion' pathway found in animals, is by rearrangement of the molecule via L-galactono-γ-lactone; the other is via D-glucasone. Both lack evidence of key enzymes. The peroxide formed from dismutation of singlet oxygen by SOD is reduced to water by ascorbate in a reaction catalyzed by ascorbate peroxidase, yielding monodehydroascorbate (MDA). Ascorbate peroxidase contains heme; a two-electron oxidized intermediate is formed which oxidizes ascorbate producing two molecules of MDA, and the enzyme is reduced again. There is about one molecule (a monomer) of ascorbate peroxidase per PSII RC, with one isoform (38 kDa mass) on the thylakoid and one (32 kDa) in the stroma. It is important to note that ascorbate peroxidase is rapidly denatured by very small concentrations of hydrogen peroxide if ascorbate is not present, as the pool is consumed; at the damaging concentration several proteins of the PCR cycle are inactivated. This further decreases the use of electrons for CO_2 assimilation and thus exacerbates the damage. Photosystem I is also inhibited by active oxygen, so it is essential that the ascorbate cycle and other processes function to remove any active oxygen species formed. In recycling, reduction of MDA uses reduced ferredoxin in thylakoids and competes very effectively with reduction of $NADP^+$. MDA is also reduced by NAD(P)H to ascorbate in the presence of MDA reductase: the enzyme in chloroplasts is about 55 kDa mass and there is one molecule per five PSI, located near the thylakoid membrane. It is very specific for MDA, and uses NADH (K_m 5 µM) rather than NADPH (K_m 22–200 µM). As reduced ferredoxin is a more effective reductant than NAD, MDA reductase is probably located away from ferredoxin on the membrane, where NAD(P)H is available. Another route for the recycling of MDA is spontaneous disproportionation (two MDA give one ascorbate and one dehydroascorbate); ascorbate is recycled, and dehydroascorbate (DHA) is converted by reduced glutathione in the presence of DHA reductase to ascorbate. The reaction is rapid under physiological conditions. MDA is also produced in the lumen, when it is acid, from ascorbate in the de-epoxidation of violaxanthin to zeaxanthin, and by reduction of ascorbate by PSII when electron flow from that photosystem is blocked. The MDA then disproportionates as there is no reduced ferredoxin or NAD(P)H in the lumen, and the DHA diffuses across the membrane. Dehydroascorbate is reduced to ascorbate by reduced glutathione, a reaction catalyzed by DHA reductase and essential if the reduced ascorbate pool is to be maintained. Mutants lacking this enzyme suffer photobleaching in bright light. DHA reductase is a 23 kDa protein with SH-groups at the active site. The oxidized glutathione is reduced:

$$\text{Dehydroascorbate} + \text{glutathione}_{red} \longrightarrow \text{glutathione}_{ox} + \text{ascorbate}$$

$$\text{glutathione}_{ox} + \text{NADPH} \xrightarrow{\frac{\text{Glutathione}}{\text{reductase}}} \text{glutathione}_{red} + \text{NADP}^+ \qquad (5.15)$$

by the appropriate reductase, which is at low concentration (0.02 molecules/PSI RC), and NADPH. Glutathione, a tripeptide (γ-glu-cys-glu; there are variants), is important in sulfur metabolism as well as in energy regulation and protection against active oxygen and xenobiotics (e.g. heavy metals). In chloroplasts it may be at concentrations of 1–5 mM. Synthesis is by a common pathway in all organisms and is stimulated by stress conditions; the rate of synthesis probably determines the content in plants. Glutamate from nitrogen metabolism and cysteine from sulfur metabolism are combined (reaction catalyzed by γ-glutamylcysteine synthetase and using ATP), giving γ-glutamylcysteine which reacts with glycine (possibly from the photorespiratory glycollate pathway) to give glutathione; the reaction is catalyzed by glutathione synthetase. The system is regulated by substrate and product control of the enzymes and there is much interest in modifying glutathione metabolism to increase the protection of plants to stresses, for example by increasing the activity of enzymes by genetic manipulation, overexpressing them to increase the concentrations of ascorbate and glutathione. In some cases, protection is enhanced under different stresses but not in others, suggesting that the systems are complex both in the way that the content of glutathione is regulated and its role in plants under specific conditions. Bright light damages mutant plants lacking SOD, catalase or other enzymes which control the oxidation–reduction state of the chloroplast, or when adverse conditions cause stomatal closure and reduce the consumption of NADPH, allowing high-energy states and reducing compounds to accumulate in the light-harvesting apparatus. Better understanding of the pathways of ascorbate synthesis, and of the regulation of ascorbate and glutathione regulation is required, as they have far-reaching effects on the cellular energy (redox) balance under a very wide range of environmental conditions.

5.5 Hydrogen production during photosynthesis

Gaseous hydrogen (H_2) is produced by heterocysts in blue-green algae, some photosynthetic bacteria and primitive eukaryotic algae. Anaerobic environments are essential for the hydrogenase function. Nitrogenase also possesses hydrogenase activity and H_2 can be used as a source of electrons for N_2 reduction via ferredoxin. Electrons from PSI are supplied to H^+:

$$2\,H^+ + 2\,e^- \xrightarrow[\text{Hydrogenase}]{\text{Light}} H_2 \tag{5.16}$$

No ATP is required and electrons come from reduced organic substrates, such as NADH or glucose.

5.6 Chlorophyll fluorescence

Even at large flux, most of the photons incident on the light-harvesting antenna of the photosystems are captured by pigments and some of the energy is transferred via excitation in the pigment bed to the RCs and thence to the chemical reactions. However, not all of the energy is used for photochemistry; thermal deactivation and excitation transfer to nonfluorescent pigments, for example to the PSI antenna, contribute to the use of energy. One of the energy dissipation routes is by fluorescence, which is emitted mainly at 685 and 740 nm at room temperature. Chlorophyll *a* in solution emits some 30% of the light absorbed as fluorescence but under physiological conditions the maximum is about 3% and the minimum about 0.6%. Most emission at normal temperatures is from chl *a* of PSII, with five major components, at 680 nm produced by LHCII, 685 nm from CP43, 695 nm from CP47, 720 nm from the PSI core and 740 nm from LCHI. PSI only produces significant fluorescence at low temperature (77K) because P700 is relatively stable and can still trap excitation energy and dissipate it as heat, but $P680^+$ rapidly returns to the ground state. The rate of fluorescence emission, F, is proportional to the light absorbed, I, to the fraction, β, of the energy reaching the PSII with chl *a* concentration [chl PSII] and quantum yield, φ_f [fluorescence relative to radiationless dissipation (*d*) transfer to other molecules (*t*) and photochemistry (*p*)]. The relation is:

$$F = I \times \beta \times [\text{chl PSII}] \times \varphi_f \tag{5.17}$$

The rate constant K (units s^{-1}) for each process expresses the rate of decay; the greater the constant the shorter the intrinsic lifetime of a pigment molecule, $T = 1/K_f + K_d + K_p + K_t$.

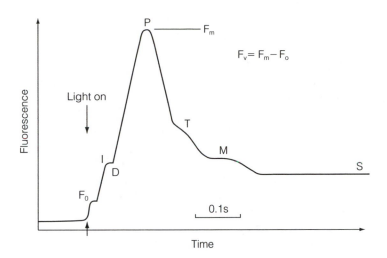

Figure 5.10. A typical Kautsky curve – the chl *a* fluorescence transients induced by illumination of a darkened leaf. See text for details

With β and [chl PSII] constant during fluorescence changes, the quantum yield of fluorescence, φ_f = intensity of fluorescence/intensity of absorbed light or $K_f/(K_f + K_d + K_p + K_t.P)$, where P is the fraction of PSII RCs that can accept energy from chlorophyll.

The time course of fluorescence emission is shown in *Figure 5.10*. At F_0 on the induction curve the PSII RCs are fully oxidized, $P = 1$ and the quantum yield of fluorescence is given by $K_f/(K_f + K_d + K_p + K_t)$. At maximum fluorescence, F_m, in the presence of DCMU or after a saturating light pulse, Q_A is blocked so the maximum yield of fluorescence, φ_{fm} is obtained. The variable fluorescence, F_v, is $F_m - F_0$, which is also the difference between the respective quantum yields and shows the yield of photochemistry. The maximum yield of photochemistry is given by the ratio of K_p to the sum of the other rate constants.

Variable fluorescence, F_v, is a very important component derived from the fluorescence signals as it reflects the balance of energy between that in the antenna and reaction center and that used by the photochemical and other processes. It is related to the efficiency with which energy is used and therefore to the pathways of chemical processes downstream of the photosystems. Fluorescence provides a signal, from events at the heart of photosynthesis, that can be measured easily and continuously with suitable photodetectors using relatively simple amplifiers and filters to remove interfering light. It is a very important technique for understanding the links and dynamics between the light reactions and metabolism (see *Figure 5.4* and Chapter 12).

In the sequence of reduction of electron chain intermediates each step has a different time constant; the initial charge separation of P680 and reduction of pheo is very fast (3 ps) whereas pheo to Q_A is much slower (100–200 µs) with open RCs and 300 s^{-1} when closed. Overall reduction of PQ requires 1–2 ms. Consequently the fluorescence emission from a population of photosystems after a period in darkness is very dynamic, the so-called fluorescence transient, or Kautsky curve (*Figure 5.10*). In darkness, all the photosystem components and acceptors are uncharged but when light is absorbed by the antenna P680 is oxidized to P680$^+$. Fluorescence increases rapidly from F_0 to an 'inflection' point called I, caused probably by reduction of Q_A molecules linked to PSII but not connected to Q_B so they are rapidly reduced; in spinach up to 30% of PSII may be in this state. As Q_A is progressively reduced in the other, attached, PSII particles, so the fluorescence rises very rapidly. The dip 'D' reflects the imbalance between fluorescence from attached and unattached Q_A. When Q_A is fully reduced and point P is reached, fluorescence is at the maximum F_m. Further reduction of Q_B and PQ uses electrons from Q_A and so the fluorescence signal decreases. During the decrease in fluorescence from P, transients T and M often appear before steady-state fluorescence, S, is reached. These reflect a number of processes that are poorly understood, involving feedback control of energy transfer by such processes as build-up of the proton gradient in the thylakoids, rearrangement of light-harvesting complexes and the induction of carbon and nitrate metabolism.

If the electron transport from Q_A to Q_B is blocked by the herbicide DCMU, which binds to the Q_B sites on the D1 protein in the PSII RC, then no further transfer of electrons is possible and fluorescence is maximal. Also with a very brief pulse (2 s) of light, which saturates all photosystems, Q_A is fully reduced and F_m is achieved. The rate of Q_A reduction is proportional to the photon flux, cooperation between PSII units and their heterogeneity, the size of the PQ pool and its rate of re-oxidation and the 'downstream' electron transport. However, during this phase F_v is smaller than expected from the state of Q_A reduction because excitation is transfered from closed to open centers. There are also different types of PSII units – PSIIα, PSIIβ and PSIIβ-Q_B^-. The PSIIα is the active PSII complex core associated with light harvesting in appressed grana and is linked to the electron transport chain. PSIIβ is in nonappressed grana, has a smaller antenna than the granal PSII, and is active in electron transport. However, it differs from the other forms in the redox potential and ability to oxidize water. PSIIβ-Q_B^- is inactive.

Fluorescence is quenched, that is decreased, by a number of factors, as we have seen, and quenching coefficients with values of 0–1 may be assigned to the processes. Thus quenching, q, of variable fluorescence is described by:

$$q = (F_v - F_v')/F_v \text{ or } F_v'/F_v = 1 - q \tag{5.18}$$

where F_v' is in the quenched state and provides valuable information on the causes of the changed fluorescence. The main physiological mechanism of quenching is by photochemical quenching q_P (or q_Q), which depends on oxidized Q_A and shows the proportion of excitons that are converted to chemical energy in PSII. Other mechanisms are nonphotochemical quenching, q_N (or q_{NP}), resulting from energy-dependent quenching (q_E) or state-transition quenching (q_T) or photoinhibitory quenching (q_I). Energy-dependent quenching derives from the development of the pH gradient in the thylakoids and has a half-life of about 1 min. As pH builds up it affects the membrane's capacity to allow electron flow, so there is a strong dependence of q_E on pH. The mechanism depends on the action of the violaxanthin cycle in the antenna, increasing the rate constant and k_D, although other models of the mechanism, for example alternative cyclic photochemical processes or quenching in the antenna, have been suggested. The state transition quenching comes from the energy transfer from PSII to PSI during state 1 to state 2 transitions, which is a shift from high to low fluorescence. This mechanism is thought of as a change in the exciton distribution in LHCII from PSII to PSI. In the light LHCII becomes phosphorylated as light activates (via electron flow and reduced PQ) a protein kinase located on the stromal surface. This kinase phosphorylates threonine residues on the exposed LHCII, using ATP (plus Mg^{2+}) and requires about 10 min, a time-scale similar to the state 1 to state 2 changes. It was once considered that this modification enabled the protein to migrate from granal to stromal regions and donate energy to PSI. However, a number of alternative possibilities have been suggested, for example that a complex of PSIIβ units, LHCII in the phosphorylated form and PSI is formed, allowing energy to 'spill over'. Further discussion of fluorescence applied to leaves is in Section 12.5.

5.7 Photoinhibition

The photoinhibitory quenching, q_I, is related to damage caused by excessive light energy and has a half-life of about 40 min. This quenching results from increased nonphotochemical processes and is detected as a decrease in F_v/F_m following exposure of leaves to bright light. It correlates with a loss of quantum yield of photosynthesis. Probably, q_I is related to damage to the D1 protein of PSII caused when the energy load on the system is excessive. The oxidation–reduction status of the primary electron acceptor Q_A controls photodamage under a wide range of environmental conditions, depending on the physiological state of the plant, for example when CO_2 (and O_2) are not available to act as substrates for electrons. Dissipation of excitation energy by charge-recombination at PSII, in the O_2-saturated environment produced by water splitting, generates singlet oxygen. Excited P680 will reduce pheophytin, but if Q_A is already reduced (as it will be when the electron transport chain cannot pass electrons to terminal acceptors) then the oxidized P680 and reduced pheophytin form a long-lived (tens of nanoseconds) charge-recombination intermediate. This has a large probability of forming P680 triplet. The P680 triplet is efficiently quenched by $^3O_2$ (triplet oxygen) forming singlet oxygen ($^1O_2^{\cdot-}$), which has a lifetime of 10–30 µs in the lipid environment of the thylakoid membrane. Singlet oxygen may then react with organic matter in the environment, including lipids and proteins, specifically D1, covalently changing the reaction center. However, it may act on peptide bonds indirectly, by altering the protein structure and allowing attack by proteases. With increasing photon capture, and no increase in the chemical utilization of energy, the probability of photodamage increases. In addition, ultraviolet (UV-B) radiation, which is about 1% of the radiation reaching vegetation, causes damage, particularly to thylakoids, similar to the effects of very large photon flux. There is controversy over what causes photoinhibition: is the magnitude of the photon flux (photons/unit time) important or the absolute number of photons captured and damaging D1? If the flux is the essential factor, then the rate of damage should increase nonlinearly with increasing photon flux above a threshold. If the absolute number of 'hits' is important then damage should be a linear function of photon flux. The latter is most likely, with a simple probability of photodamage; as light intensity increases, so does excitation transfer to the reaction center, and the rate constant for damage rises. The larger the size of the antenna, the greater the number of excitations of PSII and the greater photodamage. The more electrons from the electron transport chain that are used in CO_2 assimilation, the Mehler reaction or in photorespiration, the less likely is photodamage. Suboptimal temperatures for plants also increase photoinhibition, probably because the biochemical steps which ultimately use electrons are slowed, increasing the reduction state of Q_A. It is this which is crucial, with a larger probability of photodamage when Q_A is reduced and a linear increase in the rate constant for damage. Thus, any conditions which alter the relative rates of excitation reaching PSII and using electrons from PSII will change the probability of damage to D1 and PSII. The more active oxygen formed the greater the damage, and the site depends on whether radicals are formed at the acceptor or donor side. If on the acceptor side degradation fragments (23 kDa N-terminal and 10 kDa C-terminal) of D1 are formed; donor-side photoinhibition gives a 24 kDa C-terminal fragment, as if the loop between the A and B helices were broken. UV-B gives different fragments suggesting a different site of damage, near the B helix of the D1 protein close to the Q_A; perhaps UV-B excites the reduced form, which is why UV has greater effects in combination with photosynthetically active radiation. Large photon flux and small CO_2 assimilation, particularly at low temperature, which also slows consumption of electrons by alternative 'sinks', are classic conditions for photoinhibition.

The redox state of Q_A acts as a signal which operates by a transduction pathway to trigger expression of nuclear genes for specific components of the photosynthetic system (and also of other aspects of metabolism) which are protective; plants may be susceptible to cold and freezing temperatures unless 'hardened' or 'acclimated'. Acclimation involves altered composition and structure of the photosystems, increased content of protective systems such as the violaxanthin cycle, and greater protein activity associated with increased CO_2 assimilation (e.g. Rubisco in rye) which consumes electrons under adverse conditions. This results in accumulation of sucrose and other compounds with cryo-protectant functions. Such changes enable plants to survive and be productive in what would have been otherwise damaging environments. Recovery from photoinhibition involves transfer of damaged PSII complexes to the stromal membranes, where they are exposed to the stroma, the damaged D1 is dephosphorylated (otherwise the repair is stopped) and degraded, before a new D1 is inserted and the complex is re-inserted into the membrane. Relatively little is known of the mechanisms of these processes. There is a dynamic balance between damage to D1 and its degradation and replacement, which is rate-limiting.

References and further reading

Anderson, J.M., Park, Y.I. and Soon, W.S. (1998) Unifying model for the photoinactivation of photosystem II in vivo under steady-state photosynthesis. *Photosynth. Res.* **56**: 1–13.

Andersson, B. and Barber, J. (1994) Composition, organization, and dynamics of thylakoid membranes. In *Molecular Processes of Photosynthesis*, Vol. 10, *Advances in Molecular and Cell Biology* (ed. J. Barber). JAI Press, Greenwich, pp. 1–53.

Aro, E.-M., Virgin, I. and Andersson, B. (1993) Photoinhibition of photosystem II. Inactivation, protein damage and turn-over. *Biochim. Biophys. Acta* **1143**: 113–134.

Asada, K. (1999) The water–water cycle in chloroplasts: Scavenging of active oxygens and dissipation of excess photons. *Annu. Rev. Plant Physiol. Plant Mol. Biol.* **50:** 601–639.

Baker, N.R. and Bowyer, J.R. (eds.) (1994) *Photoinhibition of Photosynthesis: from Molecular Mechanisms to the Field.* BIOS Scientific Publishers, Oxford.

Barr, R. and Crane, F.L. (1997) Chloroplast electron transport inhibitors. In *Handbook of Photosynthesis* (ed. M. Pessarakli). Marcel Dekker, New York, pp. 95–112.

Bennett, J. (1991) Protein phosphorylation in green plant chloroplasts. *Annu. Rev. Plant Physiol. Plant Mol. Biol.* **42:** 281–311.

Böger, P. and Sandmann, G. (1998) Action of modern herbicides. In *Photosynthesis: A Comprehensive Treatise* (ed. A.S. Raghavendra). Cambridge University Press, Cambridge, pp. 337–351.

Bricker, T.M. and Ghanotakis, D.F. (1996) Introduction to oxygen evolution and the oxygen-evolving complex. In *Oxygenic photosynthesis: the Light Reactions.* (eds D.R. Ort and C.F. Yocum). Kluwer Academic, Dordrecht, pp. 113–136.

Cogdell, R.J., Fyfe, P.K., Barrett, S.J., Prince, S., Freer, A.A., Isaacs, N.W. and Hunter, C.N. (1996) The purple bacterial photosynthetic unit. *Photosynth. Res.* **48:** 55–63.

Cramer, W.A., Soriano, G.M., Ponomarev, M., Huang, D., Zhang, H., Martinez, S.E. and Smith J.L. (1996) Some new structural aspects and old controversies concerning the cytochrome b_6f complex of oxygenic photosynthesis. *Annu. Rev. Plant Physiol. Mol. Biol.* **47:** 477–508.

Deisenhofer, J., Epp, O., Mild, K., Huber, R. and Michel, H. (1985) Structure of the protein subunits in the photosynthetic reaction center of *Rhodopseudomonas viridis* at 3A resolution. *Nature.* **318:** 618–624.

Demmig-Adams, B. and Adams, W.W. (1992) Photoprotection and other responses of plants to high light stress. *Ann. Rev. Plant Physiol. Mol. Biol.* **43:** 599–626.

Dutton, P.L. (1986) Energy transduction in anoxygenic photosynthesis. In *Encyclopedia of Plant Physiology (N.S.), Vol. 19, Photosynthesis III. Photosynthetic Membranes and Light Harvesting Systems.* (eds L.A. Staehelin and C.J. Arntzen). Springer-Verlag, Berlin.

Foyer, C.H. and Mullineaux, P.M. (eds.) (1994) *Causes of Photooxidative Stress and Amelioration of Defense Systems in Plants.* CRC Press, Boca Raton, FL.

Govindjee and Coleman, W.J. (1990) How plants make oxygen. *Sci. Am.* **262,** 42–51.

Hankamer, B., Barber, J. and Boekema, E.J. (1997) Structure and membrane organization of photosystem II in green plants. *Ann. Rev. Plant Physiol. Plant Mol. Biol.* **48:** 641–671.

Huner, N.P.A, Öquist, G. and Sarhan, F. (1998) Energy balance and acclimation to light and cold. *Trends Plant Sci.* **3**: 224–230.

Kok, B., Forbush, B. and McGloin, M. (1970) Co-operation of charges in photosynthetic oxygen evolution. 1. A linear four step mechanism. *Photochem. Photobiol.* **11**: 457–475.

Krause, G.H. and Weis, E. (1991) Chlorophyll fluorescence and photosynthesis: the basics. *Annu. Rev. Plant Physiol. Plant Mol. Biol.* **42**: 313–349.

Kushner, D.J. (1985) The *Halobacteriaceae*. In *The Bacteria*, Vol. VIII, *The Archaebacteria* (eds C.R. Woose and R.S. Wolfe). Academic Press, Orlando, FL, pp. 171–214.

Lanyi, J. (1996) Proton translocation mechanism and energetics in the light-driven pump bacteriorhodopsin. *Biochim. Biophys. Acta* **1183**: 241–261.

Lavergne, J. and Briantais, J.-M. (1996) Photosystem-II heterogeneity. In *Oxygenic Photosynthesis: the Light Reactions* (eds D.R. Ort and C.F. Yocum). Kluwer, Dordrecht, pp. 265–287.

Melis, A. (1999) Photosystem-II damage and repair cycle in chloroplasts: what modulates the rate of photodamage *in vivo*. *Trends Plant Sci.* **4**: 130–135.

Merchant, S. and Dreyfuss, B.W. (1998) Post-translational assembly of photosynthetic metalloproteins. *Annu. Rev. Plant Physiol. Plant Mol. Biol.* **49**: 25–51.

Michel, H. and Deisenhofer, J. (1986) X-ray diffraction studies on a crystalline bacterial photosynthetic reaction center. A progress report and conclusions on the structure of photosystem II reaction centers. In *Encyclopedia of Plant Physiology (ICS)*, Vol. 19, *Photosynthesis 111. Photosynthetic Membranes and Light Harvesting Systems* (eds L.A. Staehelin and C.J. Arntzen). Springer, Berlin, pp. 371–387.

Mitchell, P. (1975) The protonmotive Q cycle: a general formulation. *FEBS Lett.* **59**: 137–139.

Nishida, I. and Murata, N. (1996) Chilling sensitivity in plants and cyanobacteria: the crucial contribution of membrane lipids. *Annu. Rev. Plant Physiol. Plant Mol. Biol.* **47**: 541–568.

Noctor, G. and Foyer, C.H. (1998) Ascorbate and glutathione: keeping active oxygen under control. *Annu. Rev. Plant Physiol. Mol. Biol.* **49**: 249–279.

Oakamura, M.V. and Feher, G. (1992) Proton transfer in reaction centers from photosynthetic bacteria. *Annu. Rev. Biochem.* **61**: 861–896.

Oettmeier, W. (1992) Herbicides of photosystem II. In *Topics in Photosynthesis*, Vol. 11,

The Photosystems: Structure, Function and Molecular Biology (ed. J. Barber). Elsevier, Amsterdam, pp. 349–408.

Ohad, I. *et al.* (1994) Light-induced degradation of photosystem-II reaction center D 1 protein *in vivo*: an integrative approach. In *Photoinhibition of Photosynthesis: from Molecular Mechanisms to the Field* (eds N. Baker and J.R. Bowyer). BIOS Scientific Publishers, Oxford, pp. 161–177.

Park, Y.I. *et al.* (1996) Electron transport to oxygen mitigates against the photoinactivation of photosystem II *in vivo*. *Photosynth. Res.* **50**: 23–32.

Polle, A. (1996) Mehler reaction—friend or foe in photosynthesis? *Bot. Acta* **1**(109): 84–89.

Rees, D.C., Komiya, A., Yeates, T.O., Allen, J.P. and Feher, G. (1989) The bacterial photosynthetic reaction center as a model for membrane proteins. *Annu. Rev. Biochem.* **58**: 607–633.

Renger, G. (1997) Mechanistic and structural aspects of photosynthetic water oxidation. *Physiol. Plant.* **100**: 817–827.

Rögner, M., Boekema, E.J.H. and Barber, J. (1996). How does photosystem 2 split water? The structural basis of efficient energy conversion. *Trends Biol. Sci.* **21**: 44–49.

Sauer, K. and Debreczeny, M. (1995) Fluorescence. In *Advances in Photosynthesis*, Vol. 3, *Biophysical Techniques in Photosynthesis* (eds J. Amesz and A.J. Hoff). Kluwer Academic, Dordrecht, pp. 41–61.

Scheller, H.V., Naver, H. and Møller, B.L. (1997) Molecular aspects of photosystem I. *Physiol. Plant.* **100**: 842–851.

Schreiber, U., Kühl, M., Klimant, I. and Reising, H. (1996) Measurement of chlorophyll fluorescence within leaves using a modified PAM fluorometer with fiber-optic microprobe. *Photosynth. Res.* **47**: 103–109.

van Gorkom, H.J. and Gast, P. (1995) Measurement of photosynthetic oxygen evolution. In *Advances in Photosynthesis*, Vol. 3, *Biophysical Techniques in Photosynthesis* (eds J. Amesz, and A.J. Hoff). Kluwer Academic, Dordrecht, pp. 391–405.

Van Kooten, O. and Snel, J.F.H. (1990) The use of chlorophyll fluorescence nomenclature in plant stress physiology. *Photosynth. Res.* **25**: 147–150.

Whitmarsh, J. and Pakrasi, H.B. (1996) Form and function of cytochrome *b*-559. In *Oxygenic Photosynthesis: the Light Reactions* (eds D.R. Ort and C.F. Yocum). Kluwer Academic, Dordrecht, pp. 249–264.

Synthesis of ATP: photophosphorylation

6.1 Introduction

Adenosine triphosphate is central to the functioning of all biological systems. Many metabolic processes, such as the assimilation of carbon dioxide, protein synthesis and ion pumping, require ATP. ATP functions in many different ways, by providing energy for some processes, by phosphorylating metabolites and so enabling them to enter into complex biochemical reactions, and by regulating reactions, for example as an effector of enzyme reactions. These functions of ATP and the mechanisms are considered in other chapters. The way in which ATP is synthesized has been the subject of detailed biochemical analysis over many years. More recently, structural analysis of the enzyme responsible has provided much additional insight, so that the process and the mechanism are now well understood at the molecular level. The mechanism of ATP synthesis is remarkably similar in organisms as disparate as bacteria, animals and plants, with a concentration gradient of protons providing energy and the flow of protons through an enzyme complex providing the physical mechanism for ATP synthesis. In plants, the energy of photons captured in the photosynthetic pigments and converted to chemical energy in the form of electrons of high redox potential passing along an electron transport chain in thylakoids in the chloroplasts generates the proton gradient (see Chapter 5) for synthesis of ATP. In mitochondria of nonphotosynthetic (e.g. animals) and photosynthetic organisms (e.g. plants) in darkness and in light, respiration of preformed substrates, products of photosynthesis such as carbohydrates, coupled to oxidative electron transport and generation of a proton gradient, is the main mechanism for synthesis of ATP for most of the biological world. Aerobic bacteria use such mechanisms for ATP synthesis. Despite the different sources of energy for photo- and respiratory ('oxidative') phosphorylation, both involve the use of proton gradient to 'drive' the enzyme reactions that synthesize ATP synthesis; this chapter considers the mechanism of photophosphorylation in chloroplasts.

6.2 The metabolic role of ATP

ATP has two anhydride (pyrophosphate) bonds, and is chemically stable under physiological temperatures, even in aqueous solution. Hydrolysis of ATP to ADP yields inorganic phosphate (HPO_3^- or P_i) and -31 kJ mol^{-1} of energy, and is achieved by cleavage of ATP by the enzyme ATPase, which transfers the P_i group to water:

$$ATP + H_2O \xrightarrow{\text{ATPase}} ADP + P_i + \text{energy} \tag{6.1a}$$

ADP may also be hydrolyzed to AMP, releasing ATP.

$$ADP + H_2O \rightarrow AMP + P_i + \text{energy} \tag{6.1b}$$

The energy released by hydrolysis comes from the electrostatic repulsion of negative charges on the phosphate groups, and from resonance stabilization and the large enthalpies of solvation of the reaction products. The Gibbs free energy released in the reactions is 'coupled' by biochemical mechanisms to do work, for example, the ion ATPases couple the hydrolysis of ATP with the transport of ions. In many metabolic reactions hydrolysis of ATP is stoichiometrically linked to the chemical transformations. Under cellular conditions the energy made available for biochemical reactions from the hydrolysis of ATP reaction may be much greater, 50–60 kJ mol^{-1}, and is coupled to biochemical reactions by transferring the phosphate group to other compounds in the presence of suitable enzymes. ATP may be used directly in many cellular reactions, or other phosphorylated nucleotides may be used, for example guanosine and uridine nucleotides (GTP and UTP), to which ATP transfers P_i groups:

$$ATP + GDP \xrightarrow{\text{Nucleotide diphosphokinase}} ATP + GTP \tag{6.2}$$

ATP is required in metabolic pathways where phosphorylated intermediates are interconverted, for example the photosynthetic carbon reduction cycle (Chapter 7). There are basically three different types of mechanisms in which ATP is involved: (1) the enzyme may itself be phosphorylated (e.g. ion ATPase), which alters the conformation and thus drives the process; (2) no covalent bond forms between enzyme and ATP [e.g. adenylate kinase, equation (6.3)], but the conformational changes allow the reaction to take place; and (3) the enzyme forms phosphorylated intermediates in the course of interchanging groups (e.g. glutamine synthetase, which uses ATP in transferring NH_3 to glutamate to form glutamine; Section 7.8.1). When ATP donates phosphate groups to compounds, it increases their reactivity, for example glucose is phosphorylated to glucose-6-phosphate by ATP and hexokinase before consumption in glycolysis and respiration. The anhydride bonds of ATP are not, as often said, 'richer in energy' or of 'higher energy' than those of many other compounds and do not therefore 'drive' metabolism simply by providing energy. The ability of a chemical reaction to do work is related to the state of equilibrium of the reaction: the further from equilibrium the more energy is available. The hydrolysis of ATP [equation (6.1a)] is important because, under conditions within the cell (which are far from those of reactions *in vitro*), it is displaced from equilibrium and the Gibbs free energy change, ΔG, is favorable for doing metabolic work. ATP has an intermediate phosphate group transfer potential, as defined by the free energy of hydrolysis, and can function as a phosphate group carrier. In an analogous way $NADP^+$ and NAD^+ act as electron and H^+ carriers in metabolism. ATP forms complexes with ions, particularly magnesium, and many reactions require Mg-ATP rather than ATP alone; this is related to the energetics and structure of the reaction and its constituents. The central role of ATP in cellular metabolism is considered by Westheimer (1987); its stability at normal temperatures and near-neutral pH in the cell, the range of enzyme reactions in which it is involved, its role as an almost universal donor and acceptor of phosphate groups to other molecules, activating them for biochemical reactions, all show its unique characteristics and importance.

In actively metabolizing cells of eukaryotes, ATP concentrations are not the same in different cell compartments because of differences in rates of synthesis or import of ATP and rates of consumption in reactions, and the different types of reactions, so ATP will be in different equilibrium states. Exchange of phosphorylated compounds (often not ATP directly) takes place between compartments where reactions consume or produce ATP, thus regulating the ATP pools in different parts of the cell. Cell metabolism is balanced with respect to the energy available for ATP synthesis, the requirements for ATP in the particular reactions, and the supply of substrates. Turnover of ATP in cells is rapid: the total 'pool' in leaf cells may be broken down and resynthesized within 500 ms; therefore metabolism responds quickly to the supply of, and demand for, ATP. If synthesis slows, metabolism also slows and as different pathways require different amounts of ATP or are differentially regulated by the concentration of ATP or ADP, so the response of the system is modified. Because of the rapid consumption of ATP in almost all cellular activities, the total synthesis of ATP by organisms is very large and consumes most of the energy used by organisms. Indeed, the total mass of ATP 'turned-over' in a day is very large, often comparable to the total mass of the organism.

The proportions of the phosphorylated nucleotides, ATP, ADP and AMP, in cellular compartments, under different metabolic conditions (which change the requirement for nucleotides), is controlled by adenylate kinase:

$$ATP + AMP \xrightarrow{\text{Adenylate kinase}} 2\,ADP \qquad (6.3)$$

This means that the forms of adenylate are regulated close to an optimum for the many processes involved in metabolism. The phosphorylation state in tissues is expressed by the energy charge (AEC) (Atkinson, 1977):

$$AEC = \frac{[ATP] + \frac{1}{2}[ADP]}{[AMP] + [ADP] + [ATP]} \qquad (6.4)$$

An AEC of 1 is a condition of all ATP, and an AEC of 0 all AMP. Atkinson suggested that the ATP-regenerating enzymes have minimum velocity at large AEC and maximum at small, whereas enzymes consuming ATP act in reverse. So in cells with rapid synthesis of ATP compared to demand for ATP, EC is large and ATP synthesis is slow. Conversely with little ATP synthesis and large demand, AEC is small and ATP synthesis is rapid, given the required conditions. Equilibrium is attained between supply and demand because of the response of enzyme systems to AEC. However, the rate of reactions is not linear with AEC but changes most rapidly as AEC decreases from approximately 0.9 to 0.6 and only slowly with a further decrease. Metabolism is therefore very sensitive to small changes in AEC and is closely regulated by the supply of, and demand for, ATP. Because of the difficulty of measuring ATP and other nucleotides in cells and the sensitivity of metabolism to relatively small changes in the concentrations, the role and importance of the adenylates are often underestimated. Another measure of the role played by adenylates in metabolism is the phosphorylation potential, $P = (\text{ATP})/(\text{ADP})$ (P_i), which takes account of the influence of inorganic phosphate but does not involve the adenylate kinase system, as does AEC.

Control of enzyme reactions is not only by AEC or the availability of ATP, ADP or AMP as substrate but by these molecules acting as allosteric effectors of enzyme reactions. Such effectors bind away from the reaction site, where the conversion of substrates to products occurs, and so do not enter into the reaction *per se* but change the catalytic behavior of the enzyme. Effectors may be positive or negative, increasing or decreasing the rate of reaction and acting in a regulatory way, and linking different aspects of metabolism within the cell or whole organism. If ATP is a positive effector of an enzyme, then in the light – when ATP concentrations are large – the reaction will be increased, but it will be slowed or stop in darkness when ATP concentration falls. An important photosynthetic example is the effect of ATP on 3-phosphoglycerate kinase of chloroplasts: the enzyme uses ATP in formation of 1,3-diphosphoglycerate and is stimulated at high AEC. It also generates ATP in the reverse reaction, which is inhibited by ATP and high AEC (0.9–1.0) and stimulated by low AEC (0.7). Both forward and reverse reactions are inhibited by AMP. Such complex control based on phosphorylated adenylates provides for a very subtle balance between processes and is an essential feature of cellular metabolism and maintenance of cellular homeostasis.

6.3 Measurement of ATP and other adenylates

Because of the rapid turnover of ATP and other nucleotide phosphates in actively metabolizing cells, accurate measurement of amounts requires that metabolism of tissues or cells must be stopped quickly (within milliseconds) on sampling, and that any further metabolic reactions during extraction and measurement must be stopped if the true state of the system is to be determined. This may be done by plunging tissues into very cold solvents, for example, pentane at $-20°C$, or by 'freeze-clamping' them between the jaws of metal tongs at liquid nitrogen temperature ($-196°C$), which effectively stops metabolism, but may not permanently 'kill' enzymes if the temperature rises. To stop reactions, the enzymes are denatured, for example by grinding the tissue with trichloracetic acid, followed by extraction of adenylates in very cold solvents. The concentration of ATP may be measured by one of several methods, for example by chromatographic separation from other nucleotides on ion exchange columns, often using high-pressure liquid chromatography (HPLC) and detection of ATP with ultraviolet light after separation and elution. Enzymatic methods are frequently used to measure the ATP in extracts, for example hexokinase converts glucose to glucose-6-phosphate using ATP, the glucose-6-P is oxidized by $NADP^+$ and the resultant NADPH is detected spectrophotometrically, as 1 mol ATP consumed produces 1 mol NADPH. Another sensitive method for ATP measures the bioluminescence produced when an extract from the light organs (lanterns) of fireflies, containing luciferin and the enzyme luciferase, reacts with ATP. The emitted photons are measured with a sensitive photometer.

The state of nucleotide phosphates in tissues can also be measured in living tissues by ESR, which allows the amounts to be determined and also compartmentation; however, the conditions under which measurements are made are often rather far from the normal physiological state.

6.4 Synthesis of ATP

6.4.1 Energy transducing mechanisms in phosphorylation

ATP is synthesized from ADP and P_i, by reversal of equation (6.1a) and, as mentioned, the mechanism is very similar in the widest range of organisms. The enzyme, ATP synthase, and a proton gradient across the membrane in which part of the ATP synthase is embedded are required. To achieve the gradient, an intact membrane vesicle enclosing a space is necessary. In bacteria the gradient forms across the cell membrane, in which the ATP synthase is positioned, from ambient solution at the exterior of the cell to the inside of the cell (see Chapter 1). The gradient of proton concentration is from the outside of the cell into the cell's interior or cytosol (see *Figure 1.2*). In the case of the chloroplast, the H^+ gradient is from the lumen into the stroma (see *Figures 1.2 and 6.1*). ATP synthase, in all organisms, has a basal part, called F_0, in the membrane and a head part, F_1, attached to F_0 and protruding into the cytosol. It is the gradient of protons which causes the proton flux through F_0 and drives ATP synthesis. In chloroplasts CF_0 is in the thylakoid membrane and CF_1 is in the chloroplast stroma (*Figure 6.1*). As discussed (Section 5.2.3) the flux of H^+ from one side of the membrane to the other through the proton conducting basal part (F_0) provides the energy for ATP synthesis by F_1, and generates ATP. In mitochondria the enzyme is called MF_0F_1 and is on the inner (matrix) surface of the inner membrane so the ATP is produced within the space formed between the two membranes. In photosynthetic bacteria, ATP synthase is called BF_0F_1 and ATP is synthesized inside the cellular membrane.

6.4.2 Photophosphorylation and the chemiosmotic theory

The mechanism by which the redox potential energy of electron transport in membranes of ATP-synthesizing organelles is coupled to the synthesis of the anhydride bond of ATP was a matter for heated debate for many years. Several hypotheses were advanced to account for experimental observations that electron transport was coupled at three sites (in mitochondria) or two (in chloroplasts) to ATP synthesis and that the demand for ATP and the supply of ADP and P_i could regulate electron transport. The relationship between ATP synthesis and electron transport was measured by determining ATP production or P_i consumption in relation to O_2 uptake by mitochondria or to O_2 evolution in chloroplasts. Any hypothesis also had to account for ion accumulation (e.g. calcium in mitochondria) which was known to be related to the electron transport processes and could be driven by ATP hydrolysis in the absence of electron transport. Another phenomenon requiring explanation was how artificial compounds of very diverse type,

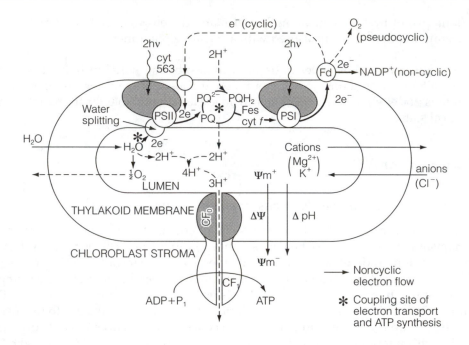

Figure 6.1. Schematic relationship between electron transport driven by light, proton accumulation in the thylakoid lumen and ATP synthesis by the ATP synthase (coupling factor) enzyme complex CF_0–CF_1; $\Delta\psi$ and ΔpH are, respectively, the electrical potential difference and H^+ gradient providing the proton motive force and the 57 kJ of energy required for synthesis of a mole of ATP

applied to mitochondria or chloroplasts, could uncouple ATP synthesis from electron transport. Electron transport was thought to produce a high-energy form of an energy-transducing intermediate, such as a particular structure of a protein, which could provide energy for ATP synthesis and related processes. Uncouplers, it was suggested, prevented the formation of, or destroyed, this high-energy form. If a component in the membrane, electron carrier or protein for example, altered its configuration as the redox potential changed, allowing ADP and P_i to bind and form the anhydride bond, it should have been detectable spectroscopically or by other means. Despite much experimentation, no 'high-energy' chemical intermediate was identified. No direct coupling between the electron transport chain and ATP synthesis could be shown, as electron transport could proceed without ATP synthesis.

The chemiosmotic hypothesis, developed by Mitchell (1966) and now universally accepted, provided a general mechanism for coupling the energy in electron transport to phosphorylation. It combined into a coherent scheme many experimental observations related to ATP synthesis, for example, the need for intact organelles (e.g. mitochondria) or membrane vesicles (e.g. the thylakoids in chloroplasts), the requirement for coupling factor attached to the membrane and the need for electron transport. It also accounted for the observed increase in alkalinity of the chloroplast stroma in the light, in addition to the other features mentioned previously. The hypothesis, that photophosphorylation in chloroplasts and in photosynthetic bacteria, and respiratory chain phosphorylation in mitochondria, employs the same basic energy-transducing mechanism driven by a proton flux through the ATP synthase enzyme, is now accepted. Energy from electron transport is conserved as a high concentration of protons on one side of the vesicle membrane and a low concentration on the other. The high concentration is outside the cell membrane in the photosynthetic bacteria, between the inner and outer mitochondrial membranes, and inside the thylakoid lumen in chloroplasts. In photosynthesis, the energy of electrons activated by light reactions is conserved by coupling electron transport in thylakoid membranes (*Figure 6.1*) to H^+ transport across the membrane into the thylakoid. The capture of four photons leads to accumulation of two H^+ from oxidation of one molecule of water, and two H^+ are transported from the stroma (which becomes alkaline) by the plastoquinone pump. The difference in H^+ concentration across the membrane is equivalent to a gradient in pH, ΔpH. Electron transport and proton pumping also cause an electrical potential difference, $\Delta\psi$, to develop across the membrane, which in chloroplasts is positive inside the thylakoid lumen and in mitochondria and negative in the matrix. Under the influence of $\Delta\psi$ and of ΔpH acting together, protons in the thylakoid lumen will tend to move from the lumen, across the membrane, to the stroma to preserve electrical neutrality and decrease the gradient of H^+. This force is called the proton motive force, pmf (also called the proton electrochemical potential, $\Delta\mu_H^+$). The two components $\Delta\psi$ and ΔpH are energetically equivalent (in thermodynamic terms) but the mechanism is not the same; there must be a mechanism whereby $\Delta\psi$ is converted into proton movement; the mechanisms are discussed later. Thus:

$$\text{pmf} = \Delta\mu_H^+ = \Delta\psi + \Delta\text{pH} \tag{6.5}$$

The Gibbs energy change for the transfer of 1 mol of H^+ down a gradient of H^+ between the inside, i, and outside, o, of the membrane vesicle in the absence of an electrical potential, is:

$$\Delta G = 2.3\ RT \log_{10} \frac{[H^+]o}{[H^+]i} \tag{6.6}$$

The Gibbs energy change for the transfer of 1 mol of H^+ ions down the electrical potential gradient $\Delta\psi$ (in millivolts) is:

$$\Delta G = -mF\Delta\psi \tag{6.7}$$

where m is the number of charges (for protons, $m = 1$), and F is the Faraday constant ($9.65 \times 10^4\,C\,mol^{-1}$).

As Gibbs energy differences are additive, the total Gibbs energy difference or pmf resulting from the transfer of 1 mol of ions down an electrical potential gradient of $\Delta\psi$ (mV) and an H^+ concentration gradient is:

$$\Delta G = -mF\Delta\psi + 2.3\ RT \log_{10} \frac{[H^+]o}{[H^+]i} \tag{6.8}$$

In units of electrical potential the combined energy for the two sources of proton flow, with $m = 1$, is:

$$\Delta\mu_{H+} = pmf = \Delta\psi - \frac{2.3RT}{F}\Delta pH \qquad (6.9)$$

which at 30°C becomes, with units of millivolts:

$$\Delta\mu_{H+} = \Delta\psi - 60\,\Delta pH \qquad (6.10)$$

Lipid membranes have very low permeability to H^+. The protons, however, can flow in a controlled manner through the CF_0–CF_1 enzyme complex, producing the change in enzyme configuration required for ATP synthesis. A certain gradient ('pressure') of H^+ is needed to 'drive' the mechanism. The flow of protons is called protonicity, by analogy with electricity, and it is a property of the flow of H^+ through CF_0–CF_1 which synthesizes ATP. The chemiosmotic hypothesis does not provide a description of the molecular mechanism of energy transduction at the level of enzyme sites (enzymological and more recently crystallographic studies have provided this many years after the chemiosmotic theory was formulated), but rather a mechanism of coupling ATP production with electron transport.

The magnitudes of $\Delta\psi$ and ΔpH have been measured by several techniques. Synthetic lipophilic ions (e.g. tetraphenyl-phosphonium), called Skulachev ions after the discoverer, penetrate lipid membranes due to the extensive π-orbital system (Section 2.1.2) even though charged. Their absorption spectra change with conditions, so that, with careful calibration, an estimate of $\Delta\psi$ can be made optically. The ΔpH has been measured from fluorescence quenching of 9-aminoacridine in chloroplasts. Microelectrodes have also been employed to measure pH and ion concentrations. Important indicators of $\Delta\psi$ are carotenoids, bound in the thylakoid (and other membranes); they respond to the large electrical field (3×10^5 V cm^{-1}) which develops, for example, on illumination, by a rapid (nanosecond) change in absorption towards longer wavelengths. This electrochromic shift is readily measured without disturbing conditions within the system and may be calibrated to provide a measure of $\Delta\psi$; however it reflects the $\Delta\psi$ only in the immediate vicinity of the carotenoid in the membrane, not necessarily at the membrane surface.

Under steady illumination $\Delta\psi$ is 10–50 mV across the thylakoid membrane, whereas ΔpH is 3 units, equivalent to 180 mV, and therefore the most important component of pmf in thylakoids; in mitochondria $\Delta\psi$ is the most important component. However, in suddenly illuminated chloroplasts $\Delta\psi$ may also be more important than ΔpH in developing pmf; $\Delta\psi$ develops within 10^{-8} s due to electron transport but ion transport is much slower. The positive charge in the thylakoid lumen causes anions to move in from the stroma to balance the electrical charge. Chloride ions (particularly *in vitro*) enter to join the H^+ in the lumen; in Jagendorf's instructive phrase 'thylakoids pump hydrochloric acid into themselves in the light'. Cations, Mg^{2+} and K^+, move into the stroma and with time $\Delta\psi$ decreases as H^+ accumulates. Up to 1 mmol of H^+ accumulates per mg of chlorophyll and increases the membrane potential. As the lumen is small (or 2.3×10^{-6}

$m^3 g^{-1}$ chlorophyll), a small change in H^+ (1 mmol) would greatly alter pH and could damage the membrane. However, most of the protons (99%) are buffered by proteins in the lumen, so local areas of the luminal space possibly provide the protons for specific coupling factor complexes. The buffering capacity is probably important in regulation of the flow of protons in rapidly fluctuating light. Large $\Delta\psi$ is most important with the onset of illumination and during fluctuations in intensity, allowing a pmf to develop before H^+ accumulates appreciably and enabling ATP synthesis to start quickly.

A convincing demonstration of the importance of ΔpH for ATP synthesis was the complete separation of electron transport from ΔpH, achieved by subjecting intact thylakoids to an acid–base transition. Thylakoids were incubated in the dark, with a buffer solution, for example succinic acid at pH 4, which diffused into the lumen, providing a controlled internal concentration of H^+; then the external pH was raised quickly. With a gradient of four pH units, ATP was synthesized, but a gradient smaller than two pH units was ineffective. Even with electron transport inhibited by DCMU (*Table 5.2*), ATP could be synthesized, showing the pH gradient to be the driving force, not electron flow.

6.4.3 Relationship between ATP synthesis and proton and electron flux

The H^+/ATP stoichiometry, the number of protons crossing the membrane per ATP molecule synthesized, is central to understanding the mechanism of coupling factor. Two basic methods of determining the H^+/ATP ratio are measurement of proton flow through the complex, together with ATP synthesis under steady state, or by calculations from the thermodynamics of phosphorylation and the energy gradient, at given pH, ATP, ADP and P_i concentrations. Both methods are demanding, so there is still uncertainty about the ratio. The measurements of proton flow linked to ATP synthesis is subject to errors in measuring the proton flux as the H^+/e^- ratio is complicated by the Q-cycle activity, and to nonspecific proton leakage across the membranes. Calculations depend on accurate estimation of $\Delta\mu_{H+}$ by using probes which may have secondary effects, lack of true equilibrium, *et cetera*. The estimated values of H^+/ATP have increased from 3 to 4 with time, as better methods have been used. However, there is still uncertainty about the 'true' value, because there is a basal rate (nonphosphorylating) of H^+ flux and the phosphorylation per H^+ depends on conditions, particularly ΔpH. When ΔpH increases from 2 to 2.5 the K_m for ADP decreases, that is the enzyme becomes more efficient; from 2.5 to 3 the K_m for ADP stabilizes and then increases as pH rises further.

Mitchell hypothesized that two H^+ were required for synthesis of one ATP, but currently it is considered to be four in chloroplasts (thylakoids) but only three in mitochondria (because the latter have an ADP/ATP transporter which results in one proton entering per ATP synthesized). The ratio is expected to be related to the number of c subunits in the F_0 part of the ATP synthase complex (see below), which is 12 in CF_0, and the number of β-subunits in F_1, which is 3, as release of three ATP from F_1 requires one rotation of the complex which results from one rotation of F_0, that is c/β of 12/3 requires 4 H^+/ATP. However, if only nine c subunits are involved, as is possible in mitochondria, then c/β is

9/3, that is 3 H^+/ATP. The ratio is not yet established (see Haraux and de Kouchkovsky, 1998). From the discussion of electron transport, eight photons give four e^- transported to $NADP^+$ and eight H^+. Thus, with 4 H^+/ATP, two ATP would be made, an ATP/2 e^- ratio of two per ATP made by noncyclic electron transport. With 3 H^+ required per ATP synthesized, the ATP/2 e^- ratio is 1.33. The energy required, calculated from equation 6.10 is about 240 mV, ΔpH and $\Delta\psi$ of 20 mV, a total pmf of 260 mV. To make ATP requires about 30 kJ per mole or (from $\Delta G = -nF\Delta E'$, where E' is the difference in redox potential, F is Faraday's constant and n the number of reducing equivalents) about 150 mV. At the small ATP concentration in the stroma more energy may be required, perhaps as great as 230 mV, within the energy available for ATP synthesis using 4 H^+ but not with fewer protons.

6.4.4 *Paths of electron flow and coupling to phosphorylation*

Three paths of photosynthetic electron flow linked to ATP synthesis have been recognized and are shown in *Figure 6.1*. ATP synthesis coupled to a 'linear' flow of electrons from water to $NADP^+$ is called noncyclic photophosphorylation because the electrons are transferred to an acceptor, which passes them as reductant to metabolic reactions, and the electrons do not return to the electron transport chain. In cyclic photophosphorylation, electrons are cycled from PSI back to the electron transport chain. Pseudocyclic photophosphorylation involves electron flow to O_2, and to H^+, thus synthesizing water, rather than to NADP; it is a variant of noncyclic electron transport, but with a different electron acceptor. It is, of course, the electron movement which is cyclic or noncyclic not phosphorylation. However, the inexact but historical term remains.

Non-cyclic photophosphorylation. With adequate substrates available to oxidize NADPH, primarily CO_2, the linear flow of electrons is coupled to ATP synthesis. Both PSII and PSI are needed. The action spectra for both CO_2 assimilation and ATP synthesis are very similar and they saturate at similar light intensities. Measured ATP/2 e^- ratios for non-cyclic photophosphorylation are between 1.5 and 2; higher ratios may be obtained if allowance for the basal rate of electron transport is made. However, it is not established if the basal rate occurs when the processes are coupled or only when uncoupled. It is important to know the *in vivo* rates of ATP formation per H^+ because at least 1.5 ATP must be synthesized for each 2 e^- if most of the ATP required for CO_2 fixation is produced noncyclically, as is probable.

Cyclic photophosphorylation. When non-cyclic electron flow is prevented, electrons from PSI or more probably ferredoxin (shown by the greater sensitivity to antimycin, an Fe-S inhibitor), pass back to plastoquinone via cytochrome 563, a b type cytochrome also called cyt b_6 (*Figure 6.1*). Cytochrome involvement is shown by spectral changes. The ratio of cytochrome to PSI is 2:1, so two cytochromes take one electron each from ferredoxin fed by two different PSI centers and reduce plastoquinone. Cytochrome b_{563} has a potential of -0.18 V and donates e^- to plastoquinone at about zero potential. Coupling with ATP synthesis is associated with the transfer of H^+ from plastoquinone to the thylakoid lumen. Only energy from PSI is used, that is, it is driven by light above 680 nm. With no net transport of e^-, water is not split and no O_2 is evolved. However, when the acceptors of PSI are fully reduced the e^- flow cannot start; a slow flux of e^- from PSII to PSI maintains the correct redox potentials ('poises' the system) so that electrons can move. DCMU, which stops electron flow from PSII to PSI, inhibits cyclic photophosphorylation by interfering with 'poising'. The importance of cyclic photophosphorylation *in vivo* is not clear; it may be most important in physiologically intact tissues when noncyclic electron transport is slowed by lack of CO_2 and O_2, and in dim light. Cyclic and noncyclic photophosphorylation may cooperate; the former poises the system and the latter generates ATP. Reduced $NADP^+$ and ferredoxin (when nearly all reduced) also regulate electron flow in cyclic photophosphorylation.

Pseudocyclic photophosphorylation. This requires both photosystems, like non-cyclic photophosphorylation but with ferredoxin reducing an 'oxygen reducing factor', which passes electrons to molecular oxygen as the terminal electron acceptor. The two-step O_2 reduction forms the superoxide radical O_2^- and then, by the action of superoxide dismutase, (Section 5.4.3) hydrogen peroxide. Further metabolism of hydrogen peroxide by catalase yields water and O_2 is released, so the cycle is a way of consuming electrons and energy without generating reductant but with synthesis of ATP. Electrons can also be donated from the reduced acceptors of PSI to H^+ giving H_2, the reaction being catalyzed by hydrogenase; this occurs in some algae. Pseudocyclic photophosphorylation is measured by the uptake of O_2 caused by light and by the effect of ADP and P_i on it. Electrons go from water to O_2 back to water, so the process is not cyclic as the same electrons are not recycled. The rate is greater at high O_2 concentration than low and when CO_2 fixation is slow; it saturates at higher light intensity than cyclic photophosphorylation and has the same $P/2\ e^-$ ratio as noncyclic photophosphorylation, because the same sites of coupling are employed.

6.4.5 ATP hydrolysis by coupling factor

When CF_0–CF_1 is removed from the thylakoid membrane it will hydrolyze ATP in a reaction which is essentially a reversal of the mechanism of synthesis but is not necessarily by a reversal of the mechanism. Of course, there are many ATPase enzymes which function normally to couple the energy from hydrolysis of ATP to transport, for example, of ions such as sodium; here the mechanism is a reversal of the proton-driven ATP synthesis with Na^+ moving instead of H^+. Different amino acid composition of the ion channel in the enzyme complex is responsible for the selectivity towards the ion. Coupling factor may also operate in reverse to pump protons into the thylakoid in darkness or in very dim light *in vivo*. This may have a physiologically important role in maintaining the correct pH for metabolic processes and giving the correct conditions (poising) to enable electron transport and ATP synthesis to start quickly upon illumination, that is regulating the pmf and proton gradient. If hydrolysis of ATP were to occur without regulation between darkness and light, then the system would become unbalanced with cytosolic and stromal ATP concentrations not adjusted to other metabolic needs. Also, ATP synthesized by the mitochondria would be consumed in a futile cycle which would waste energy and be inefficient. In darkness, the enzyme is largely inactivated by oxidation of one of its protein components (the γ-subunit), but this is achieved relatively slowly, although ATP synthesis stops within milliseconds of cessation of illumination.

Regulation is related to the redox state of disulfide links on the γ-subunit which is determined by thioredoxin f (Section 7.3).

6.5 Enzyme mechanism of phosphorylation

6.5.1 The structure of coupling factor and mechanism

In chloroplasts CF_0–CF_1 spans the thylakoid membrane with CF_0 in the lipid bilayer and CF_1 projecting into the chloroplast stroma (*Figure 6.2*). CF_1 was first detected by Avron in 1963, as a protein which, when removed from the thylakoid, uncoupled

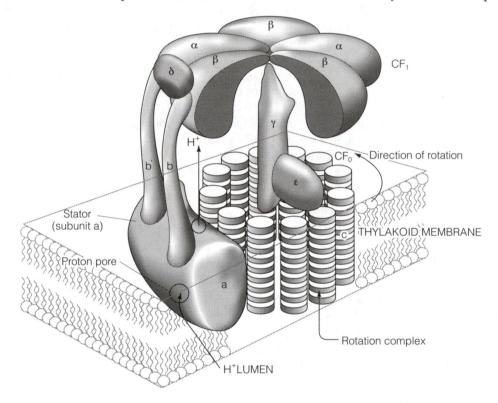

Figure 6.2. Idealized structure of chloroplast coupling factor, CF_1–CF_0, illustrating the stator (fixed position) complex of the a subunit embedded and probably anchored in the thylakoid membrane, with the b and b' units connecting it to CF_1 with three α and β subunits forming the enzymatic complex) with the δ anchoring the b and b' to an α–β dimer. The rotor complex of CF_0 is made of subunits c, forming a ring, in close connection, but not joined to it. The c-ring is linked to the asymmetric γ-subunit by ϵ. The ring rotates in the lipid membrane as a consequence of the entry of H^+, which alters the charge on the c-ring which moves forward (anti-clockwise) so turning γ in the central cavity of the CF_1 enzyme complex. As the asymmetric polypeptide γ passes each enzyme site on the β-subunit the configuration is changed, leading to synthesis of ATP. With 12 c-subunits four charges are required to move the γ from one active site to the other, giving one ATP per four H^+ per ATP and 12 H^+ for a complete rotation of CF_0

photophosphorylation but when replaced, restored ATP synthesis. CF_1 is only loosely attached and is readily removed if the cation concentration (particularly Mg^{2+}) is low, for example, after treatment with EDTA, a cation-chelating chemical. CF_1 is thought to move around on the membrane under the influence of ionic and electrical charges. Analysis of the composition of CF_1 has been done by many methods including disrupting the isolated complex with urea, which breaks hydrogen bonds between subunits, followed by polyacrylamide gel electrophoresis in the presence of the detergent sodium dodecyl sulfate (SDS). The CF_1 is composed of five types of protein (polypeptide) subunits α, β, γ, δ, ε in the ratio 3:3:1:1:1 and has a molecular mass of 325 kDa. The structure of mitochondrial ATP synthase, MF_1, from bovine heart muscle has been determined by X-ray diffraction at 0.28 nm resolution (see Abrahams *et al.*, 1994). Understanding of the molecular arrangement within the subunits and between them has confirmed and clarified what was long suspected about both the structure of the CF_1 complex and the mode of catalysis, and has thrown new light on the unique characteristics of this most important of enzymes. It has been described by Boyer (1997) as a 'splendid molecular machine', a fitting description of what is a nano-sized rotary motor, driven by protons and coupled to the active catalytic sites of ATP synthesis. Coupling factors are of very similar subunit structure in all organisms, with extensive homology in the amino acids of the polypeptides of the subunits. Clearly, the nature of coupling factor and the mode of catalysis were developed early in evolution and, despite the great changes that organisms have undergone since, the enzyme has not changed radically.

The form of CF_1 is a flattened sphere about 8 nm high and 10 nm across, formed of three elongated (approximately 10 nm long axis) α-subunits of 55–56 kDa mass and three elliptical (5–8 nm) β-subunits (52–54 kDa) symmetrically arranged alternately, like the segments of an orange, and forming a central cavity (*Figure 6.2*). The two different subunits are almost identical, consisting of an N-terminal six-stranded β-barrel, a central domain from the two subunits containing the nucleotide binding sites and a C-terminal domain of six β and seven α helices. The nucleotide binding domain contains a nine-stranded β-sheet with nine α-helices associated. The catalytic sites communicate to the external solution via a water-filled conical tunnel allowing the phosphate and ribose moieties of ADP and ATP access, and permitting diffusion of the substrates and products. The β-barrel domains of the subunits form the top of the complex, probably serving as structural elements holding the hexamer together. The α- and β-subunits are asymmetrical, reflecting the differences in the catalytic sites and the nucleotides in them and on the position of the γ-subunit which fills the central cavity of the hexamer and protrudes out of the hole in the base. Formation of the hexamer is not dependent on the presence of γ, but active synthesis of ATP is dependent on γ. When the hexamer from *Thermobacillus* is disrupted in the presence of ATP and magnesium, α–β-dimers are formed with only weak ATPase activity, suggesting close association between pairs of the heterodimer subunits. Although apparently identical, the α-subunits have different characteristics, for example lucifer yellow vinyl sulfone covalently binds to a lysine residue (Lys378) on only one of the three α-subunits, indicating a role in binding perhaps with components of the F_1 part of the complex (b and b'). Subunit β binds the nucleotides (there is some uncertainty about the number of binding sites with weak and strong binding); there are possibly two strong binding sites which can accommodate a number of inhibitory analogs of ATP. There are important amino acid sequences which form the binding sites:

a lysine (162) in a loop composed of many glycine residues near a phosphoryl residue is most important. Tyrosine at 345 in the β-subunits is closely related to the catalytic site for ATP synthesis but tyrosine 368 in the β-subunit is more involved with the noncatalytic site activity. The catalytic nucleotide binding sites are at the interfaces between the α- and β-subunits, although with most of a site composed of amino acid residues from the β-subunits and side chains of the α contributing to the site structure, for example an arginine α-373 is present in the catalytic site in the β-subunit. The γ-polypeptide (37 kDa) is contained asymmetrically within the cavity of the α- and β-subunits and greatly affects the structure of the catalytic sites: this was seen from electron microscopy combined with monoclonal antibody and immunogold labeling. Also, small angle X-ray diffraction analysis with time resolution showed extensive structural changes associated with catalysis. Electron spin resonance showed that ATP and Mg^{2+} had large effects on the structure of the nucleotide binding sites. The C-terminal of γ is twisted and sits symmetrically within the 1.5 nm deep 'dimple' formed by the terminal residues of the inner side of the α- and β-subunit complex. The amino acids of the surfaces within the α–β cavity below the β-barrels at the top and of α are very hydrophobic and so are considered to form a smooth bearing (i.e. with no hydrogen bonds), allowing rotation of the γ-subunit inside the complex. The lower part of the γ-subunit is composed of a left-handed antiparallel coiled coil with a second helix made of amino acids 1–45, which sticks out about 3 nm from the main part of γ. Thus, the lower part of γ is displaced 0.7 nm from the long axis between the α and β-subunits so that the γ-subunit is strongly asymmetrical. The unique feature of ATP synthase is the rotary mechanism for catalysis, which was suggested on the basis of the enzymatic reaction sequence before the structure was known. The rotary motion is now largely accepted (for a contrary view, see McCarty *et al.*, 2000). The asymmetry and the 'low friction bearing' in the α–β-subunit hexamer are related to the rotation of γ, which turns in the cavity and acts as a cam, opening and closing the catalytic sites between α- and β-subunits and thus driving the catalytic sequence of ATP synthesis. The enzyme turns very rapidly, 1000 times s^{-1}, achieving very fast ATP synthesis.

Although less well defined, the other subunits of the F_1 complex clearly have important roles. The ε-subunit is associated with the lower part of γ and probably links it to the polar loops of the c-subunits of CF_0 in the lipid membrane; removing it stimulates ATPase activity and it regulates H^+ leakage and thus photophosphorylation. The δ-polypeptide is thought to bind the CF_1 to the b- and b'-subunits associated with the CF_0. Removing δ has little effect on ATPase activity but is required to block H^+ leakage and thus to regulate photophosphorylation.

The CF_0 complex is composed of polypeptides called I (18–19 kDa), II (16 kDa), or b and b' in prokaryotes, IV (27 kDa), called a in prokaryotes, and III (c in prokaryotes) in the ratio of 1:1:9–12. The prokaryote terminology will be used. As *Figure 6.2* shows, the c-subunits, also called the 'proteolipid', form a ring within the lipid membrane, and are involved in proton translocation. The c has two curved membrane spanning helices, connected by a short hydrophilic loop (18 amino acids) which faces the stroma and may connect to amino acid residue 31 of the ε-subunit of CF_1. The membrane helices cross at an angle of about 30°. One of the helices contains an aspartate or glutamyl amino acid residue (asp61 in *E. coli*) positioned in the center of the membrane in a very hydrophobic environment; this is essential for function. The residue binds dicyclohexylcarbodiimide

(DCCD), thus blocking proton translocation. CF_0 allows very rapid and specific H^+ transport: 2×10^5 H^+ per CF_0 per second with μ_H^+ equivalent to 30 mV. This is 10^3 times faster than that required for ATP synthesis. The CF_1 complex interacts with the CF_0, slowing the rate of proton movement during ATP synthesis. It is possible that the c-subunits are paired.

Understanding of the role of subunit a is less well developed. The single copy is very hydrophobic, with six helices spanning the membrane; the proton pore is possibly formed by one of these helices, which has many charged amino acid residues. Arginine 210, glutamate 219 and histidine 245 (in *E. coli*) are of great importance, with arginine 210 absolutely conserved in the polypeptide from all sources; it can be moved but with loss of activity. The general view of the position of a is that it is outside the ring of c-subunits and anchored in the membrane. The two b-subunits are elongated, with a hydrophilic helical head region which is associated with F_1. Cross-linking studies show that the hydrophilic head associates with an α–β-dimer from the F_1 hexamer, probably anchored by the δ polypeptide. This role for δ is shown by its location on the outer surface of CF_1, by covalent attachment of it to the complex which does not destroy catalytic competence, and by the requirement for it in reconstruction of the enzyme using subunits obtained from different sources, for example mitochondria and bacteria. The N-terminal regions of the b and b' polypeptides are hydrophobic and it is thought that this links to the a subunit in the membrane. Thus the a, b and b' and F_1 parts of the complex are joined, forming a fixed (if elastically coupled) 'stator' as in an electric motor, where the poles of a magnet are fixed and the rotor turns within it.

The ring of c-subunits, considered to be the rotor of the coupling factor complex, turns in the thylakoid membrane lipids; MGDG is required for effective function of coupling factor, as the lipid bilayer provides the electrostatically neutral environment in which the rotation of the c-subunit ring takes place. There is good evidence that the c-subunits cooperate and act as a single unit which is attached to γ by ε. Thus, as the ring of c-subunits rotates so does γ, turning anti-clockwise during ATP synthesis within the cavity formed by the α- and β-subunits. Because of the asymmetry of α this alters the structure of the catalytic sites sequentially. For this rotary mechanism to operate, CF_1 must be anchored to the membrane.

Evidence for the rotational mechanism has been obtained by several ingenious methods, including attaching eosin to the γ-subunit and fixing the α- and β-subunit complex to ion exchange resin. Using a technique called polarized absorption recovery after photobleaching (PARAP), applied under conditions of ATP hydrolysis, it was possible to detect the relaxation of the signal with a relaxation time of about 100 ms, which is similar to the 70 ms for the breakdown of ATP. Another method was to attach a fluorescent actin filament, up to 2.5 μm long, to the γ-subunit to act as a macroscopic label, and at the same time fix the CF_1 complex to a support. When ATP was supplied the movement of the actin filament was observed in a direction expected for hydrolysis of ATP. A three-step rotation was observed, in agreement with the analysis by PARAP. Thus, a rotary mechanism is now accepted for the synthesis of ATP by CF_0–CF_1. The rotation of the central unit is driven by the flux of protons from the lumen to the stroma. However, the mechanism of this is not yet clear. One H^+ enters subunit a in the membrane and passes to a c-subunit which is protonated on the acidic amino acid residues. At the same time, another protonated c-subunit loses its H^+ via a to the stroma at another point on the circle. Each protonation results in a positive charge to a c-subunit and because of the electrostatic interactions between proteins (a and c polypeptides) the protonated c can only exist within the lipid bilayer around F_0. This electrostatic interaction results in repulsion and movement of the rotor. It seems likely that a single rotation cycle requires 12 H^+ in chloroplast coupling factor but only nine in mitochondria. Different models for the way that rotation is driven have been suggested. Junge *et al.* (1997) suggest that the c-subunits are always protonated and electroneutral when facing the lipid membrane but lose H^+ when facing the a polypeptide and thus alter in charge. If the channel for protons from a entering c is at a different position for the channel by which the proton is lost to the stroma via a, direction will be imparted to the rotation. Thermally induced movements of the ring (Brownian motion) are converted into directional movement by the charge on c related to the charge on a.

The role of the a-subunit in proton translocation is of considerable importance, yet still incompletely understood. It is thought to have a proton channel, divided into two sections connected at the center of the lipid membrane, one half receiving H^+ from the high concentration (low pH) lumen compartment and the other half connected to the low concentration, high-pH stromal compartment. At the junction between these, protons are exchanged with acidic residues in α-helices of two c-subunits. These residues are blocked by DCCD, or by modifying the acidic residues genetically, which prevents ATP synthesis. Other amino acid residues, for example an arginine and glutamate or histidine, are associated with the proton transfer site. The pH difference between the two halves of the proton channel is probably responsible for the interconversion of $\Delta\psi$ and ΔpH, allowing both 'driving forces' to operate although they are of different type (Section 6.4.2), with strict coupling between protonation and deprotonation. This type of mechanism may provide the structure for the concept of a 'proton well' which was suggested to explain many of the observations concerning ATP synthesis and the energetics.

6.5.2 Mode of catalysis

The sequence of events and conditions required for ATP synthesis were known (see Boyer, 1993, 1997) long before the detailed structure of the coupling factor complex was elucidated, and is shown in *Figure 6.3*. Details of the active sites of coupling factor, how the nucleotides and P_i are bound in them and the changes in their conformation and of the electronic forces which operate during catalysis, according to the part of the reaction cycle that is considered, are now becoming clear as a consequence of the detailed structural analysis. The articles by Abrahams *et al.* (1994) and Boyer (1997) provide details of the structural and enzymatic processes involved in the first and smallest rotary mechanism known in biology. The 'binding change mechanism' involves cooperation with the release of ATP from one site, depending on the binding of substrates ADP and P_i to another (without which proton translocation cannot occur and conformation cannot change), with the three sites on CF_1 acting in a sequence and going through the same cycle. In other words the sites are equivalent. It is not clear if all three sites must be occupied or only two; measurement of the K_m values for substrates suggests that two sites must be filled with substrates, with different affinities. ATP is synthesized by indirect changes in conformation of the active sites on the subunits. The synthesis of ATP

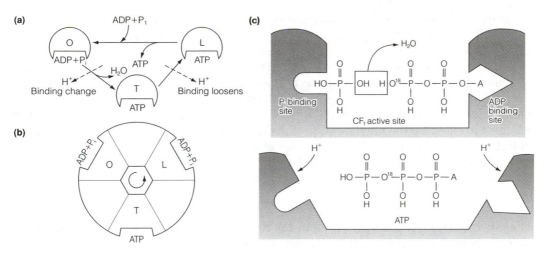

Figure 6.3. An illustration of the binding change mechanism and sequence for the synthesis of ATP on CF$_1$. (a) ATP is formed when the ADP and P$_i$ which bind in the 'open' site (O) undergo the enzymatic reaction caused by rotation of γ which converts the site, in a low energy requiring structural change in the active site, to a tight binding site (T) containing ATP. Further rotation of γ results in release of ATP, in the main energy requiring step, and formation of an open site to which ADP and P$_i$ bind, forming the L active site structure once again. (b) The sequence for an individual CF$_1$ complex (from Boyer, 1993). The active site structure is shown, greatly simplified in (c) the positions of ATP, Mg^{2+} and a water molecule involved in providing the required energy mechanism for electron and proton exchange between the substrates are indicated. The structure of the site is mainly formed by the β-subunit but the presence of residues from the α-subunit is essential for catalysis (based on Abrahams *et al.*, 1994). (c) The spontaneous reaction with H$^+$ removal from ADP terminal phosphate, shown by incorporation of ^{18}O from ADP into ATP, and removal of OH from P$_i$

involves considerable cooperation at the molecular level between the constituent amino acids of the active site, formed between the polypeptides of the α- and β-subunits of CF$_1$, and also between the polypeptides of the complex as a whole to achieve conversion of energy in the form of the proton gradient across the thylakoid membrane into chemical energy in the form of ATP. A model of the active site (physically identified as on the β-subunit but with a major contribution from α) existing in three distinct conformations is now accepted (*Figure 6.2*). On each α–β-hexamer are three active enzyme sites, the catalytic sites in the β-subunits and the nucleotide binding sites between the two types of subunit (*Figure 6.3*). The sites are called open (O), loose (L) and tight (T). One complete rotation of the α-subunit within the water-filled α–β-hexamer results in synthesis of three ATP molecule. Consider the state of the 3 sites at any time: O has low affinity for ADP and P$_i$ which attach but the site is catalytically inactive. This converts O to L with ADP and P$_i$ attached to the binding sites: ADP spontaneously loses H$^+$ from the terminal phosphate group (shown by incorporation of ^{18}O from ADP into ATP) and OH is removed from P$_i$, a water molecule, forming the anhydride bond. Simultaneously L converts to a T site, with ATP fixed within it. Then ATP is released by the main energy-requiring step in catalysis; the energy required is about equivalent to hydrolysis of ATP. Some of the energy may come from the previous change in conformation. Binding of ADP and P_i to CF_1 results in changed structure and water content (shown by the altered tritium exchange), demonstrating that molecular arrangement of subunits and conformation of active sites change with conditions and are involved in catalysis. The interconversion of the sites occurs as a consequence of the rotation of γ. Although the main requirement for energy is in the conversion of the T to an L site, the L to O change also requires energy provided by the passage of protons through CF_0 and rotation of γ. Thus, each individual site undergoes three binding changes (and conformational changes) in synthesis of one ATP, so, as three ATP are made per rotation under steady-state synthesis and 12 H^+ are needed to drive the cycle by one full turn of γ, four H^+ produce one ATP.

There is considerable uncertainty about the role that nucleotide binding at noncatalytic sites plays in the structure of the catalytic site for ATP synthesis. Adenylates other than the substrates bind to three sites rather tightly. They are close to catalytic sites, but the significance of this is unknown. The average distance between the adjacent sites is 2.7 nm; ADP and P_i bind at particular positions in the β-subunits tightly at two sites, about 90 μm of binding sites per chloroplast. A CF_1–ADP and P_i complex may be formed, which is inhibited by arsenate. Studies with ^{18}O-labeled ATP, P_i and H_2O show that as protons pass through CF_0 they remove an oxygen atom on the phosphate group, forming water (*Figure 6.3*). The electron on the phosphate group moves to the ADP terminal phosphate, forming the anhydride bond. This is thought to require little change in energy. Protons in CF alter the conformation of the peptide chains and change the binding energy between ATP and the enzyme complex, allowing ATP to escape into the stroma. ADP + P_i then bind again to the enzyme reaction site in its energized conformation, which then returns to its original form, producing the anhydride bond. Release of ATP allows ADP to bind and phosphorylation occurs spontaneously, the release of ATP requiring energy. Inhibition of ATPase by ADP may reflect 'clogging' of the sites. Removal of subunits regulating the conformational change then prevents phosphorylation but leaves the enzyme in a state capable of hydrolysing ATP.

6.5.3 Activation of coupling factor and regulation of activity

Not only does the electrochemical proton gradient provide the energy for ATP synthesis but it also activates the enzyme complex in chloroplasts. The mitochondrial enzyme does not require activation. Without activation ATP is not made and ATP cannot be hydrolyzed. Upon activation there are changes in conformation and exchange of water with the medium, as was elegantly shown by incubating CF_1 with tritiated water, removing the enzyme, denaturing it in urea and measuring incorporation of radioactivity. When the complex was energized by illumination or a pH gradient, there was exchange of tritium from the medium to hydrogen on the protein. Under de-energizing conditions, approximately 100 H atoms per CFI remained hidden in the proteins, possibly associated with conformational changes. Proton exchange was essential for ATP synthesis. Reduction of the complex is also a requirement; this is achieved chemically with thioreducing agents such as dithiothreitol (DTT), which reduces a specific disulfide bridge on the γ-subunit (measured with N-ethyl maleimide, which reacts with -SH groups). This decreases the $\Delta\mu_{H^+}$ required for activation, increases the period for which the enzyme remains active and increases the efficiency. Reduction of two cysteine residues on γ (Cys199 and Cys205) in a domain of some 40 amino acids found only in chloroplast coupling factor γ is essential for activity. Another two cysteines are also involved in activation under some conditions. One is accessible to maleimide when the membrane is de-energized (γCys322) and the other when $\Delta\mu_{H^+}$ is produced in the light. In darkness, ATPase is inactive but is activated by DTT reacting with sulfhydryl groups on the γ-subunit which are hidden within the complex in the dark but become exposed in the light (γCys89). The γCys322 may be at the end of the C-terminal α-helix and the γCys89 in another α-helix about 3–7 nm apart. Evidence from binding of different size maleimides suggests that there are large structural (conformational) changes in γ, with periodic movement of the subunit up and down within the cavity of CF_1, depending on the energy state of the thylakoids and related to the passage of protons. Such changes in shape and position may be important in the mechanism by which rotation of γ alters the conformation of the active sites. Reagents which bind to -SH groups (e.g. dithiothreitol) increase ATPase, suggesting that sulfhydryl groups regulate ATP synthesis. Light activates CF_0–CF_1, not only via ΔpH (see above) but by the thioredoxin system which regulates other chloroplast enzymes (Section 7.3) changing the sulfhydryl groups on the γ-subunit. In the unactivated state a much greater ΔpH is needed to drive photophosphorylation. CF_1 activity is regulated by several factors, including H^+ and the concentration of substrates ADP and P_i. ATP inhibits ATP synthesis and ADP and P_i inhibit ATP hydrolysis so that the complex is allosterically controlled. Nucleotides bind to multiple sites, not active in ATP synthesis, depending on the energized state of CFI (indicating conformational changes in CFI proteins), but their role in catalysis is not understood. Energy from the proton gradient is essential for releasing ATP by changing conformation of CFI; this stage may be the most energy demanding. Activation of coupling factor *in vivo* has been shown by measuring the electrochromic shift at 515 nm; it requires only limited light and is rapid on illumination after darkness, whereas the deactivation is slow so the activation state of the ATP synthesizing system is probably sensitive to the onset of light but not very sensitive to rapid changes in photon flux during the day.

In illuminated thylakoids, ATP synthesis is the main function of CF_0–CF_1 but this complex also catalyzes hydrolysis of ATP. Treatment of isolated CFI with the protein-digesting enzyme trypsin, heat, or with sulfhydryl reagents (e.g. dithiothreitol) stimulates ATPase activity and Ca^{2+} is required. The α-subunit controls ATPase activity. This may have a physiological significance, allowing ATP to drive proton accumulation, providing control over the ionic balance of thylakoids, for example in darkness, when regulation of the state of the membranes and ionic concentration creates the conditions needed for rapid synthesis of ATP on illumination.

6.6 Uncouplers of ATP synthesis

A gradient of proton concentration is required for ATP synthesis by coupling factor, so intact vesicles with a bounding membrane impermeable to H^+, as in the thylakoid system, are needed for H^+ accumulation; the lipid membrane is an effective barrier. Illumination of intact thylakoids without ADP or P_i causes H^+ accumulation from water-splitting and a 'back-pressure' on the PQ pump, slowing electron transport to a basal rate, corresponding to leakage. When the substrates ADP and P_i are available, H^+ flows through CF_1–CF_0, ATP is synthesized, the concentration of H^+ drops and the electron transport rate increases until equilibrium is attained. Treatments causing loss of ΔpH or $\Delta\psi$ uncouple electron transport and ATP synthesis, and the coupling factor itself may be inhibited by analogs of substrates and products. There are, accordingly, many different compounds which interfere with the mechanisms. Inhibition of electron transport will prevent the proton gradient developing. If only some CF complex is removed by EDTA, which chelates Mg^{2+}, then H^+ leaks through CF_0 and no ATP is made. Inhibition of ATP synthesis by arsenate or thiophosphate allows H^+ flow through CF_1 and destroys ΔpH. Phlorizin (a glucoside from roots) blocks ATP synthesis but prevents H^+ flow. Membranes are also made 'leaky' by lipid-soluble proton ionophores, H^+ transporting compounds, such as carbonylcyanide p-trifluoromethoxyphenyl hydrazone (FCCP). The negatively charged molecule moves along the $\Delta\psi$ gradient into the lumen, where it is protonated, then moves back to the medium, where it is deprotonated and recycles, collapsing ΔpH. Dinitrophenol (DNP) carries anions and cations (H^+) in response to $\Delta\psi$, and is a very effective uncoupler in mitochondria but not in thylakoids. Other ionophores carry ions; for example, valinomycin, a depsipeptide (with alternating hydroxy and amino acids) antibiotic from bacteria, dissolves in the membrane and transports K^+ out of the lumen, changes $\Delta\psi$ and uncouples. Valinomycin and DNP together very effectively uncouple chloroplasts as they remove H^+ and K^+ from the lumen, collapsing ΔpH and $\Delta\psi$. Nigericin also acts in this way. Gramicidin is a very effective uncoupling peptide, forming a pore or channel in the membrane allowing very rapid efflux (10^{-7} s^{-1}) of monovalent ions. If ammonia enters the lumen and achieves a concentration above 10^{-3} molar (below this and the protons buffer the pH), it is protonated to ammonium ions (NH_4^+) and destroys ΔpH; anions also enter to balance the excess charge increasing the osmotic potential, leading to chloroplast swelling. Ammonia is produced in many metabolic processes so it is a physiological uncoupler under some conditions.

6.7 Phosphorylation and physiological control

Photosynthesis is a process which is extremely well regulated, to maintain function under a range of environmental conditions, such as light, temperature and water supply, but also in response to conditions within the cell and plant, for example P_i availability and carbohydrate production. At the core of the process, the synthesis of ATP and the balance between ATP and NADPH synthesis and concentrations are critical. Despite much attention, these processes are not well quantified, and there is no model which can account for or simulate the processes. However, a general description is possible. In dim light, electron transport is slow and so is the rate of synthesis of NADPH and ATP; thus the supply limits metabolism. Synthesis of ATP may be more restricted than that of NADPH, so more electrons must flow in cyclic electron transport to generate the ΔpH for ATP synthesis. Activation of coupling factor occurs in dim light so this is not a limitation.

With increasing photon flux, the synthesis of ATP and NADPH increases, so that the activated metabolic processes can proceed rapidly. If imbalance occurs between the supply and use of the products then regulatory mechanisms operate to balance the system. Excess reductant leads to increased ATP synthesis via the recycling of electrons and proton pumping. Such regulatory mechanisms are particularly important, for example, if the supply of CO_2 is limited and thus also the consumption of ATP and NADPH, when they are being rapidly synthesized in bright light; then ΔpH rises and activates the xanthophyll cycle of energy dissipation by increasing nonphotochemical quenching in the thylakoids, so removing the excess energy (or part of it, see Section 3.4.4), and secondary mechanisms of removing reductant (such as transfer of electrons to oxygen forming superoxide which is then broken down enzymatically) come into play. Pseudocyclic electron flow may also permit ATP synthesis when $NADP^+$ is not available, perhaps in very intense light. Low demand for ATP, which slows the rate of supply of ADP to CF_1 or inadequate P_i supply, inhibits phosphorylation. Photosynthesis by C3 plants (Section 4.3.1) requires an $ATP/2\ e^-$ ratio of at least 1.5 for CO_2 reduction; as other processes also consume ATP, particularly in the light, either the ratio *in vivo* is greater than that *in vitro* or additional ATP is synthesized in other ways. Regulation of the ATP/NADPH ratio is important for photosynthesis but poorly understood. Affinity of components of the e^- transport chain for substrates may be expected to determine the relative flux of e^- into parts of metabolism, e.g. ferredoxin-$NADP^+$ reductase has a much greater affinity for $NADP^+$ than NAD^+ (a K_m of 10^{-5} M compared to 3×10^{-3} M), but under unphysiological conditions it catalyzes H^+ transfer between NADPH and NAD^+ and also oxidation of NADPH by several electron acceptors. Also, the e^- flux will depend on light, which determines the saturation of the e^- transport chain, relative to CO_2, O_2 and NO_3^- supply.

The evolutionary history of species affects their ability to adjust to conditions. Many C3 plants have considerable adaptability to the light environment, because the genetic mechanisms for synthesis of photosynthetic components are able to adjust the structure of the system to light conditions during development of the leaves, and by regulation of ATP and NADPH synthesis (amongst other processes). However, other C3 plants are permanently adapted to deep shade (e.g. with many grana) and have a very limited capacity to adjust the structure of photosynthetic system and are therefore relatively inefficient in bright light. The balance of the processes between different tissues within plants is determined by the composition of the system, for example some C4 plants (such as *Zea mays*) have agranal bundle sheath but granal mesophyll chloroplasts, which determines the relative capacities for ATP synthesis and electron transport and water splitting. This sets a balance between ATP and NADPH synthesis, resulting in relative inefficiency in dim light, but efficiency in bright light. Granal number and size may regulate the area of membrane not only for light harvesting and electron flow between photosystems, but also H^+ flux through coupling factor. Coupling factor is almost certainly controlled by feedback mechanisms, because of its extremely important position in metabolism and because of the very dynamic nature of thylakoid energetics. It is a challenge to understand the nature of the regulation in varying environments because of the very central role that it plays in cellular metabolism.

References and further reading

Abrahams, J.P., Leslie, A.G.W., Lutter, R. and Walker, J.E. (1994) Structure at 2.8 Å resolution of F_1-ATPase from bovine heart mitochondria. *Nature* **370**: 621–628.

Atkinson, D.E. (1977) *Cellular Energy Metabolism and its Regulation.* Academic Press, New York.

Boyer, P.D. (1993) The binding change mechanism for ATP synthesis – some probabilities and possibilities. *Biochem. Biophys. Acta* **1140**: 215–250.

Boyer, P.D. (1997) The ATP synthase – a splendid molecular machine. *Annu. Rev. Biochem.* **66**: 717–749.

Haraux, F. and de Kouchkovsky, Y. (1998) Energy coupling and ATP synthase. *Photosynth. Res.* **57**: 231–251.

Jagendorf, A.T., McCarty, R.E. and Robertson, D. (1991) Coupling factor components: structure and function. In *The Photosynthetic Apparatus: Molecular Biology and Operation* (eds L. Bogorad and I.K. Vasil). Academic Press, San Diego, CA.

Junge, W., Lill, H. and Engelbrecht, S. (1997) ATP synthase: an electrochemical transducer with rotary mechanics. *TIBS* **22**: 420–423.

McCarty, R.E. (1996) An overview of the function, composition and structure of the chloroplast ATP synthase. In *Oxygenic Photosynthesis: the Light Reactions* (eds D.R. Ort and C.F. Yocum). Kluwer Academic, Dordrecht, pp 439–451.

McCarty, R.E., Evron, Y. and Johnson, E.A. (2000) The chloroplast ATP synthase: a rotary enzyme? *Annu. Rev. Plant Physol. Plant Mol. Biol.* **51**: 83–109.

Mitchell, P. (1966) Chemiosmotic coupling in oxidative and photosynthetic phosphorylation. *Biol. Rev.* **41**: 445–502.

Ort, D.R. and Oxborough, K. (1992). *In situ* regulation of chloroplast coupling factor activity. *Annu. Rev. Plant Physiol. Plant Mol. Biol.* **43**: 269–291.

Sabbert, D. and Junge, W. (1997) Stepped versus continuous rotatory motors at the molecular scale. *Proc. Natl Acad. Sci. USA* **94**: 2312–2317.

Westheimer, F.H. (1987) Why nature chose phosphates. *Science* **235**: 1173–1178.

Chapter

7

The chemistry of photosynthesis

7.1 Introduction

All dry matter produced by photosynthetic organisms comes from the use of reductant (ferredoxin and NADPH) and adenylate energy, ATP, from the light reactions, to synthesize chemical products from simple organic starting materials, basically CO_2 and nitrate and sulfate ions. These synthetic reactions are fundamental to plant function and are common to prokaryotes and eukaryotes. The mechanisms of higher plants evolved from common origins, with a prokaryotic oxygen-evolving cyanobacterial cell entering into symbiosis with a eukaryotic cell, and integrating with its functions. The interactions in the eukaryotic photosynthetic cell between the processes of light transduction and chemical synthesis and the components responsible for them are substantial, as expected for a highly regulated system which has evolved over millions of years.

Here, the general mechanisms are described in some detail; differences in mechanisms and the regulatory processes between organisms, which are important but less well understood, will not be considered further. The chemical reactions of carbon dioxide, nitrate and sulfate reduction are very tightly interconnected, in terms of material and energy flows and their regulation within cells, so are considered here in one chapter, although separated for convenience. Carbon dioxide is the major energy consuming process in photosynthesis; nitrate and sulfate reduction (Section 7.8) use 10% or less of total energy in plants, but are vital aspects of photosynthetic metabolism. Understanding of the basic mechanisms helps to explain how cells, leaves, plants and ecosystems function in particular environments.

Photosynthesis of assimilates of carbon and nitrogen in higher plants takes place primarily in the chloroplast stroma, which contains many different types of proteins, including the enzymes of CO_2, nitrate and sulfate assimilation and metabolites. These complex reactions require substrates produced inside the chloroplast, principally ATP and NADPH, and imported from the cytosol (e.g. inorganic phosphate), or from mitochondria (e.g. organic acids). Enzymes are affected by conditions within the chloroplast, for example by concentrations of inorganic phosphate or the metabolites produced by the reactions which they catalyze or from other enzyme reactions; such products act as effectors, which alter enzyme activity but are not consumed in the reactions. The rate of CO_2 fixation is determined by enzyme activity, the supply of substrates and the conditions in the chloroplast. Also, there is substantial interaction between respiration, in mitochondria, and photosynthesis, with interchange of intermediates. Utilization of assimilated carbon depends on transport of products out of the chloroplast, via translocator systems, and on the demand from growth and for maintenance of the plant, for which respiration (and hence production of CO_2) is essential. Carbon assimilation is regulated by highly integrated mechanisms which allow the photosynthetic system to maintain its activity at rates appropriate to the demands of, and changing conditions within, the plant. All these processes are, directly or indirectly, complex functions of environmental conditions which are very variable and change, often rapidly, in complex ways. This chapter examines the basic mechanism of CO_2 assimilation and the associated processes, and how the characteristics of different components interact to achieve such a highly regulated and effective system of assimilate production.

7.2 Carbon dioxide assimilation

Carbon dioxide assimilation is a cyclic, autocatalytic, process (*Figure 7.1*), variously called the Calvin cycle (after M. Calvin who described the system in the 1950s), reductive pentose phosphate pathway, photosynthetic cycle or photosynthetic carbon reduction cycle (PCR cycle). PCR cycle emphasizes the carbon, reduction and cyclic aspects, so the term is used here. The PCR cycle is the fundamental CO_2 assimilatory process in all photosynthetic organisms, including prokaryotes; it appears to have developed early in evolution and to have retained its characteristics, a theme noted earlier during discussions of ATP synthesis and electron transport. The mechanism of CO_2 assimilation by the PCR cycle is called C3 photosynthesis, as the first stable product is a 3-carbon compound. Additional processes for accumulating CO_2 have arisen which do not replace the PCR cycle, but rather add to it; they are the C4 and CAM mechanisms (see Chapter 9).

7.2.1 Mechanisms of the PCR cycle

Assimilation of CO_2 is described by:

$$3\,CO_2 + 9\,ATP + 6\,NADPH + 5\,H^+ \rightarrow C_3H_5O_3P +$$
$$9\,ADP + 8\,P_i + 6\,NADP^+ + 3\,H_2O + 468\,kJ\,mol^{-1} \tag{7.1}$$

This summary greatly simplifies the complexity of the process. There are 11 enzymes involved in the 13-step carboxylation cycle (*Figure 7.1*) and the individual reactions are identified by numbers which correspond to the enzymes listed in *Table 7.1*. In the carboxylation reaction catalyzed by the enzyme RuBP carboxylase-oxygenase (Rubisco), an acceptor molecule, ribulose bisphosphate (RuBP) combines with CO_2, producing 3-phosphoglyceric acid (3PGA). The acceptor is then regenerated from 3PGA in reactions consuming NADPH and ATP; for each CO_2 assimilated a minimum of three ATP and two NADPH + H$^+$ are needed. They are produced, as discussed in Chapters 5 and 6, by light-driven thylakoid electron and proton transport. The role of the PCR cycle is to regenerate RuBP for further CO_2 assimilation. If more carbon is fixed than is used to regenerate RuBP, it is exported from the chloroplast as triose phosphates, or starch is synthesized in the chloroplast. Control of export and consumption of PCR cycle intermediates is

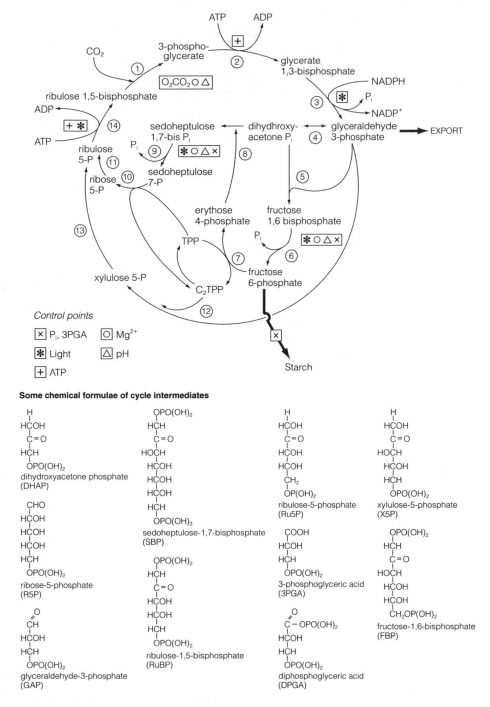

Control points

⊠ P_i, 3PGA	⊙ Mg^{2+}
✳ Light	△ pH
⊞ ATP	

Some chemical formulae of cycle intermediates

H
|
HCOH
|
C=O
|
HCH
|
OPO(OH)$_2$
dihydroxyacetone phosphate
(DHAP)

CHO
|
HCOH
|
HCOH
|
HCOH
|
HCH
|
OPO(OH)$_2$
ribose-5-phosphate
(R5P)

O
‖
CH
|
HCOH
|
HCH
|
OPO(OH)$_2$
glyceraldehyde-3-phosphate
(GAP)

OPO(OH)$_2$
|
HCH
|
C=O
|
HOCH
|
HCOH
|
HCOH
|
HCOH
|
HCH
|
OPO(OH)$_2$
sedoheptulose-1,7-bisphosphate
(SBP)

OPO(OH)$_2$
|
HCH
|
C=O
|
HCOH
|
HCOH
|
HCH
|
OPO(OH)$_2$
ribulose-1,5-bisphosphate
(RuBP)

H
|
HCOH
|
C=O
|
HCOH
|
HCOH
|
CH$_2$
|
OP(OH)$_2$
ribulose-5-phosphate
(Ru5P)

COOH
|
HCOH
|
HCH
|
OPO(OH)$_2$
3-phosphoglyceric acid
(3PGA)

O
‖
C–OPO(OH)$_2$
|
HCOH
|
HCH
|
OPO(OH)$_2$
diphosphoglyceric acid
(DPGA)

H
|
HCOH
|
C=O
|
HOCH
|
HCH
|
OPO(OH)$_2$
xylulose-5-phosphate
(X5P)

OPO(OH)$_2$
|
HCH
|
C=O
|
HOCH
|
HCOH
|
HCOH
|
CH$_2$OP(OH)$_2$
fructose-1,6-bisphosphate
(FBP)

Figure 7.1. The photosynthetic carbon reduction cycle, with numbered reactions corresponding to the enzymes listed in *Table 7.1*. The upper part of the figure includes the carboxylation and reduction steps; the lower part shows the regeneration of the CO_2 acceptor ribulose-l,5-bisphosphate (RuBP)

Table 7.1. Enzymes of the photosynthetic carbon reduction cycle (reactions are numbered as in *Figure 7.1*) with their approximate mass, specific activity (SA = μmol min^{-1} mg chlorophyll^{-1}) and Michaelis constant (K_m). The free energy change ΔG^s, at the steady-state physiological (i.e. stromal) concentrations of substrates is for *Chlorella* in 40 Pa CO_2 and 21 kPa O_2. Large negative ΔG^s indicates a probable control reaction

Reaction number	Enzyme	SA	Percentage of total in chloroplast	Total mass (kDa)	Number of subunits	ΔG^s	Increased by	Decreased by	K_m
1	Ribulose bisphosphate carboxylase/oxygenase	10	100	550	8	−41	High pH, CO_2, Mg^{2+}, FBP	Gluconate, SBP?	CO_2 12 μM, O_2 250 μM, RuBP 40 M
2	Phosphoglycerate kinase	900	90	44	1	+16	3PGA, ATP, Mg^{2+}	DGPA, ADP	3PGA, 0.5 μM, ATP 0.1 μM
3	Triosephosphate dehydrogenase (NADP glyceraldehyde P dehyd.)	100	—	140	4	−6.7	Light, NADPH, ATP	P_i? DHAP	DPGA 1 μM, NADPH 4 μM
4	Triosephosphate isomerase	200	—	53	2	−7.5	Alkaline pH	PEP, RuBP, FBP glycolate P, PGA	DHAP 1 mM, GAP 0.4 mM
5, 8	Aldolase	100	—	140	4	−1.6	High GAP/DHAP ratio, pH 8	RuBP, ADP, PGA	FBP 20 μM, GAP 0.3 mM, DHAP 0.4 mM, SBP 20 μM
6	Fructose bisphosphatase	100	90	140	4	−27	High pH, Mg^{2+}, ATP, reductant, light	P_i	FBP 0.2 mM, 800 μM activated
7, 10, 12	Transketolase	150	—	140	2	−5.9	Mg^{2+}, high pH		Xu5P 100 μM
9	Sedoheptulose bisphosphatase	0.5	—	70	2	−29.7	Light, pH, Mg^{2+}	P_i	SBP 13 μM (reduced)
11	Ribose-5-phosphate isomerase	10	—	54	2	−0.5	Freely reversible, pH 8		R5P 0.5 mM, ATP 0.1 mM
13	Ribulose-5-phosphate 3 epimerase	30	—	46	2	−0.6	Freely reversible		X5P 0.5 mM
14	Ribulose-5-phosphate kinase	320	—	83	2	−15.9	Reductant, ATP, energy charge?, pH > 7	ADP	R5P 2.5–70 μM, ATP 60 μM

essential to prevent the cycle from 'running down', and is critical for maintenance of such an autocatalytic process. If export of metabolites from the chloroplast is slow then phosphorylated intermediates accumulate and the chloroplast becomes depleted of inorganic phosphate, both of which decrease rates of enzyme reactions, so the cycle cannot function efficiently in CO_2 fixation. Rate of cycle turnover in the steady state depends on the rates of NADPH and ATP synthesis, on enzyme activities, on the export of carbon to the cytosol and, of course, on the availability of CO_2.

Starting with carboxylation of RuBP, Rubisco produces two molecules of 3PGA [reaction (1)], a reaction unique to the PCR cycle. Because of the great importance of this reaction, its mechanisms are discussed in Section 7.2.2. Reaction (2): phosphoglycerate kinase uses ATP to phosphorylate 3PGA to a more reactive state in 1,3-diphosphoglyceric acid with two acid anhydride bonds. In reaction (3) NADP glyceraldehyde-3-phosphate dehydrogenase substitutes H^+ for the phosphate group in 1,3-diphosphoglycerate. The enzyme uses $NADP^+$ in the light but NAD^+ in darkness during catabolic ATP synthesis; the respiratory (mitochondrial) enzyme requires NAD^+. This is the only reduction reaction in the PCR cycle and is therefore of great importance. Glyceraldehyde-3-phosphate (GAP) is converted by triose phosphate isomerase (reaction 4) to dihydroxyacetone phosphate (DHAP); these two compounds are used in reactions (8) and (12), where 3-carbon units are converted, by fructose-1,6 bisphosphate/sedoheptulose-1,7-bisphosphate aldolase, to yield 6- and 7-carbon compounds in the sequence of reactions leading to regeneration of RuBP. Triose phosphates are condensed [reaction (5)] to the 6-carbon compound fructose bisphosphate; the aldolase has maximum activity in alkaline conditions, which are found in the chloroplast stroma during illumination. Dephosphorylation of FBP by fructose bisphosphatase gives fructose-6-phosphate (F6P). The reaction has a free energy change of -25 kJ mol^{-1}, so it is not reversible and is considered a control point in the cycle (for more detailed discussion of regulation of the PCR cycle see Section 7.4).

Regeneration of RuBP is achieved by interconversion of 3-, 4-, 5- and 6-carbon compounds. Transketolase (steps 7, 10 and 12) removes 2-carbon fragments (glycoaldehyde) from F6P and sedoheptulose-7-phosphate (S7P), attached to the thiamine pyrophosphate (TPP) co-factor of the enzyme. Erythrose-4-phosphate (E4P) reacts with DHAP in the aldolase reaction (8) giving sedoheptulose-1,7-bisphosphate. This is dephosphorylated [reaction (9)] by sedoheptulose bisphosphatase, only found in the PCR cycle. Ribose-5-phosphate (Ru5P) is made from S7P [reaction (10) catalyzed by transketolase], and converted to ribulose-5-phosphate [reaction (11)] by ribose-5-phosphate isomerase. Another 5-carbon sugar, xylulose-5-phosphate, is formed by transketolase [reaction (12)] from glycoaldehyde and glyceraldehyde-3-phosphate and is converted to Ru5P [reaction (13)] by ribulose-5-phosphate 3-epimerase. The 'final step' [reaction (14)] is the phosphorylation of Ru5P to the substrate RuBP by phosphoribulokinase (also called ribulose-5-phosphate kinase), another enzyme unique to the PCR cycle, with a free energy change of -15 kJ mol^{-1}, the fourth most negative in the cycle.

7.2.2 Enzymes of the PCR cycle

Characteristics of individual enzymes are discussed because they determine how the PCR cycle uses the ATP and NADPH from the light reactions. Regulation of the cycle is complex, with changes in concentrations of substrates and products of one step affecting the behavior of others, and with some enzymes regulated by light-induced signals, or by metabolic effectors, which modify the flux of carbon through the cycle. Regulation is considered in detail later.

Ribulose-1,5-bisphosphate carboxylase/oxygenase (Rubisco).

Rubisco (EC 4.1.1.39), the carboxylating enzyme, occurs in all photosynthetic organisms; it comprises up to 50% of the soluble protein (6 mg per mg chl) in leaves of higher plants such as wheat supplied with large amounts of nitrogen fertilizer, and is probably the most abundant protein on earth. Much is known of its enzymology and structure: Roy and Andrews (2000) provide a detailed, current view. Its importance in photosynthesis was recognized very early from correlations between the amount of protein and photosynthesis, and later the characteristics of the enzyme provided a mechanism for many features of CO_2 and O_2 sensitivity of the process and of photorespiration. In higher plants the enzyme is a hexadecameric complex of two dissimilar subunits in equal number, eight large and eight small subunits (called LSU and SSU) of 55 and 15 kDa, respectively, a total mass of 550 kDa; these are arranged as shown in *Figure 7.2*. This complex structure is an L_8S_8 type of Rubisco which is found in most higher prokaryotes and eukaryotes. There is one catalytic site on each large subunit, that is eight catalytic sites per complete complex (holoenzyme) for the higher plant enzyme. However, this complex multimeric enzyme structure is not essential for CO_2 assimilation as Rubisco from the photosynthetic bacterium *Rhodospirillum rubrum* has only two LSUs (molecular mass 52 kDa) and other bacteria have L_4S_4 and other combinations of units. However, Rubisco from higher eukaryotes has a 10-fold higher efficiency in assimilating CO_2 but much slower catalysis than the *R. rubrum* L_2; the reasons for this are not known. LSUs are very similar in amino acid structure in all organisms but the SSUs differ widely. The LSUs are synthesized in the chloroplast stroma and encoded as a single gene in chloroplast DNA whereas the SSUs are encoded by a family of nuclear genes and made in the cytosol with a 40–60 amino acid long N-terminal extension. This targets the protein translocating

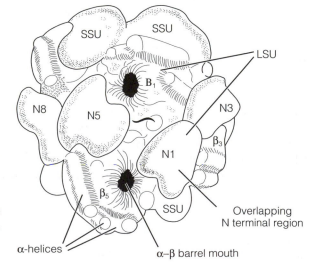

Figure 7.2. Illustration of the subunit structure of ribulose bisphosphate carboxylase–oxygenase enzyme. Detailed structure based on X-ray crystallography, showing the small subunits (SSU), the N terminal and barrel (B) domains of the large subunits (LSU) numbered 1–8. The α-helices around the B domains are shown and also the 'mouths' of two α/β barrels. N domains of one LSU normally cover the mouths of a barrel on another LSU, e.g. N5 covers the B1 barrel.

system in the chloroplast envelope (see Chapter 10) so the SSU is transported into the stroma where the extension is removed by a protease and the S-subunit is bound to L-subunits synthesized there. Folding of the L-subunit requires a molecular chaperone (once called Rubisco binding protein) to make the active tertiary structure (see Chapter 10). It is not yet possible to express active higher plant Rubisco in prokaryotes or vice versa by genetic transformation (although the protein is made) because of the absence of the specific folding machinery.

Analysis of Rubisco by X-ray diffraction has revealed details of the structure, particularly of the active catalytic site. The LSU has an N-terminal domain (amino acid residues 5–134 in spinach), a connector region (residues 135–168), a main domain called the barrel (residues 169–432) and the C-terminal domains (residues 433–477). The N-terminal is made up of four antiparallel β-sheets with four surrounding α-helices; the connector region has a short helix which blocks the end of the barrel. The barrel is made of eight-stranded α-helices and eight β-sheets, arranged in a complex structure. These β-sheets are in parallel, forming the tube-like barrel; the α-helices surround the β-sheets on the outside and the large structure seems to be necessary to maintain the active site coordinates for effective catalysis. The Rubisco active site is, however, not on one LSU only; it is at the junction of the N-terminal of one LSU where its fifth β-strand interacts with the pair of β-strands in loop 6 of the C-terminal domain of the other. The active site, at which carboxylation and oxygenation take place, has three lysine and one glutamate amino acid residues, and is close to the mouth of the barrel. Lysine residues are on one large subunit C-domain but the glutamate is on the N-domain of the neighboring large subunit. The phosphate binding site is on loop 5 of the C-terminal domain. Thus, the active site of Rubisco is a pocket, formed at the interface between two subunits. Both the α/β-barrels and cooperation between subunits in forming an active site occur in other enzymes and are not unique to Rubisco; these enzymes may have very little homology between amino acid sequences, yet the active sites are always at the C-end of β-strands. The amino acid structure of the enzyme is most highly conserved in the β-sheet close to the active site at the barrel opening, probably because electronic configuration at the site of catalysis is very critical for carboxylation. Indeed, altering single amino acids at very different positions in the large subunit nearly destroys carboxylation. Such sensitivity to the configuration may make it difficult to reduce the oxygenase activity of the enzyme (see later discussion) by manipulation of the site although some progress is being made with modifications to loop 6 of the C-terminal domain.

Small subunits of Rubisco consist of an antiparallel sheet of β-strands with flanking α-helices. A long C-terminal extension reaches down to the back of the barrel on an LSU and two SSU extensions contact the barrel of an LSU. This, together with the extensive contact between LSUs, suggests cooperation of LSU and SSU subunits in catalysis but no clear explanation for the structure role of the SSU has been advanced. Bacterial (prokaryotic) and dinoflagellate (eukaryotic) Rubisco without SSUs is functional, suggesting that SSUs are not essential for catalysis. However, Rubisco from cyanobacteria and plants was 99% inactivated when the SSUs were removed; re-addition of them restored activity, and even adding SSUs from other sources, such as spinach, did so. The SSUs seem to improve the rate of CO_2 assimilation. Another role for them is in holding the LSUs together. The four SSUs at each pole of the complex interact and also with the LSUs: one SSU is connected with two others and with three LSUs, possibly with helix 8 of the LSU barrel domain. This arrangement may alter the binding energy and active site structure and energy distribution. This would explain the 100-fold increase in reaction rate, and improved specificity and carbamylation. SSUs may also stabilize Rubisco under the different conditions within the chloroplast.

Rubisco reactions and mechanisms.   Rubisco catalyzes two principal reactions with the substrate RuBP, CO_2 assimilation or carboxylation, considered above,

$$RuBP + CO_2 = 2 \times 3\text{-PGA} \qquad (7.2a)$$

and O_2 assimilation or oxygenation

$$RuBP + O_2 = \text{phosphoglycollate} + 3\text{-PGA} \qquad (7.2b)$$

For these reactions to proceed Rubisco must be activated by addition of CO_2 and Mg^{2+}. A scheme summarizing the reactions is:

$$ (7.3) $$

where E is the enzyme, C is CO_2, O is O_2, M is Mg^{2+} and R is RuBP.

E + R is the enzyme site which is not catalytically active and binds RuBP; it is converted to the 'open' but still inactive site E by the activity of a protein, Rubisco activase. The open site is then activated by addition of CO_2, giving E + C, and then magnesium, forming a catalytically competent enzyme complex E + C + M. This binds RuBP, forming E + C + M + R, and this complex then reacts with CO_2 giving E + C + M + R + C, or O_2 giving E + C + M + R + O. Details are discussed in the following sections.

Rubisco: activation.   Rubisco requires activation by addition of CO_2, that is the enzyme must be carbamylated. This CO_2 does not take part in the reaction between RuBP and CO_2 in photosynthetic CO_2 reduction. The carbamylating CO_2 binds to a lysine residue (number 201 in the spinach enzyme) situated in the catalytic site, and coverts one of the carbamino oxygen groups on the lysine 201 side chain from a positive to a negative charge and two protons are released. The carbamylation reaction is slow but once the site is carbamylated Mg^{2+} binds rapidly to it; two other ligands (carboxyl O-atoms from acidic groups of Glu 204 and Asp 203) also link with the magnesium, which is then coordinated, that is held but not covalently bound in the active site, giving the correct energy distribution in the active site for effective catalysis. The enzyme is thereby activated, allowing the substrates to bind and the carboxylation reaction to proceed. Conditions in the chloroplast which are necessary for the activation (e.g. the CO_2 and Mg^{2+} concentrations) are not well established, but carbamylation occurs at small CO_2 concentrations. Activation is required for the formation of the correct electronic configuration of the active site, and incidentally permits activity only when CO_2 and magnesium availability is sufficient. The uncarbamylated site, without magnesium, may be occupied by three water molecules or, more importantly, by RuBP which binds three orders of magnitude more tightly to E (giving E + R in the scheme above) than to the activated site, E + C + M + R.

The way that activation affects the active site of the enzyme and the binding of the substrates has been shown by the structure of the spinach enzyme, resolved to 0.24 nm with a transition state analog of RuBP, 2-carboxyarabinitol-1,5-bisphosphate (CABP), bound to the active site. CABP and related compounds are effective inhibitors because they bind tightly to, and thus block, the catalytic site. This characteristic can be used to measure the number of sites on the protein by using ^{14}C-labeled CABP, which binds very tightly to the sites: the number of sites is proportional to the amount of radioactivity which can be easily measured. Structural studies and analysis of Rubisco altered by site-directed mutagenesis, show that lysine residue 201 is absolutely essential for enzyme function, and is in a strongly conserved sequence of amino acids forming the active site. The amino acids have particular functions at stages in catalysis and establish the correct electronic configuration of the active site into which RuBP binds by its phosphate groups (*Figure 7.3*). RuBP binds before CO_2 or O_2, to basic residues, possibly on lysine. No sites for the gases exist unless RuBP is bound first. When RuBP binds, a tautomeric change in its electronic configuration occurs. The C2 carbonyl of RuBP is normally slightly positive (electron attracting) because electrons are pulled towards the oxygen in the 2' carboxyl of the molecule. This and the C2 hydroxl group of the bisphosphate bind to the magnesium. The phosphate groups are at different distances from the magnesium, altering the electronic structure and allowing carboxylation. Loss of a proton from C3 results from polarization of the carbonyl by magnesium and a base nearby must also help in the removal of the proton, which is an important process, but not fully understood. Formation of a keto-enol equilibrium produces a nucleophilic (proton attracting) enediol in *cis*-configuration, allowing the CO_2 to react at the negatively charged C2 of RuBP. Alternatively, the structure may be a carbanion, especially in the hexadecameric forms, and contribute to the greater efficiency compared to other Rubisco types. The amino acids have specific roles, for example Asp 203 is involved in enediol formation but not hydrolysis, whereas Glu 204 participates in both reactions. A threonine (Thr 173) which is close to the C2 of the RuBP substrate is possibly related to carbanion formation; it is present in L_8S_8 forms but not others which have Ile 164 instead. The position of the CO_2, Mg^{2+} and carboxyl group of RuBP are contiguous, as shown by ^{13}C NMR. Details of the chemistry of the reactions are given by Gutteridge and Gatenby (1995) and Roy and Andrews (2000).

Inhibitors of Rubisco and the role of Rubisco activase.

Although it may appear strange, it is a characteristic of Rubisco that RuBP binds tightly to the uncarbamylated form of Rubisco, preventing the activation steps and inhibiting CO_2 assimilation. Also, the reactions of Rubisco produce compounds, analogs of RuBP and intermediates of the reaction, which also inhibit the enzyme by binding tightly to the active sites. Such

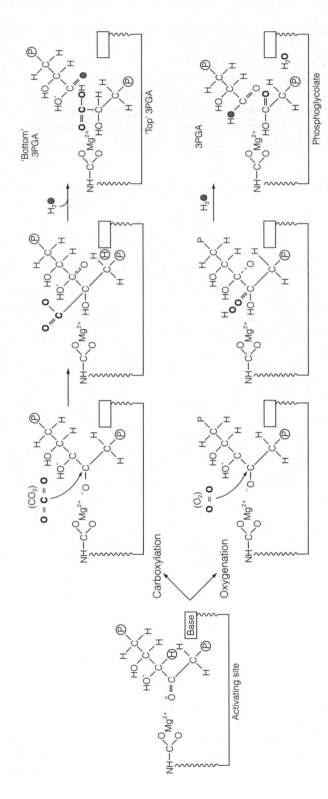

Figure 7.3. Schematic of the reaction site of ribulose bisphosphate carboxylase and the oxygenase showing molecular events in carboxylation and oxygenation of RuBP to give 3-phosphoglyceric acid (3PGA) or phosphoglycolate and 3PGA, respectively

'misfire' products are made about once in every 400 turnovers of the active site and result from imperfect partitioning of the enediol. For example, D-xylulose 1,5-bisphosphate is made by incorrect protonation of the C3 intermediate and 3-keto-D-arabinitol 1,5-bisphosphate by proton attack on C2 of the enediolate intermediate; both these compounds bind tightly to the catalytic site. In addition, regulation of Rubisco activity *in vivo* involves inhibitors. One tight-binding inhibitor produced from carbohydrate metabolism in chloroplasts under very low light or in darkness is 2'-carboxy-D-arabinitol 1-phosphate (CA1P). Specifically, carbon from the fructose-1,6-bisphosphate of the PCR cycle is enzymatically rearranged to hamamelose bisphosphate, which is dephosphorylated to hamamelose. This is converted to carboxyarabanitol and phosphorylated to CA1P. CA1P binds to sites not occupied by RuBP, thereby blocking activity. CA1P is a normal nocturnal inhibitor of Rubisco in many higher plants (e.g. *Phaseolus vulgaris*, and tobacco), but not in others (e.g. wheat and spinach). CA1P may reach 5 mM concentration, greater than that of Rubisco active sites. CA1P possibly prevents the consumption of RuBP and intermediates of the PCR cycle in darkness and prevents binding by metabolites from the many other reactions which occur in darkness. A more important function may be protection of the enzyme from proteolysis. In the light CA1P, once released from Rubisco, is metabolized to 2'-carboxy arabinitol by a relatively specific phosphatase, CA1P phosphatase, activity of which increases in the presence of photosynthetic product (e.g. 3PGA, RuBP and NADPH), suggesting stimulation of breakdown when conditions are correct for photosynthesis, but it probably is not light activated. The propensity of Rubisco to form inactive complexes with a range of metabolites means that it is less efficient in CO_2 assimilation: the Rubisco activase reaction serves to overcome this problem.

Rubisco activase.   The inhibition of Rubisco by RuBP binding and the action of other tight-binding inhibitors is reversed by a protein, called Rubisco activase [see the reaction sequence for Rubisco activation, equation (7.3)] which was discovered in an *Arabidopsis* mutant lacking active Rubisco. Activity of Rubisco decreases under small photon flux and increases in large flux as inhibitors are formed and bind or are removed, respectively. The 'deactivation' and 'activation', as they are termed, require several minutes and are also related to the removal and addition, respectively, of both Mg^{2+} and CO_2 at the active sites of Rubisco. A very important step in activation is removal, in a reversible reaction, of RuBP bound very tightly to the decarbamylated (inactive) enzymatic site (E+R). This is done by Rubisco activase, a tetrameric protein of 47 kDa subunits, which may be 1–2% (tobacco) or 5% (*Arabidopsis*) of soluble protein in leaves, giving a stoichiometry of one activase tetramer to 80 Rubisco catalytic sites in tobacco and one to *ca* 25 sites in *Arabidopsis*. Activase binds to blocked sites, altering the closed-loop conformation and thereby changing the binding affinity for different substrates, so increasing the rate at which inhibitors dissociate from decarbamylated Rubisco. Activase may recognize the different forms of Rubisco and reacts much faster with the E + R than E + C + M + R form (which is catalytically competent). Thus, RuBP and other inhibitors are removed, allowing carbamylation and magnesium attachment to form the catalytically competent enzyme sites, followed by RuBP binding, thus permitting CO_2 assimilation.

Activase may also maintain a balance between light-dependent RuBP regeneration and the rate of carboxylation, decreasing fluctuations in the pools of PCR cycle intermediates, and the flux of carbon out of the chloroplast and to secondary metabolism. Rubisco activase requires ATP hydrolysis for the energy required to alter the active sites, suggesting that it is a form of molecular chaperone. Rubisco activase is one of the most abundant chloroplast proteins because it is relatively slow; activation increases exponentially, and may limit the rate of response of CO_2. Decreasing the amount of activase by antisense methods shows that it is very important in regulating the rate of CO_2 assimilation, for example during induction as photon flux increases. Under these conditions, activase may control photosynthesis (control coefficient of 1, see Section 7.4.2). However, a large part of the activity in the normal ('wild-type') plants must be removed before maximum Rubisco activity is affected. The effects are greater under conditions which place large demands on the capacity of Rubisco, particularly when there is a need for rapid activation, for example in fluctuating light with large changes in photon flux. Activase thus stabilizes metabolism as well as regulating CO_2 flux when RuBP supply is already large. Probably, an optimum for CO_2 assimilation exists between activase and Rubisco amounts and activity which depends on light, with the ratio of activase to Rubisco greater in fluctuating light than in relatively constant bright light (see Mott and Woodrow, 2000). Rubisco activase is nuclear encoded; there are two isoforms of 41 and 46 kDa, which differ in thermal stability, but increasing the proportion of the larger more stable protein did not improve the heat stability. The structure has been analyzed, showing how the protein links to Rubisco and changes the active site. Activase may be regulated by magnesium, ATP/ADP ratio, sugar phosphates and light via the thioredoxin system. Such complex regulation may be expected for an essential site (Rubisco) in the PCR cycle.

Once Rubisco is activated, RuBP must bind to the active site. The number of RuBP molecules per active site required to give maximum rates of assimilation is still controversial; probably 1.5–2 RuBP per site is required. It may be necessary to ensure that all sites are filled with RuBP to prevent tight binding inhibitors (which are byproducts of carbohydrate metabolism) and even other bisphosphates and isomers of RuBP, from blocking them and inhibiting photosynthesis. This is probably important when the supply of CO_2 or RuBP is limited, so linking energy transduction to carbon assimilation.

Rubisco: carboxylation reaction. The CO_2 reacts [equation (7.2a)] directly and electrophilically with the C2 of the tautomeric complex of RuBP, forming the 6-carbon intermediate 2-carboxy-3-ketoarabinitol-1,5-bisphosphate. Water is essential but its relation to the reaction must be carefully regulated. Water donates OH^- to the C3 carbonyl group producing a *gem*-diol which is deprotonated, causing the C2–C3 bond to break, giving a 3PGA molecule (D-stereoisomer). Then a further protonation of the other half of the molecule releases the second D-isomer of 3PGA, that is two molecules of the D-stereoisomer of 3PGA are produced in total per molecule of RuBP plus CO_2 complex. The removal of the C3 proton from RuBP involves Asp 203, Glu 204 and Lys 175, which is in the loop 1 of the C-terminal segment of the barrel. However, if hydrolysis of the 6-carbon intermediate proceeds non-enzymatically, both the L- and D-isomers are formed. The enzyme controls the stereochemistry of the reaction. The carbanion intermediate may also lose the phosphate group (once in every 100 reactions – see earlier discussion of 'misfire' reactions), thus forming pyruvate. The specificity of different Rubiscos is influenced by Glu 60, Thr 65 and Asn 123 residues in the N-terminal region of one L-subunit, with Lys 334 in loop 6 which stabilizes the intermediates; Thr 342 and Val 331 determine the flexibility of the loop. All loop amino acids seem important both for flexibility and correct position of the amino groups, relative to the position of RuBP and intermediates, indeed loop 6 plays an essential role in catalysis, moving into the active site during enediol formation. It is interesting that there are differences in catalytic efficiency: sunflower has a Rubisco some 16% more efficient at CO_2 fixation than spinach, yet with the same essential active site amino acids, so differences in amino acids outside the site, indeed distant from it or in the S-subunit, must determine the activity. More detailed analysis of the catalytic competence of engineered Rubisco combined with structural studies is clarifying the details of catalysis, and showing how very subtle the process is. Indeed the differences in efficiency of the bacterial and higher plant enzymes, although substantial, involve only the equivalent of the energy of a weak hydrogen bond (*ca* 4 kJ mol^{-1}). If distributed across the whole protein this would be difficult to detect in terms of the structure, although if related to a specific group it may be more evident. Genetically engineering Rubisco to increase efficiency is, therefore, likely to be more difficult than once envisaged, although better understanding of the molecular structure during catalysis will aid progress in that direction.

Rubisco: oxygenation reaction.

Rubisco is a bifunctional enzyme, catalyzing the reaction [equation (7.2b)] of molecular oxygen with RuBP at the same catalytic site as the carboxylation and producing phosphoglycolate, a 2-carbon phosphorylated compound, and 3-PGA. The mechanism of oxygenation by the enzyme is still not well understood. The oxygen molecule is 0.012 nm long, is relatively stable, uncharged and not polarized. This contrasts with CO_2, which is 0.0232 nm, stable but polarized (although neutral overall). Other types of oxygenase enzyme contain a transition metal (e.g. Fe) or a redox prosthetic group, which transfers electrons, but Rubisco has no such group and there is no comparable monooxygenase enzyme. Theoretically, the addition of ground-state triplet O_2 to the enediol in the reaction is not allowed because of spin restrictions and there are no mechanisms allowing the addition to occur. However, it does occur! It is not clear how the oxygen is activated, perhaps by coordination to the Mg^{2+} in the presence of the enediol of RuBP and formation of the C2-carbanion which reacts with O_2 giving a 2'-hydroperoxy 3-keto arabinitol 1,5-bisphosphate intermediate (*Figure 7.3*). This is subsequently hydrolyzed by water bound to the metal. The products are one molecule of the D-streoisomer of 3-PGA and one of 2-phosphoglycolate. The second protonation step of the carboxylation reaction is not involved in the oxygenase reaction. If the RuBP oxygenase reaction is performed in the presence of $^{18}O_2$, one ^{18}O joins to C2 of RuBP to produce phosphoglycolate labeled in the carboxyl group, and the other is released as $H_2{}^{18}O$. An O from H_2O reacts with C3 of RuBP to give 3PGA. Possibly the enediol form of RuBP can attract the electrons in C of CO_2 or of O_2; the required electronic changes in RuBP are produced by the coordinated Mg^{2+} and bound activating CO_2. That oxygenation and carboxylation occur at the same site is shown by inhibition of both functions by CABP, other intermediates of the PCR cycle and pyridoxal phosphate. Also, both functions are activated to the same degree by CO_2 and Mg^{2+}. However, the two reactions are differentially affected by pH, and by temperature which changes the O_2/CO_2 solubility ratios in solution. If manganese ions replace magnesium ions in the activation of the isolated enzyme, the ratio of oxygenase to carboxylase activity increases. As with carboxylation, catalysis involves distinct steps associated with changes in enzyme structure and movement of parts of the Rubisco complex, particularly loop six, that changes in position, covering and opening the site. The reaction sequence then involves different amino acids entering the active site, altering the electronic structures and the acid–base characteristics of the site. This affects the binding of RuBP, and the electronic structure, thus allowing the reactions with CO_2 or O_2 and then with water. The dynamic molecular structure now emerging explains the production of RuBP analogs and the reasons for their inhibitory action.

Competition between RuBP carboxylase and oxygenase reactions. Oxygen and CO_2 compete for RuBP at the catalytic site of Rubisco; they are mutually competitive inhibitors. In the reaction sequences for Rubisco [equation (7.3)], the interconversion of the forms of enzyme complex sites and their rate constants (k) provides the basis for understanding the competition between CO_2 and O_2 (see von Caemmerer, 2000, Chapter 12 for details). The Michaelis–Menten constants for the carboxylase (K_c) reaction is given by $K_c = (k_{11} + k_{10})/k_9$ and oxygenase (K_o) by $K_o = (k_{14} + k_{13})/k_{12}$. The dissociation of RuBP $K_{ir} = k_8/k_7$. The rate of turnover of the carboxylation site is $k_{ccat} = k_{11}$ and of the oxygenation site is $k_{ocat} = k_{14}$. The rate, V_c, of carboxylation is given by $k_{ccat} E + C + M + R + C$, and the maximal rate, which depends on the total concentration of enzyme sites (E_t, the sum of all forms of E), is given by $V_{cmax} = k_{ccat}E_t$. At saturating RuBP concentration, often used to measure the reaction rates *in vitro*, V_c (denoted W_c) in the presence of competitive inhibition by O_2 is given by:

$$V_c = W_c = \frac{V_{cmax} \times C}{C + K_c(1 + O/K_o)} \tag{7.4}$$

where V_{cmax} is the maximum rate, C and O are the partial pressures of CO_2 and O_2 in equilibrium with the dissolved gases in the chloroplast stroma, and K_c and K_o are the Michaelis–Menten constants for CO_2 and O_2, respectively. The rate, V_o, of the oxygenase reaction is given by $k_{ocat} E + C + M + R + Ot$ and the maximal rate by $k_{ocat}E_t$. With saturating RuBP, $V_o = W_o$, and for competitive inhibition by CO_2 is:

$$V_o = W_o = \frac{V_{omax} \times O}{O + K_o(1 + C/K_c)} \tag{7.5}$$

where V_{omax} is the maximum rate of oxygenation. These equations describe the responses of Rubisco to substrates with competitive inhibition. The K_m for CO_2 in the oxygenation reaction equals the CO_2 inhibition constant for the oxygenase and O_2 has the same K_m in the oxygenase reactions as K_i in the carboxylase reaction. Oxygen competes inefficiently with CO_2 for RuBP at the catalytic site and only at high molar ratio of O_2 to CO_2 is the oxygenase reaction significant, as discussed earlier.

Specificity factor. The carboxylation and oxygenation rates may be expressed as $V_c/V_o = (V_{cmax} K_o/K_c V_{omax})C/O$, where the term in brackets is the relative specificity (specificity factor) of the enzyme for CO_2 compared to O_2 when the gases are at equal partial pressures or concentrations: $S_{c/o} = \tau = (V_{cmax} K_o)/(V_{omax} K_c)$ or $(V_c/K_c)(V_o/K_o)$. Specificity expresses the intrinsic carboxylation capacity of Rubisco relative to oxygenation and is used as a measure of the efficiency of Rubisco from different organisms. It determines the CO_2 compensation point in C3 leaves. There is a large (20 times) range: $S_{c/o}$ is 77 and 94 mol CO_2 mol O_2^{-1} in tobacco and wheat at 25°C, the extremes of the range for C3 plants, 58–82 mol mol^{-1} for the C4 plants *Setaria* and *Amaranthus*, decreasing to 50–60 mol mol^{-1} in green algae, 35–48 for cyanobacteria and 10–60 for photosynthetic bacteria (*Rhodospirillum* is about 10), so enzyme complexity appears to be associated with greater specificity.

The two reactions may also be expressed as a ratio of oxygenase to carboxylase, α (*Figure 7.4*) which increases as the ratio of O_2 to CO_2 increases:

$$\alpha = \frac{V_o}{V_c} = \frac{V_{omax}}{V_{cmax}} \times \frac{OK_c}{CK_o} \tag{7.6}$$

Hence, at small CO_2 partial pressure α increases strongly with increasing oxygen.

The reason for the differences in specificity may be that the reaction mechanism is not by formation of a Michaelis complex, but depends on the rate of reaction of CO_2 and O_2 with the enediol. This in turn depends on the differences in free energy of the gases and intermediates in the reactions, which are very small, equivalent to a hydrogen bond in the entire molecule. Such a large difference in specificity if related to such a small energy difference suggests that the greater efficiency in C3 Rubiscos compared to others is the result of very small changes in quaternary structure. Hence, substituting Mn for Mg and altering amino acids well away from the active site, alters the catalysis, generally decreasing specificity, but sometimes giving a small increase.

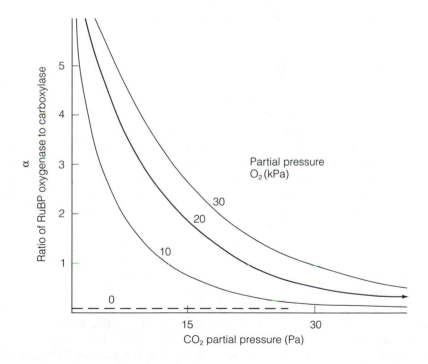

Figure 7.4. Ratio of RuBP oxygenase to RuBP carboxylase activity, α, in relation to CO_2 and O_2 concentrations for extracted enzyme

Significance of carboxylation and oxygenation: photorespiration.
The oxygenase activity of RuBP carboxylase is important because the PG formed in chloroplasts is transported or diffuses out and is metabolized by the glycolate pathway (Section 8.7), involving reactions in the peroxisomes and mitochondria, and CO_2 release in the mitochondria. This CO_2 is thus a form of respiration, which occurs in the light as photosynthesis proceeds and is distinct from tricarboxylic acid CO_2 production. It is, therefore, called photorespiration (Section 8.7). It offsets CO_2 assimilation by the PCR cycle and thus decreases the efficiency of CO_2 assimilation. When the ratio of O_2 to CO_2 in the chloroplast stroma is small, α approaches zero and photorespiration is negligible. However in air, with the partial pressure of O_2 1000-fold greater than CO_2 (Chapter 11), α is about 0.4; photorespiration is large and 20–30% of the fixed carbon is lost. When α is 2, carbon lost by photorespiration equals gross CO_2 assimilation (see Chapter 12) and there is no net gain of carbon. Some types of plants have developed mechanisms which avoid or minimize loss of photorespired carbon, viz. C4 and CAM plants; those mechanisms are considered in Chapter 9. Crop plants with large α would be more efficient at CO_2 assimilation but attempts to select this desirable trait or to modify the enzyme chemically or by genetic manipulation have, so far, produced no marked improvement, rather the opposite – selection of more efficient enzyme remains a major challenge.

7.2.3   Other PCR cycle enzymes

Rubisco has been discussed in detail and here the other enzymes are considered. The first four have been considered rate-limiting, based on the reaction and amounts of enzyme activity. More detailed discussion of the concepts and assessment of rate limitation is given in Section 7.4.

Fructose-1,6-bisphosphatase and sedoheptulose-1,7-bisphosphatase.
Fructose-1,6-bisphosphatase (FBPase, EC 3.1.3.11) and sedoheptulose-1,7-bisphosphatase (SBPase, EC 3.1.3.37) in chloroplasts of higher plants are separate enzymes, with very strong specificity for their respective substrates, in contrast to prokaryotes, where one enzyme has both functions. FBPase is a homodimer of 40 kDa subunits; there are two isoenzymes, one in the chloroplast functioning in the PCR cycle, and the cytosolic form, regulating flux of assimilates through gluconeogenesis. The two isoforms have very different regulatory characteristics and structural differences. Chloroplastic FBPase catalyzes an irreversible reaction under physiological conditions and is regarded as a major regulatory site of the PCR cycle. The enzyme is almost inactive in the dark, when it is oxidized, and is activated by light via the ferredoxin–thioredoxin f mechanism (Section 7.3), attaining maximum activity at alkaline pH and with high Mg^{2+}, which is characteristic of a chloroplast enzyme. The FBPase activity is also small with low concentrations of FBP but increases greatly above a threshold concentration. This characteristic provides varying control of the PCR cycle and of processes leading to and from FBP. Synthesis of FBP is dependent on the production of triose phosphate; when this is rapid FBP accumulates, stimulating FBPase. If FBP decreases below the threshold the reaction slows, allowing the cycle to attain a new equilibrium. FBPase is rapidly activated by the thioredoxin system. If FBPase activity is decreased by an antisense reduction of the protein, CO_2 assimilation falls substantially with 60% loss of protein, but before the flux falls there is substantial

compensation resulting from substrate accumulation, with FBP activating FBPase. This suggests that SBPase is normally close to limiting assimilation in bright light. Also, FBPase in the chloroplast is not affected by fructose-2,6-bisphosphate, which is a very strong allosteric effector for the cytosolic FBPase. SBPase in plants is very specific for its substrate, is light activated via the thioredoxin system and is close to limiting the cycle activity, especially in bright light. In low light, the amount of enzyme required is much smaller or the activity decreases as the system adjusts.

3-phosphoglycerate kinase.

3-Phosphoglycerate kinase (PGK, EC 2.7.2.3) is a monomer of 44 kDa, which undergoes large conformational change as the substrates bind before transfer of the γ-phosphate of ATP to the carboxyl group of 3PGA, giving a reactive bisphosphate for the following reduction step in the cycle. About 90% of PGK is in the chloroplast of higher plants and catalyzes a reaction with large positive free energy change. It is therefore controlled by the end products, the only PCR cycle enzyme so regulated. ADP and a low ATP/ADP ratio slow the reaction, as does accumulation of glyceraldehyde-3-phosphate, regulating the cycle in relation to ATP synthesis and consumption. The enzyme is not regulated by light or allosteric effectors.

Triose phosphate dehydrogenase (NADP glyceraldehyde 3-phosphate dehydrogenase).

Glyceraldehyde-3-phosphate dehydrogenase (GAPDH, EC 1.2.1.13) occurs in higher plant chloroplasts as a heterotetramer; in other organisms it is a homotetramer. The enzyme has a cysteine at the active site to which the bisphosphoglycerate binds, allowing reduction of the carbonyl group by hydride transfer from NADPH. The resulting hemithioacetal bond is cleaved and GAPDH is released together with P_i. GAPDH can use both NADPH and NADH but the K_m for NADH is so small that it is not important during photosynthesis. The reaction is complex, depending on the activation state of the enzyme, which is stimulated by reduced ferredoxin and the thioredoxin system. Control of two steps in 3-phosphoglycerate metabolism by products of the light reactions and the PCR cycle provides coordination of cycle function in light and dark, preventing large fluctuations in intermediates or, more importantly, their depletion. As there is so much 3PGA kinase in chloroplasts, it is not limiting, nor is triose phosphate isomerase. However, GAPDH is a limiting reaction, as demonstrated in antisense plants, where assimilation was affected by decreasing the amount of enzyme to less than 35% of wild type. There is strong compensation by the increasing ATP/ADP ratio in the transgenic plants.

Phosphoribulokinase.

In the regeneration of RuBP, phosphoribulokinase (PRK, EC 2.7.1.19) transfers the γ-phosphate of ATP to the hydroxyl group C1 of R5P. It is restricted to photosynthetic organisms and is a homodimer of 40 kDa subunits in green algae and higher plants but a hexamer or octamer of 32 kDa mass in prokaryotes. It catalyzes a very exergonic, irreversible reaction, is stimulated by ATP and is regulated by adenylate energy charge and the ferredoxin–thioredoxin system. It does not regulate or limit the photosynthetic rate until substantially decreased (85% or more of wild type) in amount, as increased ribulose 5-phosphate concentration and ATP/ADP ratio maintain the flux of carbon through the enzyme, indicating the advantage of substantial overcapacity of the enzyme. However, in very bright light, and substantial loss of enzyme, PRK becomes limiting.

Nonregulatory enzymes. A number of the PCR cycle enzymes have been regarded as having no regulatory functions, but recent analyses have shifted perceptions (Section 7.4.1). Triose phosphate isomerase (EC 5.3.1.1; TPI) is a homodimer in the cytosol and chloroplast of higher plants. Interestingly, the chloroplast TPI is probably of mitochondrial origin and was transferred to the chloroplast. It catalyzes a reversible ketose–aldose isomerization of dihydroxyacetone phosphate and glyceraldehyde 3-phosphate. Fructose-1,6-bisphosphate/sedoheptulose-1,7-bisphosphate aldolase (EC 4.12.1.13) exists in two distinct forms, with complicated distribution in prokaryotes and eukaryotes and different mechanisms of action and structure without any sequence similarity but of similar mass – 40 kDa per subunit. Probably this reflects convergent evolution of enzyme function. Transketolase (EC 2.2.1.1; TK), as the name suggests, catalyzes a reversible transfer of a two carbon ketol group from F6P to glyceraldehyde-3-phosphate, giving xylulose-5-phosphate, or from S7P to glyceraldehyde-3-phosphate, forming erythrose-4-phosphate or ribose-5-phosphate. The reaction requires thiamine diphosphate. TK is found only in a single form and only in chloroplasts. Ribulose-5-phosphate 3-epimerase (EC 5.1.3.1) and ribose-5-phosphate isomerase (EC 5.3.1.6) both catalyze reversible reactions; they each have only a single, chloroplastic isoform.

7.3 Light activation by thioredoxin

At this point it is useful to consider how light regulates enzyme activity, including that of PCR cycle enzymes. FBPase, SBPase, $NADP^+$ glyceraldehyde phosphate dehydrogenase and phosphoribulokinase, are controlled by the redox state of proteins called thioredoxins. The thioredoxins are reduced by ferredoxin as a consequence of light-induced electron transport in the thylakoids (*Figure 7.5*) in a reaction catalyzed by ferredoxin–thioredoxin reductase. This enzyme is an Fe–S protein of 30 kDa mass, of two dissimilar subunits (heterodimer), with one subunit variable between organisms, the other conserved with an Fe–S cluster involved in electron transfer. Thioredoxins are a group of water-soluble, low mass (12–14 kDa) proteins with one S–S bond per monomer; the reaction mechanism involves a particularly striking conformational change in the molecule. It is the reactivity of the thio group which is of importance in enzyme regulation, and the thioredoxins and the enzymes they modify have very negative mid-point redox potentials of the cysteines involved in regulation. However, the reduction can occur at small electron flow, that is at low light intensities, so activation of the target enzymes occurs quickly in low light. The active site of thioredoxins is extremely reactive, with two cysteines in a Trp–Cys–Gly–Pro–Cys sequence, giving an S–S bond which is very sensitive to reductant status. The bond can be reduced to the active S–H by both the natural reductant, thioredoxin reductase, which is reduced by NADPH via FAD, and artificial reducing agents, such as dithiothreitol. The thioredoxin system is deactivated continuously by the re-oxidation of the reduced cysteines in thioredoxin by the large oxygen content of chloroplasts, by way of glutathione or ascorbate. In the light, reduction predominates over oxidation, and in darkness it is vice versa. Probably a continuous flow of e^- to O_2 oxidizes the thioredoxin so that inhibition of electron supply is accompanied by enzyme inactivation. The system is therefore not simply 'on–off' but can be continuously modulated. Such light effectors provide several points at which light can start and regulate carbon flux in the PCR cycle, although their function in intact systems, such as leaves, is still unclear.

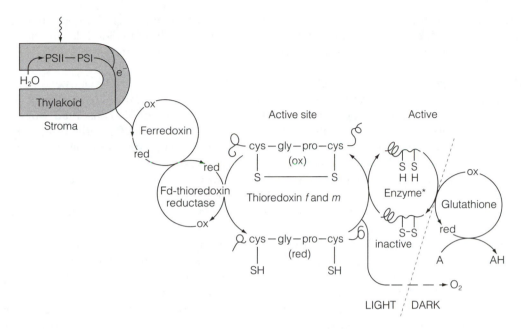

Figure 7.5. Simplified scheme of the regulation of enzyme structure and activity by the reduction of disulfide bridges with electrons from the thylakoid via thioredoxin and the mechanism of deactivation. * Activation of FBPase (by thioredoxin *f* only) NADP malate dehydrogenase, phosphoribulokinase, NADP glyceraldehyde 3P dehydrogenase. SBPase, phenylalanine ammonia lyase (thioredoxin *m* or *f*)

Thioredoxins also regulate other enzymes. NADP–malic enzyme (NADP-MDH), a C4 enzyme converting reducing power, glucose-6-phosphate dehydrogenase (G6PDH), a glycolytic enzyme, and ATP synthase, producing ATP, are examples. G6PDH is a chloroplast enzyme regulating the flow of carbon into the oxidative pentose phosphate pathway, which is deactivated by thioredoxin. This would switch off the flow of carbon to respiratory processes from photosynthesis in the light.

There are five forms of thioredoxin; of these one of *h*, four of *m*, and two of *f* occur in *Arabidopsis*. In chloroplasts, thioredoxin *m* regulates activity of malate dehydrogenase, and thioredoxin *f* activates fructose-1,6-bisphosphate. Thioredoxin *h* is probably in the cytosol (or perhaps phloem). Thioredoxins may have very specific target proteins *in vivo*. Thioredoxins *m*, *f* and *h* are nuclear encoded; the chloroplast forms have an N-terminal domain which is probably a transit peptide for transport of the protein from cytosol into the chloroplast, but *h* lacks it. The origin of the different groups of thioredoxins is complex: the *m* types are probably from cyanobacterial origin; *f* and *h* may be from the 'original' eukaryotic cell, derived from an archebacterial ancestor, into which the cyanobacterial symbiont entered. Despite similarities, thioredoxins from different organisms cannot substitute for each other. This is explained by differences in three-dimensional structure and position of the FAD and NADPH domains. There are many other proteins with amino acid sequences very similar to the thioredoxins in many plants. Some are in the chloroplast, and some are induced, for example, by water stress.

Multiplicity of thioredoxins and related proteins is a feature of plants. Thioredoxins and glutathione reductase, a very different protein containing a dinucleotide fold and active redox site, are involved in ribonucleic acid metabolism, as well as in photosynthesis.

Enzyme activation (either an increase in basal activity or a switch from inactive to active state) by thioredoxin involves, irrespective of inhibition or stimulation, reduction of S–S bonds in cysteines in the enzyme protein. These cysteines are highly conserved, but can occur in different parts of the protein. Each enzyme is regulated differently. In PRK the cysteines are not involved with catalysis but alter active site configuration, and there appears to be a similar mechanism for FBPase and perhaps SBPase, where specific cysteines have regulatory roles (mutating them removes the need for thioredoxin) and regulate site configuration. In contrast, GAPDH in the light-regulated state, uses both NADH and NADPH equally well, but the ratio depends on the state of reduction, predominantly NADH when oxidized and NADPH when reduced, and only the NADPH-dependent activity is modulated. The enzyme has two types of subunit, A (37 kDa) and B (43 kDa) of high homology; B has two cysteines in the C-terminal extension. The low-activity form of the enzyme may be a tetramer of tetramers of 600 kDa mass, dissociating when poorly activated into a form (perhaps caused by changed binding via the C-terminus of B) with two A and two B units (150 kDa), which is active, and decreasing the K_m for 3PGA. This complexity suggests the need for close regulation of processes under changing conditions. Another very important role for thioredoxin is activation of ATP synthase in light, by reduction of the γ-subunit of the CF1 complex, which changes the configuration of the rotary part of the enzyme (Chapter 6).

Only G6PDH is inactivated by thioredoxin: the K_m for G-6-P is increased but the mechanism is unclear. The NADP-MDH of the C4 chloroplast is totally inactive in the dark; thioredoxin is essential to regulate the eight cysteines in the molecule. Altering the S–S bonds changes the structure of the protein, particularly the C-terminal loop, and moves the N-terminal part away from the active site. Such a variety of modes of action suggests considerable evolution of the enzyme–thioredoxin interactions to optimize activity in relation to light–dark conditions.

7.4 **Regulation of the PCR cycle**

Increasing understanding of the cycle and its components is being used to develop quantitative analyses of the processes which can be used to predict how conditions within the plant and in the environment affect carbon assimilation. These analyses are based on simplifications of enzyme characteristics, fluxes and so on to identify the points of regulation and control, and differences between plants. The aim is to define what regulates CO_2 assimilation in the photosynthetic system as a whole, rather than the more traditional biochemical 'reductionist' approach, which considers individual parts of the system in the expectation that the particular characteristics of that subsystem will explain the whole. Such an approach has provided the basic information which is absolutely essential for understanding the PCR cycle and CO_2 fluxes. Combination of analysis of metabolite fluxes, enzyme characteristics and, more recently, effects of altering the amounts and activities of enzymes by sense and anti-sense methods, is improving understanding of the dynamics of PCR cycle regulation in relation to demand from the rest of the plant and in the face of greatly fluctuating environmental conditions. Detailed discussion is provided by Schulze (1994), Fell (1997) and Poolman *et al.* (2000). Here the regulation of the PCR cycle is addressed from the point of view of limiting factors and flux control analysis.

7.4.1 Rate limitation

With the carbon flux into the PCR cycle equal to the flux out of it, which is a state of dynamic equilibrium, the overall flux rate will depend on whichever is the slowest or rate-limiting step. In photosynthesis it may be the energy (photon flux), CO_2 supply, the rate of particular reactions within the PCR cycle or the rate of utilization of products. This is essentially a Blackman 'law of limiting factors' approach: under some conditions one set of factors dominate, under others different factors regulate the rate. However, it has long been clear that the photosynthetic system is so complex that a simple analysis of this type cannot ascribe all the effects of environment, physiology and biochemistry satisfactorily to a single step. Considering the PCR cycle enzymes, and specifically Rubisco, it is apparent from experiment that in low light it does not limit CO_2 flux – light and thus ATP and NADH do. However, with adequate light, CO_2 of 36 Pa is inadequate to fully saturate the enzyme capacity. With saturating light and CO_2, and fully activated enzyme, the slow rate of catalysis (hence the large amounts of Rubisco in chloroplasts) limits the rate of assimilation. However, it is not certain from consideration of Rubisco alone that this conclusion is true. Other enzymes in the PCR cycle may be limiting the flux and so restricting RuBP supply. Export and use of products may limit. Until an analysis can take into account all the potential factors it is not satisfactory to assume that a particular step is the limitation, however logical it appears.

The components of the system do not interact linearly, that is a change in one part does not cause a proportional increase in another. Changing the amount of protein may not result in a linear change in the activity: the activation state of the enzyme may change, the concentrations of effectors, substrates and products will alter. Hence, the flux through the system is not a simple, linear function of enzyme amount. This has been demonstrated for many PCR cycle enzymes by decreasing the protein content of chloroplasts and leaves by inhibiting the synthesis of particular enzymes with anti-sense DNA methods, which block the normal gene expression and protein production. When Rubisco content of leaves was decreased by up to 80% compared with the wild-type plants, with little change in other components, and the effects on gas exchange measured under limiting light, there was no evidence that Rubisco controlled assimilation. Only in very large photon flux and with nitrogen supply limiting the synthesis of Rubisco was there evidence of any control. Another example is provided by decreasing PRK activity: with only 10% of the wild-type activity, and under normal growing conditions, there was no effect on CO_2 assimilation. This is expected, as PRK is not regarded as rate-limiting, but in light higher than the plant experienced during growth it was rate-limiting. Activity of the enzyme increased because the substrates increased and so tended to overcome the limitation. Only in extreme conditions was the limitation sufficient to slow the PCR activity. The conclusion is that the flux of CO_2 is regulated by several processes which share control, depending on the environmental conditions, both in the short term (e.g. light during measurements of assimilation) and long term (e.g. N supply during growth which controls Rubisco amounts). The concept of a single rate-limiting step is not correct.

7.4.2 Control in metabolic systems

Reactions within the PCR cycle were characterized in the previous discussion by reference to their $\Delta G'$, the change in free energy. For the reaction:

$$A + B \xrightleftharpoons[\hspace{1.5cm}]{\text{enzyme}} C + D \qquad (7.7)$$

with substrates A and B reacting in the presence of the appropriate enzyme to give products C and D, the equation

$$\Delta G' = \Delta G^{o\prime} + RT \ln CD/AB \qquad (7.8)$$

describes how the change in free energy is related to the ratio of products to substrates; $\Delta G^{o\prime}$ is the standard free energy at pH 7, R is the universal gas constant, T the absolute temperature and ln the natural logarithm. The ratio CD/AB is the mass action ratio, Γ. At equilibrium, Γ is equal to the equilibrium constant of the reaction, K_{eq}, $\Delta G'$ is zero, and it is possible to show that $\Delta G' = RT \ln (\Gamma/K_{eq})$. The ratio Γ/K_{eq} is called the disequilibrium ratio and is a measure of the displacement of the reaction from equilibrium. A value of 1 is obtained at equilibrium, less than 1 when $\Delta G'$ is negative and the value diminishes as the free energy change becomes progressively more negative. This leads to classification of reactions as near-equilibrium (approaching 1) with very small free energy change, or nonequilibrium (disequilibrium ratio very small), having large negative free energy change and a large K_{eq}. These enzyme characteristics derive from the requirements for small metabolite concentrations (most are in the range 10^{-6}–10^{-3} M) in cells for reasons of osmotic homeostasis and rapidity of responses to changing conditions. Reactions near equilibrium catalyzed by an enzyme with capacity (specific activity × amount) in excess of the flux through the whole system were not considered rate limiting. In contrast, nonequilibrium reactions (particularly if regulated by effectors of some type, i.e. not only by substrate concentrations, or sensitive to a 'signal' and stimulating the rest of the pathway) with rates close to that of the whole pathway were regarded as limiting or regulating the rate of flux through the whole pathway. The activity of transporters and enzymes outside the chloroplast using carbon is potentially limiting, for example cytosolic FBPase, sucrose phosphate synthase and sucrose phosphatase. The concept of 'a rate-limiting step' by a particular enzyme controlling a nonequilibrium reaction determining the flux is often applied to metabolic systems.

Several models have been developed to assess the contributions of PCR cycle enzymes to control of flux. The Farquhar and von Caemmerer models based on Rubisco kinetics (see Section 12.2.3) assume that at high light RuBP is saturating and Rubisco fully activated so the enzyme is limiting and the photosynthetic rate is increased by a larger concentration of CO_2 and is responsive to O_2. At low light RuBP regeneration limits CO_2 flux. However, there is evidence that other factors limit in both low and high light. An analysis (Fridlyand et al., 1999) of the response of CO_2 flux to enzyme activities in transgenic plants based on Michaelis–Menten kinetics has the concept of 'limit of compensatory ability (LCA)' for an enzyme: if the activity falls below a particular value then the flux will decrease (*Figure 7.6*). Decreasing an enzyme exerting large control by only a small amount will decrease the flux substantially, whereas with an enzyme which has little control the activity can fall greatly before affecting the flux. If conditions change so that flux increases, the LCA will increase. This accords with much experimental data.

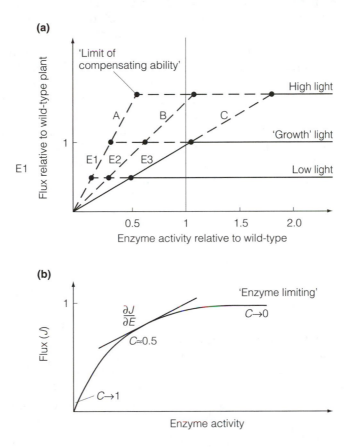

Figure 7.6. Illustration of the response of the relation between flux and enzyme activity for a metabolic pathway. (a) Dependence of flux, relative to the flux under growth conditions, on enzyme activity (relative to wild-type activity). The 'limit of compensating ability' (LCA) is indicated for three enzymes, E1, E2 and E3, differing in control over the flux. E1 has little control, E2 more and E3 full control, shown by the progressively larger LCA values. Increasing the activity would not increase the flux. In light levels below the growth conditions the enzymes' LCAs become smaller and other factors increase in importance. With much higher light intensity the LCAs increase compared to the standard condition. In the case of E3 the flux could not increase unless the activity increased (see Fell, 1997, and Fridlyand *et al.*, 1999 for details). (b) shows the response of flux through an enzyme to activity of the enzyme in a system. The flux control coefficient, $C = \partial J / \partial E$, where J is the flux and E the enzyme activity. Approximate values of C are given, showing how it differs between the theoretical limits of 0 and 1. Flux control is not generally dependent on a single step in multicomponent systems and is distributed so that the limiting values may only be obtained under extreme circumstances (see text)

There is weakness in the assumption of a rate-limiting step, for example the effects of product concentrations on reactions can greatly alter the evaluation. Also, there is strong evidence that metabolic systems regulate conditions within the cell as much as flux; homeostasis is vitally important for cell survival and has been optimized in relation to fluxes. Another very important aspect is the requirement for optimization under fluctuating CO_2 or light which places part of the reason for the PCR characteristics outside the cycle itself. Regulation and control are determined in the long term by altering amounts of enzymes (related to synthesis and degradation), by changing activities through processes such as phosphorylation (short to medium term) and in response to metabolites and signals (e.g. the thioredoxin system in PCR cycle) in the short term. Now, the use of transgenic plants has provided practical evidence of the complexity of control in the PCR cycle under a range of conditions. Description of the regulation of complex systems such as the PCR cycle therefore requires methods which provide a quantitative measure of the contribution of the individual parts to the flux through the system, under a range of conditions, which has been described as metabolic flux under homeostatic control. One approach is metabolic flux control analysis, which does not support the idea of a limiting factor, but suggests that in the PCR cycle all enzymes are potentially limiting, and that some that are not important by the nonequilibrium criterion have an important role *in vivo*.

7.4.3 Flux control analysis

Metabolic flux control analysis (MCA) (see Fell, 1997) has emerged as an important tool in analysis of metabolic, including photosynthetic, systems. The theory postulates that the control of fluxes of material through a metabolic system, in the case of the PCR cycle the rate of carbon assimilation, is not determined by an enzyme step considered regulatory on grounds of *in vitro* enzyme characteristics. Rather the method quantifies the metabolic flux for the system and assesses how it changes if activity of an enzyme changes. The extent to which control of flux may be ascribed to particular steps in the pathway is quantified by the flux control coefficient, C, the fractional change in flux, ∂J, relative to the total flux, J, resulting from a small change in the enzyme, ∂E, relative to the total amount of the enzyme, E. When flux is plotted against enzyme amount the tangent to the curve is C (*Figure 7.6a*):

$$C = (\partial J / J)/(\partial E / E) \tag{7.9}$$

The initial slope of the curve, where the change in flux is linearly related to the enzyme, has C of 1. As the curve rises and starts to flatten out, C falls progressively and on the plateau it is 0. The values of C for the different enzymes in a system add up to 1 (the summation theorem), so in a multienzyme system all enzymes contribute to the regulation of the overall flux, although in practice many will exert a negligibly small proportion of the control. This is 'distributed control', rather than a 'limiting factor control'. Thus, C of an enzyme is a property of the system not of the enzyme: that is if a particular enzyme is decreased, then C of other enzymes will increase. Similarly if conditions change and affect the fluxes and regulation, then C of each enzyme changes. An example is the increase in C as activity decreases shown by transgenic plants with decreased Rubisco or PRK, but C also changes with light, as it takes a greater proportion of the control. This means that there is no fixed value of C. Control theory shows that the effect of increasing the amount of an enzyme on the flux through the system depends on the control coefficient; with a C of 0.5, a 20-fold increase in activity will only double the flux, at a C of 0.9 the flux will increase 10-fold. Similarly, decreasing the flux requires large decrease in enzyme if C is small. The value of C is not dependent on the free energy and displacement for equilibrium. It is independent of the regulation; some enzymes have low values of C under normal conditions but are regulated (e.g. phosphogluco-isomerase of the chloroplast starch synthesis pathway). It therefore relates to changes in substrates, products and effectors (inhibitors and activators).

Flux through a pathway is determined by the metabolite concentrations as well as by enzyme activity. The effects of changes in metabolite concentration on an enzyme are expressed by an elasticity coefficient:

$$\varepsilon = (\partial V/V)/(\partial S/S) \tag{7.10}$$

where $\partial V/V$ is the fractional change in the enzyme activity caused by a fractional change in the metabolite $\partial S/S$. The ε is related to the kinetic properties of the enzyme under *in vivo* conditions at steady state. There is an elasticity coefficient for each enzyme and metabolite. The net change in activity of a particular enzyme is given by the sum of each elasticity coefficient. The value of elasticity coefficients is that they quantify the regulation of flux rate by metabolites, thus avoiding the need to ascribe control via the K_m of an enzyme and the concentration of any one metabolite.

Flux control coefficients are related to the elasticities by the connectivity theorem: this states that for different enzymes in a pathway, the sum of the flux control coefficient multiplied by its elasticity coefficient for each enzyme, is zero. That is, the dependence of flux in the system is related to that enzyme's kinetic properties and the effect of metabolites, within the context of the system characteristics. Further, a response coefficient is defined, so that changing a factor which alters the behavior of the system (but not by way of the substrates, rather by regulation) can be quantified. It is the product, for a particular enzyme, of the flux control coefficient and elasticity caused by an effector. Thus, metabolic control analysis provides a quantitative measure of the effect of the activity of an enzyme on flux in a system, the effects of changing metabolite concentrations, and the consequences of regulation with changing conditions.

MCA does not provide a set of fixed coefficients for the enzyme's contribution to the system, but quantifies the contribution within the system, under defined conditions. When conditions change, the coefficients change: flux control analysis is a way of describing a dynamic system. The single rate-limiting enzyme concept is thus not tenable, and the use of criteria such as the equilibrium state (free energy change) of an enzyme is not valid. Criticisms of the flux control analysis include the difficulties of applying it when non-PCR reactions (e.g. use of assimilates in sucrose synthesis) control the flux of CO_2, and when the light intensity limits the process (coefficient of almost 1), so that the flux control coefficients of PCR cycle enzymes are very small. Also, the responses of transgenic plants are characterized by a threshold in the relation between enzyme amount and flux; co-limitation is also questioned.

Determination of flux control and elasticity coefficients is demanding, because amounts and activities of enzymes and metabolites must be measured, amounts of enzymes changed, and a range of environmental conditions considered to understand how flux is regulated. Enzyme activity may be altered by different methods, including the use of specific inhibitors or genetic means, including mutant plants, although these often totally inhibit activity. Changing enzyme activity by altering the environment (e.g. light or nutrient supply) alters many other aspects of the system, which complicates analysis. Genetic engineering is now the chosen method for studies on plants. A range of enzyme amounts and activities is obtained by insertion of a gene, which produces RNA in the sense or antisense orientation relative to the normal RNA for that gene. Antisense studies predominate. A double-stranded RNA is formed which does not translate, decreasing production of the protein. Amounts of many of the enzymes of the PCR cycle, including those catalyzing irreversible reactions, have been altered, for example Rubisco activity has been decreased by up to 80%, using antisense to the SSU, and PRK by up to 95% compared with the untransformed (wild-type) plants. The CO_2 assimilate flux is generally unaffected by initial decrease in enzyme activity and substantial decrease (60–70% smaller than wild type) is needed before it is markedly decreased when measured under conditions at which the plants were grown. When conditions are changed, particularly to high light, the effect on flux is greater.

Rubisco.   LCA analysis shows that Rubisco activity limits the PCR cycle at low CO_2 concentrations at large photon flux, but at small flux, Rubisco it is not strongly limiting, rather electron transport limits RuBP regeneration. Increasing light will increase CO_2 flux until CO_2 supply or Rubisco becomes limiting. When Rubisco activity was decreased in transgenic tobacco, and CO_2 flux fell, RuBP increased and ATP/ADP increased, thus activating the enzyme and compensating, in part, for loss of enzyme. In this condition, Rubisco limits at smaller photon flux than in the wild type. Flux control analysis also shows that Rubisco has variable control over the flux. The control coefficient in wild-type tobacco under normal growing conditions (350 μmol m^{-2} s^{-1} light, 36 Pa CO_2) was about 0.1, but increased to 0.7 with saturating light at the same CO_2. At photon fluxes which saturate CO_2 assimilation, the flux was most sensitive to Rubisco activity (control coefficient for carboxylation almost 1, for oxygenase −0.3 as this reaction removes C from the reaction and decreases the CO_2 assimilation). Under these conditions, but with only 30% of the wild-type protein activity, the coefficient rose to almost 1 in normal light. Thus, the flux control coefficients are small for the normal plant in a low light and CO_2 environment to which it is adapted. Substantial loss of Rubisco, particularly as the environment is changed to increase the ATP, NADPH and CO_2 supply, results in a marked increase in the control exerted by the enzyme. Analysis of the effects of decreasing Rubisco activase by antisense on Rubisco activity in tobacco, showed that activase amount correlated extremely well (control coefficient of 1 – the highest observed for any system) with the rate of activation with increasing photon flux, that is under nonsteady-state photosynthesis. However, it did not correlate with loss of activity in darkness, nor did it with steady-state CO_2 flux, decreasing only when the enzyme content was less than 15% of the wild type.

Glyceraldehyde phosphate dehydrogenase.   Reduction of 3PGA by GAPDH is limiting in wild-type plants only when the CO_2 flux is very large (control coefficient of

0.2); a three-fold increase in flux gives a LCA point of about 35% (note that decreasing GAPDH to 35% of wild type caused pleiotropic effects) and decreased activity of this magnitude is required to slow the flux. There is compensation from the increased ATP/ADP ratio.

Aldolase activity.   Plants have little aldolase, which is regarded as 'near-limiting' from the LCA analysis, with a high LCA point (more than 60% of enzyme activity); reduction of the activity shows it also to be important in flux control. When activity decreased there was compensation from increasing DHAP and GAP concentrations. Limitation by transketolase may also be 'near-limiting' as the LCA analysis shows a large LCA point.

Sedoheptulose-1,7-bisphosphatase.   This enzyme catalyzes an irreversible reaction, is near-limiting in leaves and chloroplasts with a very large LCA point at high light, and at low light is in excess. Flux control showed, for plants with a wide range of SBPase activity, that this enzyme has a large flux control coefficient, approaching 1, under a range of conditions, with only a small loss of activity slowing CO_2 assimilation. In contrast the sensitivity analysis suggests that it has negligible effect on flux under high light. With only 40% of the wild-type activity there were other effects – small chlorophyll content and inhibition of growth. Analysis of plants with only 7% SBP suggests that the PCR cycle can operate without this enzyme, as the transaldolase reaction which uses erythrose-4-phosphate and forms S7P allows the flux of carbon to continue. SBPase exerts the greatest control over the flux.

Fructose 1,6-bisphosphatase.   This enzyme is not limiting, from LCA analysis, in wild-type plants at saturating light and CO_2, and thus not in subsaturating light and CO_2. However, in transgenic plants with less than 40% of wild-type activity the flux decreased, suggesting it is a 'near-limiting' enzyme. There is compensation for loss of enzyme activity by accumulation of FBP. Sensitivity analysis suggests that FBPase may be a limiting enzyme.

Phosphoribulokinase.   PRK is categorized as nonlimiting, although catalyzing an irreversible reaction, as there is a large activity in chloroplasts. Indeed, substantial decrease in activity (to less than 15% of wild type) was required before the CO_2 flux was decreased under nonsaturating photon flux and CO_2. In part, the small effect of loss of enzyme activity was due to the increased compensation by increasing Ru5P and ATP concentrations and ATP/ADP ratio. The flux control coefficient was small, so control is negligible, but as with Rubisco, control ascribed to PRK increased at high light, particularly with low enzyme activity.

Triosphosphate isomerase, phosphoglycerate kinase, ribose5-phosphate isomerase and ribulose 5-phosphate epimerase have large activities and are considered nonlimiting under virtually any conditions as 80% or more of their activities may be removed before there is any effect on flux.

Enzymes outside the PCR cycle, concerned with assimilate use, must also be considered. The triosephosphate translocator will exert control if the capacity is too small to permit assimilate efflux but estimates of flux control coefficients lack precision. Enzymes of sucrose synthesis were shown by flux control and sensitivity analyses to regulate the flux (sucrose phosphate synthase coefficient of 0.1). In contrast an enzyme engaged in chloroplast starch synthesis, ADP-glucose pyrophosphorylase (Section 8.2.3) and ATPsynthase (coupling factor) had little or no effect.

A consensus emerges about control of the PCR cycle from the different concepts and methods. There is a group of reactions which are important under particular conditions. However, the idea of a 'limiting enzyme' is not valid and the control changes with conditions. Ascribing regulation on the basis of enzyme reaction free energy is not valid. PCR cycle enzymes, such as aldolase, transketolase and GAP dehydrogenase, that catalyze equilibrium (reversible) reactions may have control capacity and affect the CO_2 flux more when slightly decreased in activity than those (e.g. PRK) catalyzing nonequilibrium (irreversible) reactions. Control exerted by a particular enzyme depends on the conditions of light and CO_2. In darkness deactivation of Rubisco ensures that the PCR cycle stops, thus avoiding depletion of intermediates and CO_2 acceptor. In low light, electron transport and ATP and NADPH synthesis determine the supply of RuBP and no PCR enzyme controls the CO_2 flux; even severe reduction in Rubisco has little effect as the sum of the control coefficients for the PCR cycle is very small compared to the light reactions. Activation of enzymes (including ATP synthase) by thioredoxin, Rubisco activation by RuBP activase and increasing Mg^{2+}, pH (from 6 to 13 mM and 6 to 8 units, respectively) and metabolite (e.g. F6P, R5P and E4P) concentrations results in sufficient capacity. Rubisco has a large capacity for activation and so can compensate to a considerable extent for any decrease in the amount of enzyme, for example as a result of decreasing the protein by antisense methods. With greater photon flux, rates of ATP, NADPH and RuBP synthesis increase, leading to faster CO_2 assimilation and the aldolase, Rubisco and SBPase capacities start to limit, probably interacting. In very bright light with limiting CO_2 Rubisco has a large control coefficient but not 1, as the other reactions and ATP synthase become limiting. If, at high light, the CO_2 concentration is large, then the Rubisco control falls drastically (to 0.1) with other reactions increasingly important. Under nonsaturating light and CO_2, that is 'normal' growth conditions, Rubisco is the main flux control, but it is not the only enzyme with control function.

Absence of a single controlling or limiting enzyme has implications for the genetic engineering of photosynthesis (e.g. the PCR cycle enzymes) to increase CO_2 assimilation. To achieve this, large increases in the capacities of several enzymes would be required, together with appropriate adjustments to the assimilate transporting and consuming systems. It also would require ATP and NADPH supply to be increased. The modification of Rubisco characteristics to remove or decrease the oxygenase would be a great improvement but, as earlier discussions showed, this is difficult to achieve given the complexity of Rubisco and its mechanisms. The PCR cycle evolved over 3.5 billion years and regulation is optimized in ways that are only now becoming apparent.

The PCR cycle is an example of a metabolic system which exploits the availability of substrates (it is opportunistic, and with respect to light-flux and CO_2 its responses are very rapid for an activated system) and also achieves stability (homeostatic flux control) in a 'robust', stable metabolic network, e.g. the relative stability of the ATP and NADPH concentrations with conditions. It controls flux at the end of a pathway by the TPT translocator and sucrose synthesis, rather than initial reactions. Also, there is considerable feedback regulation of products on enzymes earlier in the system, and covalent modification of some enzymes (e.g. sucrose phosphate synthase) which links metabolism to different metabolic signals and environmental conditions. Although there has been great progress in understanding the PCR cycle and use of the products, there is still uncertainty about how the large flux of carbon is achieved and integrated with other metabolic processes.

7.5 Induction of the PCR cycle

A delay or induction period ('lag') of several minutes occurs between illuminating leaves after a long period of darkness, and attainment of a rapid, constant rate of photosynthesis. Limitations at steady state were considered above and are further discussed in Chapter 12. What limits CO_2 flux during induction? After illumination, photochemistry and electron transport are very fast, with the electrochemical potential and ΔpH across the thylakoid membrane developing very rapidly, and $NADP^+$ is reduced within seconds, as are ferredoxin and thioredoxin. Activation of ATP synthase is rapid. These processes also occur in very low light, so in the first minutes the ATP and NADPH pools and ATP/ADP and NADPH/NADP ratios increase. Other thioredoxin-dependent enzymes are activated, slowly compared to energy transduction. However, Rubisco is activated much more slowly, requiring minutes, by Mg^{2+}, pH and Rubisco activase which exerts strong control over induction. Synthesis of inhibitors (e.g. 6-phosphogluconate) stops and as their concentrations decrease, so the rate of PCR cycle turnover increases. Concentration of RuBP in the lag phase initially rises to a large value (20 µM) as the steady-state pool accumulates, before export of carbon from the cycle, then decreases to a steady-state concentration of 5–10 µM as regulation occurs. As the steady state is attained triose phosphate is exported. During induction, as concentrations of intermediates rise and fall and enzymes in the cycle are activated at different rates, CO_2 assimilation and concentrations of intermediates oscillate as the system 'hunts' (to use an engineering term) until it approaches dynamic equilibrium, where the rate is determined by light or enzyme activities depending on the conditions. The supply of CO_2 in leaves (see Chapter 11) may restrict CO_2 assimilation compared to metabolism because stomata open slowly in the light and prolong the oscillations in CO_2 fixation, and amounts of intermediates are damped because several pools act as capacitances in the system, smoothing the demand/supply imbalance and giving only small fluctuations in the generally increasing rate of photosynthesis.

In the induction phase, every five CO_2 assimilated produce an extra RuBP, if no carbon is removed from the PCR cycle, so that the rate of CO_2 assimilation increases with each turn of the cycle. The rate of turnover is fixed by the reaction rates so, at a given temperature, it is fixed. Starting with 1 µmol RuBP per m^2 leaf turning over in unit time, after five turns of the cycle there will be 2 µmol RuBP and 5 µmol of CO_2 fixed, so the CO_2 fixation rate is 1 µmol per unit time; after two turns there will be 4 µmol RuBP and 15 µmol will have been assimilated, with the rate of CO_2 fixation 2 µmol per unit time. A further turn of the cycle gives 8 µmol RuBP and 4 µmol CO_2 assimilated per unit time. This is an exponential increase in activity which ceases when concentrations of RuBP or other metabolites or enzyme activities reach values that limit. Then the additional carbon assimilated is exported. The rate of steady-state CO_2 flux depends on the amount and activity of the enzymes, and on the size of metabolite pools. As previous discussion has shown, the system is homeostatic with respect to metabolism so the substrate/product balance and enzyme amount/activity are optimized for the environmental conditions.

7.6 Exchange of photosynthate between chloroplast and cytosol

In mature leaves products of the PCR cycle and nitrate assimilation do not accumulate in the chloroplast stroma (with the exception of 'transient' starch, Section 8.2), but are transported across the envelope to the cytosol. The fluxes of material and energy, their control and interaction with chloroplast reactions have been studied on isolated chloroplasts, with intact envelopes, free (or relatively so) of the rest of the cell substance and physiologically undamaged as far as can be established. Distribution of assimilates between these chloroplasts and defined media is measured under different conditions of light and CO_2, and with added metabolites and inhibitors. Permeability of the envelope is determined from the distribution of metabolites between the medium and the chloroplasts by, for example, using radioactively labeled substrates. How added compounds affect the rate of photosynthesis is measured by CO_2 and, particularly, O_2 exchange. The latter is rapidly determined by the polarographic oxygen electrode, an O_2-sensitive electrode which produces a voltage proportional to O_2 concentration in solution, allowing O_2 to be measured continuously as it is evolved by photosynthesis (e.g. by chloroplasts). A suspension of chloroplasts is placed on top of the electrode in a vigorously stirred small volume chamber. Light, temperature, inhibitors of photosynthesis or substrates may be changed and added to assess the effects on photosynthesis.

Methods have been developed for determining the types and rates of metabolites exchanged between cell compartments (e.g. chloroplast and cytosol). Fast, efficient separation of organelles is essential for such metabolic studies. One method employs rapid separation by centrifugation of isolated chloroplasts (or other organelles) in a reaction solution. A compound, usually radioactively labeled, to be tested for entry into the plastid is placed onto a layer of silicone oil above a denaturing solution (e.g. perchloric acid) in a centrifuge tube. After the reaction has progressed for the required time under particular conditions the tube is centrifuged, the chloroplasts are spun into the acid and killed and the radioactivity determined, allowing the amount and rate of uptake to be measured. Inclusion of metabolites and inhibitors shows the type of exchange occurring. These techniques have shown the existence of different transport systems in the membranes and the stoichiometry of the translocators in different plastids. Translocators in biological membranes are identified by saturation of the rate of transport with increasing concentration of the compound being transported. From the velocity of movement vs concentration, the K_m and V_{max} of the translocator are obtained. Analogs of the substrate with related chemical structure, which compete for the translocator sites, enable the reactions to be analyzed.

Measurements of the exchange of PCR substrates, products and effectors show that active transport processes are involved in movement between chloroplast and cytosol, rather than diffusion. Permeability of the chloroplast membrane has been demonstrated by measuring the volume of chloroplasts which can be penetrated by low molecular mass, neutral substances such as sorbitol. The space between the envelope membranes is accessible to sorbitol but the stroma and thylakoids are not (indicating that there is no direct connection between the inner membrane and thylakoid compartments). Changes of osmotic concentration in the medium cause chloroplasts to swell or shrink, altering the proportion of volume accessible to solutes and indicating that the outer membrane is freely permeable but the inner is not. Other techniques have been used to analyze chloroplast and cytosol interaction *in vivo*. Distribution and fluxes of assimilates have been measured at different times after treatment of leaves under known conditions, with radioactive CO_2 applied to label the assimilates. The tissue is then rapidly killed and organelles such as chloroplast separated from the rest of the cell using nonaqueous solvents (e.g. carbon tetrachloride), to prevent redistribution of water-soluble substances between parts of the cell.

7.6.1 Metabolite exchange

Movement of compounds across the envelope is not by simple diffusion, but CO_2 and O_2 are major exceptions. CO_2 is soluble in lipid membranes and diffuses rapidly, aided by carbonic anhydrase. Membranes have low permeability to cations, many large sugars and phosphorylated intermediates of the PCR cycle (*Figure 7.7*). Chloride ions, DHAP and P_i penetrate most rapidly; sucrose does not permeate nor does $NADP^+$ or NADPH, and ATP and ADP only enter slowly. 3PGA, which might be expected to enter readily, has only limited penetration. When plants or algae assimilate $^{14}CO_2$ the cytosol contains labeled DHAP, 3PGA, fructose- and glucose-6-phosphates, fructose bisphosphate and the nucleotide sugar uridine diphosphoglucose before labeled sucrose; there is a lag of up to several minutes. With longer time intervals, several amino acids are labeled. However, not all are made in the chloroplast and transferred unaltered to the cytosol; most are synthesized from triose phosphate (DHAP), which rapidly passes across the envelope. Fructose bisphosphate, for example, quickly appears outside the chloroplast but is synthesized from triose phosphate. Sucrose, the major translocated photosynthetic product, is made in the cytosol from DHAP (Chapter 8).

7.6.2 Transporters

The outer membrane of the chloroplast envelope has pore-forming proteins (porins) which allow substrates of up to some 10 kDa mass to diffuse freely (i.e. they do not regulate or actively carry the substances) into the intermembrane space. In contrast the inner envelope membrane is the permeability barrier between the stroma and cytosol and all products of the PCR cycle and all cytosolic requirements for the chloroplast must pass across it. Transport is generally by means of specific translocators (for details see Flügge, 2000), which are actively engaged in the processes and regulate many aspects of cell metabolism by affecting the substances transported and the rates. *Figure 7.8* indicates the types of transporters of the inner membrane of the chloroplast.

Triose phosphate, P_i transporter. One of the best understood translocation systems is the phosphate–triose phosphate–phosphoglycerate translocator, generally called the

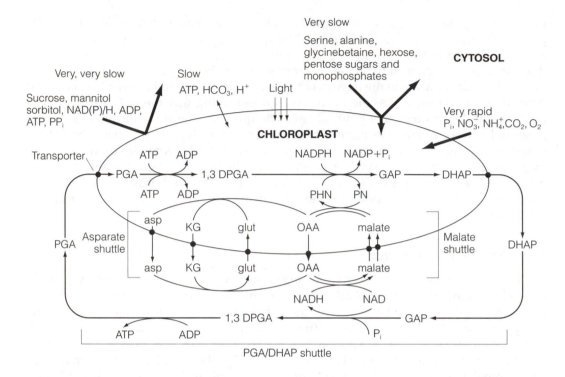

Figure 7.7. Illustration of the permeability of the intact chloroplast envelope to substances produced or consumed in the stroma and the shuttles of dihydroxyacetone phosphate and 3-phosphoglyceric acid and of dicarboxylic acids between chloroplast and cytosol. These generate ATP and NAD(P)H outside the chloroplast in the light or provide ATP and NAD(P)H to the chloroplast in darkness

triose phosphate/inorganic phosphate translocator (TPT) which controls the counter-exchange of chloroplastic triose phosphates (DHAP, GAP and 3PGA) with P_i from the cytosol. DHAP is transported rapidly [16 μmol (mg chlorophyll)$^{-1}$ h^{-1}], and 3PGA at half this rate. The TPT is the major protein of the inner chloroplast envelope and is almost restricted to photosynthetic cells. It is extracted from the lipid matrix of the membrane (which can be isolated by differential centrifugation) with nonionic detergents such as Triton X-100. The mature TPT protein (mass 61 kDa), consisting of two identical polypeptide subunits of 29 kDa (i.e. homodimers), is elliptical in shape and about 6.6 nm in length, enabling it to span the membrane. It is a very hydrophobic protein of about 330 amino acids, with only 36% of polar amino acids; hydropathicity plots indicate that each subunit has five to seven membrane-spanning α-helices (probably six, as in the equivalent mitochondrial translocator). The α-helix V has lysine and arginine residues, which form the transport path, but all 12 helices are arranged in a zig-zag fashion with the segments joining them at the membrane surfaces, forming the hydrophilic translocation channel. The channel is asymmetric, for example the cytosolic side has greater affinity for its substrates than the stromal. TPT is nuclear encoded, and is synthesized with an N-terminal peptide which directs it to, and inserts it in, the membrane (ATP is required). The protein rearranges spontaneously into the tertiary structure within the lipid bilayer

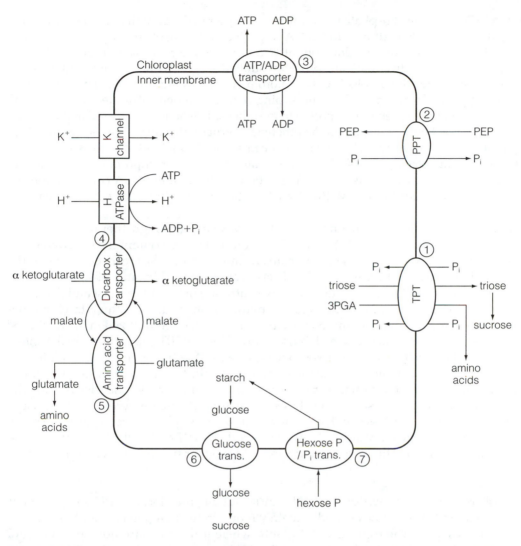

Figure 7.8. Translocator proteins in the inner membrane of the chloroplast envelope: (1) the triose phosphate inorganic phosphate translocator TPT; (2) phosphoenolypyruvate inorganic phosphate translocator; (3) the ATP/ADP translocator; (4) the dicarboxylate translocator; (5) amino acid translocator; (6) glucose translocator; (7) hexose phosphate–inorganic phosphate translocator

(see Chapter 10). There is strong homology between TPTs from different sources but not strong homology with other proteins.

The translocator is a divalent anion exchange protein with very strict specificity, binding to active site P at the end of three carbon chains but not P at the 2-position. Thus it has large affinity for DHAP (0.1 mM), P_i (K_m 0.2 mM) and for 3PGA (K_m 0.15 mM). DHAP is divalent, but 3PGA trivalent at neutral or alkaline pH; trivalent 3PGA only equilibrates slowly with the divalent form, so in illuminated chloroplasts 3PGA is retained, DHAP exported and the 3PGA/triose phosphate ratio is 10 times greater in the light than in the dark. However, the TPT has low affinity for 2PGA and glucose-6-phosphate (K_m 65 and 40 mM, respectively). It transports P_i alone but at two to three orders of magnitude less than in counter-exchange. Active sites are probably sulfhydryl groups and lysine and arginine residues (since activity is blocked by chloromercuribenzoate and benzene sulfonates, respectively). The reaction mechanism is 'ping-pong', with one substrate binding on one side of the membrane, being transported to the other side where it is released and the other substrate binds before transport to the other side. Structural changes caused by the binding of triose phosphate may alter the mechanism so that P_i can bind after the substrate has left the protein. Characteristics of the translocator are of great importance to metabolite fluxes between compartments and for regulation of DHAP and 3PGA concentrations so that 3PGA is reduced at low NADPH/NADP$^+$ ratio and low phosphorylation potential.

The TPT usually operates at only 10% of its capacity, although there is substantial exchange between cytosol and stroma as both compartments contain the substrates, but it may limit the export of triose phosphates from the chloroplast, especially when the assimilation is rapid. The amount of TPT has been decreased by antisense methods. A 30% reduction in TPT increases the 3PGA concentration in the chloroplast and decreases P_i, which stimulates activity of ADP glucose pyrophosphorylase and thus starch synthesis. Interestingly, this does not inhibit growth as the starch is remobilized in the wild-type plant more rapidly to glucose, which is transported by the GPT, particularly in darkness, so despite the much smaller sucrose concentration in the transgenic plants with very decreased TPT activity, they grow at the same rate as wild-type plants (Flügge, 1999). The smaller sucrose concentration in young leaves may be buffered by other tissues, and the glucose is metabolized to sucrose via hexokinase. If ADP glucose pyrophosphorylase is decreased by antisense, less starch is made and more triose phosphate exported to make sucrose, so growth does not suffer. However, if both the TPT and ADP glucose pyrophosphorylase are inhibited then growth is decreased as neither starch nor sucrose is available at any time.

The phosphoenol pyruvate/P_i translocator.

The P_i translocator (PPT) is present at low activity in C3 chloroplasts (and is mainly found in the nongreen tissues) but is very active in mesophyll chloroplasts of C4 plants, where it functions to move phosphoenol pyruvate (PEP) to the cytosol and import P_i. PPT is nuclear encoded, of a different type to the TPT, has similar activity with respect to Pi, but is much less active in transporting triose phosphate and 3-PGA, but much more active with 2-PGA and PEP. It is important in transporting PEP, which otherwise cannot enter plastids at all rapidly even though it is needed for metabolism, for example for aromatic amino acid synthesis by the shikimate pathway, and for fatty acid synthesis.

The hexose phosphate/P_i translocator.

This recently described translocator is less well characterized than the TPT and PPT systems and may be more important in non-photosynthetic cells. Stomatal guard cells have a glucose-6-phosphate/P_i translocator, which is important for import of carbon as hexose phosphates for starch biosynthesis as they lack fructose-1,6-bisphosphatase. Starch is metabolized to malate, which is the counter-ion for potassium exchange and central to stomatal regulation. Hexose phosphates are exchanged with P_i in non-green plastids of heterotrophic cells providing carbon for starch and fatty acid synthesis. Glucose-1-phosphate is generally the translocated hexose in starch-synthesizing amyloplasts, in contrast to other tissues importing carbohydrate, which take up glucose-6-phosphate. G6P can also be exchanged with triosephosphate via a specific translocator; it is mainly in heterotrophic tissues and probably differs from the TPT and PPT types. Glucose produced from starch breakdown is exported from darkened chloroplasts via a translocator, and used in the cytosol. This is important if plants accumulate much starch, for example as a result of inadequate nitrogen supply, low temperatures, or the TPT activity is low as in transgenics. Sucrose can be synthesized via this route. If the glucose transporter is inactivated by mutation then starch accumulates, with adverse effects on photosynthesis.

Dicarboxylic acid translocators.   Two single molecule dicarboxylic acid translocators allow the import of oxaloacetate, malate and 2-oxoglutarate into chloroplasts where they provide carbon skeletons for the assimilation of ammonium ions from nitrate reduction, and in redox exchanges. The amino acids aspartate and glutamate are also translocated by these. One transporter imports 2-oxoglutarate against malate but does not transport glutamate or aspartate. The other exports glutamate from the chloroplast for malate from the cytosol. These different transporters are essential for plant function, particularly when photorespiration is active and there are large fluxes of amino acids. The structure of the translocators is known; the inner envelope 2-oxoglutarate/malate translocator is a 45 kDa protein of 13–14 hydrophobic segments, with 12 transmembrane α-helices. Another transporter for counter-exchange of glutamine and glutamate is in chloroplasts. The oxaloacetate/malate translocator is probably a specific translocator and is very important in exchange of redox equivalents between cytosol and chloroplast. It enables the supply and concentrations of metabolites, and redox potential, to be regulated.

The very important regulation of redox state and H^+ is achieved by malate and OAA shuttles via the dicarboxylate translocator. Malate dehydrogenase (which catalyzes malate oxidation and oxaloacetate (OAA) formation) occurs on both sides of the envelope and the redox potential is determined by OAA, malate and oxidized and reduced pyridine nucleotides. Coupling the H^+ gradient with the distribution of substrate and pyridine nucleotides, and also involving the mitochondria, enables electrons to flow between compartments against a gradient of reduction potential. Malate concentration is 10^3 times greater than oxaloacetate in tissues, and this inhibits transport of OAA. However transaminations with glutamate give aspartate and 2-oxoglutarate so that the stromal NADP and cytosolic NAD systems are linked. The malate/aspartate ratio is larger in the light than in the dark in photosynthetic tissues. Together the DHAP and dicarboxylate shuttles regulate the pyridine nucleotides in both compartments. The chloroplast stroma is more reduced than the cytosol in the light, and a gradient of phosphorylation potential develops from cytosol to stroma, that is, there is more ATP outside the envelope than inside, as found experimentally by nonaqueous fractionation. Of course in the dark ATP in the stroma is at low concentration but even in the light there is relatively more ATP in the cytosol than in the chloroplast, due to the action of the shuttles, despite ATP synthesis in the stroma.

Proton pumping ATPases.   Chloroplast envelopes regulate the passage of H^+ into the medium. The pH of the darkened chloroplast stroma is smaller than that of the medium due to Donnan equilibrium of H^+ with proteins but in the light the pH of the stroma increases as H^+ is pumped into the thylakoid lumen. With acidic medium or cytosol, H^+ would enter the stroma faster than ions (Mg^{2+}) in the light and decrease the pH gradient required for ATP synthesis. However, the envelope has an ATPase and H^+ is pumped (slowly compared to the thylakoid pump) out of the stroma into the medium in exchange for K^+. Anything which penetrates the envelope and transports H^+ into the stroma inhibits photosynthesis, for example, the nitrite anion (NO_2^-) at pH 7 gives nitrous acid (HNO_2) which enters and dissociates; H^+ remains in the stroma destroying the pH gradient and NO_2^- diffuses back to the medium and recycles H^+. Other weak acid anions, such as glycolate, decrease photosynthesis in the same way.

ATP/ADP translocator. ATP and ADP counter-exchange between chloroplast and cytosol via an ATP/ADP translocator, which is a very specific (AMP, ADP-glucose and other nucleotides do not pass), 62 kDa protein with 12 transmembrane helices. The equivalent (but with no structural homology) in the mitochondrial inner membrane provides ATP for metabolism in the cytosol, particularly for sucrose synthesis from triose phosphate, which probably requires additional ATP from respiration.

Regulation of metabolism requires separation of processes and also the very rapid exchange of substrates and energy between processes in a controlled manner. The translocators provide specific regulation and mechanisms of exchange between different cell compartments and are therefore essential components of the metabolic regulatory mechanisms. Probably as the genomes of plants become better known and the comparisons with yeast and bacteria are made, the number of transport mechanisms identified will increase, as the specific regulation of metabolite and energy fluxes is of such fundamental importance in metabolism.

7.7 Shuttles of ATP and NADPH across the chloroplast envelope

Pyridine nucleotides ($NADP^+$ and NADPH in the chloroplast, NAD^+ and NADH in the cytosol) cannot pass the intact envelope, so combination of the phosphate and dicarboxylate translocators and the H^+ pump provides the mechanism for distributing assimilates and energy between compartments. A large cytosolic DHAP/PGA ratio drives the synthesis of ATP and reducing power by triose phosphate oxidation. In the chloroplast, 3PGA is metabolized to DHAP using NADPH and ATP and in the cytosol DHAP is converted back to 3PGA generating ATP and NADH. The reaction is the reverse of that in the stroma:

$$DHAP + ADP + P_i + NAD(P)^+ \rightarrow 3PGA + ATP + NAD(P)H + H^+ \qquad (7.11)$$

Exchange of ATP and pyridine nucleotides (PN) is thus directly linked to assimilate transport. Glyceraldehyde phosphate formed in the cytosol may be oxidized to 1,3-diphosphoglyceric acid without synthesis of ATP and reduce $NADP^+$ in the cytosol before NAD^+, favoring reactions using NADPH. Glyceraldehyde phosphate oxidation by this mechanism proceeds before the glycolytic enzyme is activated because the latter requires a much larger concentration of substrate. These mechanisms provide flexibility in the supply of ATP and NADPH to the cytosol for protein synthesis and ion exchange, and a mechanism to balance supply and demand within the cell. In darkness the DHAP shuttle reverses and supplies the chloroplast with ATP and NADH; starch is metabolized to DHAP which enters the chloroplast and is converted to 3PGA, forming ATP and NADPH.

C4 plants (Chapter 9) have massive flux of assimilate between the cell types during CO_2 fixation, as well as in translocation. Bundle sheath chloroplasts form 3PGA and DHAP; the latter is probably transported in exchange for P_i as in C3 plants. Large flux of malate or aspartate occurs across the cell membrane but does not involve the chloroplast envelope. Mesophyll cell chloroplasts import pyruvate and 3PGA from oxaloacetate, malate and PEP, which are exported across the envelope to the cytosol. The type of compounds formed and transported are similar to C3 plants, but the proportion and size of the fluxes probably differ greatly.

Translocators and shuttles provide a 'valve' for regulating the flows of assimilate and energy in the whole system. By coupling reactions in one compartment with those in another acting in reverse, net transport of energy and reductant is achieved without physical movement of ATP or NADPH and fine metabolic control is possible. Also specific translocators control the fluxes of triose phosphate and P_i, and organic acids, enabling close coupling of all cellular factors, under different conditions (e.g. dark–light transitions). If P_i is in short supply then starch synthesis occurs in the stroma and in darkness supplies the stroma with carbon assimilates and ATP. Over-reduction of one compartment is balanced by exchange with another. However, conditions which drive metabolism too far from equilibrium cannot then be balanced and metabolism is disrupted.

7.8 Photosynthetic assimilation of nitrogen and sulfur

An adequate supply of amino acids is essential for protein synthesis, formation of the photosynthetic and other metabolic systems and thus for plant growth and development. This is obvious, but should not be forgotten or underestimated. Nitrogen and sulfur are key components of proteins and provision of reduced forms of N and S from inorganic nitrate (NO_3^-) and sulfate (SO_4^{2-}) obtained from the environment is a major metabolic requirement. Reduction of NO_3^- and SO_4^{2-} is intimately linked with photosynthetic electron transport and carbon assimilation, which lead to amino acid synthesis.

7.8.1 Nitrate assimilation

Higher plants absorb nitrogen from their environment as the nitrate ion (NO_3^-), or as ammonia (NH_3) or ammonium (NH_4^+), but before nitrate is utilized for amino acid synthesis it is first reduced to nitrite (NO_2^-) and then NH_3 (*Figure 7.9*):

$$NO_3^- + 2\ e^- + 2\ H^+ \xrightarrow{\text{Nitrate reductase}} NO_2^- + H_2O \tag{7.12}$$

$$NO_2^- + 6\ e^- + 7\ H^+ \xrightarrow{\text{Nitrite reductase}} NH_3 + 2\ H_2O \tag{7.13}$$

175

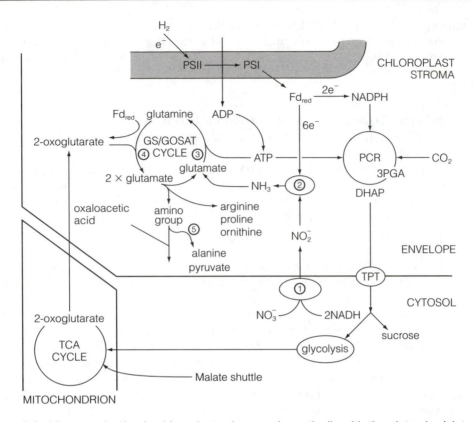

Figure 7.9. Nitrate reduction in chloroplasts shown schematically with the glutamine/glutamate cycle of ammonia assimilation, but omitting the photorespiratory nitrogen cycle. Enzymes identified in the figure are: (1) nitrate reductase; (2) nitrite reductase; (3) glutamine synthetase; (4) GOGAT; and (5) pyruvate aminotransferase

Reduction of NO_3 to NH_3 requires 8 e^-. Nitrate reductase (EC 1.6.6.1) is a 200 kDa molybdo-flavoprotein cytoplasmic enzyme which uses electrons from NADH (rather than NADPH) passing via FAD, cytochrome b_{557} and molybdenum. The electrons are probably supplied to NAD^+ in the cytosol from NADPH in the chloroplast by shuttle systems and reductant from the mitochondria may also contribute, but the extent depends on the environment and species. Reductant supply probably does not limit nitrate reduction except after long periods of darkness. Nitrate reductase, which is possibly close to the plasmalemma, is a rapidly turned over enzyme with a half-life of some 4 h. Its activity is controlled by the concentration of nitrate ions, by light (it is inactivated in darkened leaves within 20 min), CO_2 (inactivated by low CO_2), and particularly by protein phosphorylation and regulatory proteins (14–3–3 proteins). The rapid change in activity with conditions may regulate the assimilation of nitrate in relation to reductant and carbon skeletons. As concentrations of nitrite and ammonia in leaves are usually very small, they do not control the activity of the enzyme.

Nitrite is a toxic compound and therefore must be rapidly removed when formed. Reaction (7.12) proceeds with electrons supplied from ferredoxin (so it operates only in the light), not NAD(P)H, and all the intermediates are bound to the enzyme. The nitrite reductase enzyme (EC 1.7.7.1; 60 kDa) is in the stroma and contains siroheme and two additional atoms of iron and two labile sulfides per molecule. The location of nitrite reductase close to ferredoxin ensures that nitrite reduction obtains available electrons in preference to nitrate reductase, so that toxic concentrations of NO_2 are avoided. In darkness the supply of electrons from respiration probably controls conversion of NO_3^- to NO_2^- and NH_3 reduction is slower than in the light. Assimilation of NO_3 by isolated, intact chloroplasts increases the OH^- ions and changes the pH of the cell but exchange of organic acids balances this, illustrating how homeostatic balance is achieved.

Ammonia, the product of the nitrite reductase reaction, is an important assimilate which must be rapidly assimilated; it may also uncouple ATP synthesis so it is important that its concentration is kept small (in the micromolar range). Assimilation of ammonia involves formation of the amide glutamine from glutamate:

$$\text{Glutamate} + NH_3 + \text{ATP} \rightarrow \text{glutamine} + \text{ADP} + P_i + H_2O \tag{7.14}$$

The reaction is catalyzed by glutamine synthetase (EC 6.3.1.2), a protein of 360 kDa probably with eight subunits, a pH optimum of 8 and requiring Mg^{2+} for activation; approximately half the enzyme activity in higher plant leaves is found in the chloroplast and there are different isoforms. Glutamate dehydrogenase (EC 1.4.1.1; located in the mitochondria and chloroplasts) may operate at very high ammonia concentrations and mainly in catabolism:

$$\text{2-oxoglutarate} + NH_3 + \text{NAD(P)H} + H^+ \rightarrow \text{glutamate} + \text{NAD(P)}^+ + H_2O \tag{7.15}$$

Much evidence points to the glutamine synthetase reaction as the primary route for ammonia assimilation, for example, isotopic nitrogen (^{15}N or ^{13}N) from NO_3^- accumulates first in glutamine then glutamate. When [14]CO_2 is given to photosynthesizing leaves containing an analog of glutamate to block further glutamate metabolism, ^{14}C accumulates in glutamine. Glutamine synthetase is found in all plants; it is a very efficient enzyme with a high affinity (small K_m) for NH_3 (10^{-5} M), the concentration of which is therefore kept rather low. The enzyme is regulated by phosphorylation and inhibitor proteins analogous to 14–3–3 proteins.

Glutamate is synthesized from glutamine, regenerating glutamate as an acceptor of NH_3, by glutamate synthase [also called GOGAT, the acronym for the earlier name of the enzyme, glutamine (amide):2-oxoglutarate aminotransferase (oxido reductase NADP; EC 1.4.7.1)], found in chloroplasts:

$$\text{Glutamine} + \text{2-oxoglutarate} + Fd_{red} + H^+ \rightarrow 2 \text{ glutamate} + Fd_{ox} \tag{7.16}$$

The dicarboxylic acid 2-oxoglutarate is the 'carbon skeleton' for amino acid synthesis; it is supplied from the TCA cycle in mitochondria, although recent evidence suggests that mitochondria export citrate, which is decarboxylated to 2-oxoglutarate by isocitrate dehydrogenase in the cytosol (there are isoforms using NAD or NADP). An alternative route for 2-oxoglutarate from aspartate aminotransferase is less likely. Irrespective of the route, the organic acid comes from mitochondrial metabolism and demand is large The GS/GOGAT system links chloroplast C and N metabolism and peroxisomal and mitochondrial steps of the photorespiratory nitrogen 'cycles' (see *Figure 8.4*). Energy for net nitrate reduction is probably less than 10% of the energy used in CO_2 assimilation. The photorespiratory nitrogen cycle is, however, very active in C3 plants; the flux is some 10-fold greater than the net rate of nitrate reduction. Considerable reductant and ATP is consumed and the photorespiratory N and C cycles together consume over 30% of total available energy. Photorespiratory nitrogen metabolism is important because it decreases the efficiency with which light is used without increasing the net assimilation of nitrate. However, it may be important in regulating the energy supply in metabolism and protecting metabolism from high NH_3 and reductant concentrations.

Control of nitrate and ammonia assimilation. Nitrate reduction is controlled by the activity and amount of nitrate reductase, which is a rapidly synthesized and inactivated enzyme, so it is potentially sensitive to factors which affect gene transcription and translation and thus protein synthesis. Change in the amount of enzyme may provide longer-term control: increasing NR activity by enhancing constitutive expression in tobacco leaves increased nitrate assimilation, whilst decreasing NO_3^- and increasing amino acid concentrations, but it did not affect total nitrogen content or sucrose in leaves. Decreasing the expression of the NR gene by antisense methods has little effect on NR activity because regulation stimulates the activity of the protein. Thus, there is control operating on the amount and regulation of activity. Should conditions become unsuitable, then enzyme synthesis slows, as does ammonia production. Conversely, increasing sucrose and glucose in leaves stimulates the transcription of the NR gene and activation of the NR protein. So with CO_2 enrichment, plants respond by increased nitrate assimilation. Thus, there is an important interaction between sugars and nitrate supply. The activity of NR is also regulated by protein phosphorylation (as are sucrose phosphate synthase and phosphoenol pyruvate carboxylase). In the light NR is active and not phosphorylated; in darkness it is inactivated by phosphorylation of a serine residue by a protein kinase and binding of a 14–3–3 inhibitor protein (a class of ubiquitous small acidic proteins which are important regulators in plants). The inactive NR-phospho-protein is dephosphorylated by a protein phosphatase, although the reaction is slow.

A connection between NO_3^- reduction and the production of carbon skeletons would be provided by ATP or adenylate energy charge. As ATP is required for glutamine and protein synthesis, shortage of ATP would inhibit both nitrate assimilation and protein synthesis as well as the PCR cycle. Nitrate reductase synthesis is stimulated by NO_3^-, enabling the plant to respond rapidly to NO_3^- availability, and the sucrose stimulation also leads to maximum activity to exploit resources. Another control point in NO_3^- reduction is at the transport of nitrate to the assimilatory sites. Concentration of Mg^{2+} and pH and energy charge may regulate glutamine synthetase and light may stimulate enzyme activity by a mechanism similar to that of the thioredoxin system. Glutamate synthase may not be inhibited by end products of the reaction, allowing full use of NH_3 and accumulation of amino acids. Light activation and conditions in the chloroplast provide coordination between the PCR cycle, sugar supply, NO_3^- reduction, light reactions and demand for products from secondary metabolism. Regulation of all aspects of nitrate reduction is complex, presumably because the metabolic and physiological demands for a supply of amino acids for protein synthesis is of the greatest importance for plant survival and ecological and evolutionary success.

Synthesis of amino acids in photosynthesis. Chloroplasts contain enzymes for synthesis of most protein amino acids and some amino acids are rapidly labeled with ^{14}C when leaves assimilate $^{14}CO_2$. Alanine is formed by transamination of pyruvate (derived from 3PGA via phosphoenol pyruvate, possibly in the chloroplasts – the dependence of chloroplasts on mitochondria for organic acids has been discussed) with glutamate, catalyzed by pyruvate aminotransferase. Glutamate is an essential precursor for other amino acids, such as arginine. However, the last step in the arginine synthetic pathway (arginosuccinate lyase) is cytoplasmic. Aspartate is made in chloroplasts by transamination of oxaloacetic acid and metabolized in chloroplasts to lysine, threonine and homocysteine, but the final methylation of this with methionine occurs outside the chloroplast. Glycine and serine are synthesized by three routes in photosynthesis. In the phosphorylated pathway 3PGA is dephosphorylated to glycerate and this is converted to hydroxypyruvate which is transaminated to give serine. Another route is via the glycolate pathway (Section 8.7), in which glycolate is metabolized to glycine; one glycine is decarboxylated and its β-carbon added to a second glycine giving serine; the reaction is catalyzed by serine hydroxymethyl transferase in the mitochondria. A third route is via phosphoserine. Multiple routes for serine formation probably maintain its content and supply when conditions, such as CO_2 supply, vary. Amino acid synthesis requires relatively higher reductant-to-ATP ratio than does triose formation. One molecule of aspartate requires 11 NAD(P)H and 10 ATP whereas triose requires six NADPH and nine ATP. Nitrate reduction increases electron flow and accumulation of H^+ in the thylakoids, and may increase ATP synthesis, which would benefit CO_2 fixation if the PCR cycle is limited by ATP. Thus, both CO_2 and NO_3^- reduction would increase together and stimulate assimilation.

Dinitrogen fixation and photosynthesis. Higher (eukaryotic) plants cannot photosynthetically assimilate gaseous dinitrogen (N_2) because they lack a nitrogenase enzyme. However, the prokaryotic photosynthetic bacteria and blue-green algae have a nitrogenase which reduces N_2 to ammonia. As nitrogenase is inhibited or destroyed by very small O_2 concentrations, N_2 assimilation proceeds in heterocysts, cells with thick walls which are impermeable to oxygen and with only PSI activity so that no O_2 is produced. The electrons for reduction of N_2 [equation (7.14)] come from ferredoxin and are derived from sugars transported into the heterocysts and metabolized to give electrons. Cyclic photophosphorylation probably provides ATP, the requirement for which may be as high as 15 molecules per N_2 because the N_2 molecule is particularly inert.

$$N_2 + 6\,e^- + 12\,ATP + 6\,H^+ \rightarrow 2\,NH_3 + 12\,ADP + 12\,P_i \tag{7.17}$$

7.8.2 Photosynthetic sulfur metabolism

Sulfur, absorbed as the sulfate ion (SO_4^{2-}) and reduced in leaves by electrons from electron transport and ATP, is incorporated into the amino acids cysteine and methionine, and into sulfhydryl groups of coenzymes and sulfolipids. Reduction of SO_4^{2-} to sulfite

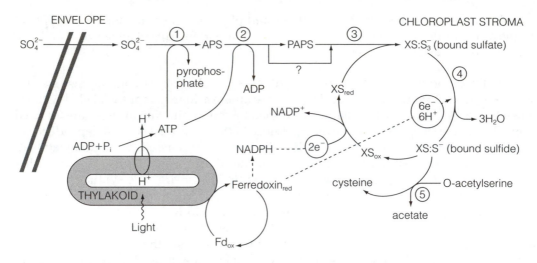

Figure 7.10. Photosynthetic assimilation of sulfate in chloroplasts, shown schematically, with enzymes: (1) ATP sulfurylase; (2) APS kinase; (3) APS sulfotransferase; (4) thiosulfonate reductase; (5) cysteine synthase. '?' denotes uncertainty of PAPS involvement in higher plant SO_4^{2-} reduction

and sulfide and incorporation into cysteine in the chloroplasts (*Figure 7.10*) is strongly stimulated by light. Activation of sulfate is required before reduction. It occurs in two stages, both requiring ATP. First SO_4^{2-} reacts with ATP giving adenosine phosphosulfate (APS). The enzyme ATP sulfate adenylyltransferase (called ATP sulfurylase), which occurs in all organisms able to reduce sulfate, has several subunits, a broad alkaline pH optimum and requires Mg^{2+} (characteristic of stromal enzymes). The sulfo- group of APS is transferred to a carrier thiol (carSH), for example glutathione in algae or a larger molecule in higher plants, such as phytochelatins, (γ-glutamyl-cysteine)$_n$-glycine, by adenosine 5′-phosphosulfate sulfotransferase:

$$APS^{2-} + carSH \rightarrow carS\text{-}SO_3^{2-} + AMP^{2-} + H^+ \tag{7.18}$$

This widely distributed enzyme is probably only in the chloroplast and has a molecular mass of 110 kDa and a K_m for APS of 10 μM; it is probably regulated by changing the amounts and is induced by the absence of SO_4^{2-} or large demand for reduced sulfur. Sulfite is reduced by sulfite reductase to free sulfide using electrons from reduced ferredoxin; energy requirements are similar to those for nitrite reduction. Sulfite reductase is of two subunits, 63 and 69 kDa, with a siroheme and one Fe_4S_4 center per subunit and a K_m for sulfite of 20 μM. In the second stage serine reacts with acetyl-S-CoA to give O-acetyl-L-serine, catalyzed by serine acetyl transferase. Then cysteine synthase reacts the O-acetyl-L-serine with sulfide (H_2S) to give L-cysteine. Cysteine synthase is a widely distributed enzyme which is abundant in chloroplasts, and which, interestingly, also catalyzes metabolism of selenide (reduced selenium found in serpentine soils) forming compounds that are toxic to animals.

Methionine is synthesized by a pathway from cysteine, the first step reacting O-phosphohomoserine and cysteine, giving cystathionine catalyzed by a specific synthase. Despite its position as the first step in the pathway it is not important in regulation. Cystathionine is cleaved by β-cystathionase to homocysteine, which is then methylated to give methionine. Glutathione (γ-glutamyl-cysteinyl-glycine, GSH) is an abundant low-molecular-mass thiol in plants, where it is very important in detoxifying active O_2 species (Section 5.4.3) and metals, and in regulation of redox states of enzymes and in transport of S. Synthesis is by γ-glutamyl-cysteine synthase in the reaction

$$Glu + Cys + ATP \rightarrow \gamma\text{-Glu–Cys} + ADP + P_i \qquad (7.19)$$

and conversion of the γ-Glu–Cys by adding glycine to the C-terminal of the dipeptide by glutathione synthetase using ATP. The enzyme is active in many different species in cytosol and chloroplast; it has a molecular mass of 85 kDa with an alkaline pH optimum and requires magnesium.

References and further reading

Appel, J. and Schulz, R. (1998) Hydrogen metabolism in organisms with oxygenic photosynthesis: hydrogenase as important regulatory devices for a proper redox poising? *J. Photochem. Photobiol. B: Biol.* **47**: 1–11.

Boichenko, V.A. and Hoffmann, P. (1994) Photosythetic hydrogen production in prokaryotes and eukaryotes: occurrence, mechanism, and functions. *Photosynthetica* **30**: 527–552.

Bryant, J.A., Burrell, M.M, and Kruger, N.J. (eds.) (1999) *Plant Carbohydrate Biochemistry*. BIOS Scientific Publishers, Oxford.

Buchanan, B.B. (1994) The ferredoxin–thioredoxin system: update on its role in the regulation of oxygenic photosynthesis. In *Molecular Processes of Photosynthesis*. Vol. 10, *Advances in Molecular and Cell Biology* (ed. J. Barber). JAI Press, Greenwich, pp. 337–354.

Calvin, M. and Bassham, J.A. (1962) *The Photosynthesis of Carbon Compounds*. Benjamin, New York.

Fell, D. (1997) *Understanding the Control of Metabolism*. Portland Press, London.

Flügge U.-I. (1999) Phosphate translocators in plastids. *Annu. Rev. Plant Physiol. Plant Mol. Biol.* **50**: 27–45.

Flügge U.-I. (2000) Metabolite transport across the chloroplast envelope of C_3-plants. In *Photosynthesis: Physiology and Metabolism* (eds R.C. Leegood, T.D. Sharkey and S. von Caemmerer). Kluwer Academic, Dordrecht, pp. 137–152.

Flügge U.-I. and Weber, A. (1994). A rapid method for measuring organelle-specific assimilate transport in homogenates of plant tissues. *Planta* **194**: 181–185.

Foyer, C.H., Ferrario-Méry, S. and Huber, S.C. (2000) Regulation of carbon fluxes in the cytosol: coordination of sucrose synthesis, nitrate reduction and organic acid and amino acid biosynthesis. In *Photosynthesis: Physiology and Metabolism* (eds R.C. Leegood, T.D. Sharkey and S. von Caemmerer). Kluwer Academic, Dordrecht, pp. 177–203.

Fridlyand, L.E., Backhausen, J.E. and Scheibe, R. (1999) Homeostatic regulation of enzyme activities in the Calvin cycle as an example for general mechanisms of flux control. What can we expect from transgenic plants? *Photosynth. Res.* **61**: 227–239.

Gálvez, S, Lancien, M. and Hodges, M. (1999) Are isocitrate dehydrogenases and 2-oxoglutarate involved in the regulation of glutamate synthesis? *Trends Plant Sci.* **4**: 484–490.

Gutteridge, S. and Gatenby, A.A. (1995) Rubisco synthesis, assembly, mechanism and regulation. *Plant Cell* **7**: 809–819.

Gutteridge, S. and Keys, A.J. (1985) The significance of ribulose-1,5-bisphosphate carboxylase in determining the effects of environment on photosynthesis and photorespiration. In *Topics in Photosynthesis*, Vol. 6. (eds J. Barber and N.R. Baker). Elsevier, Amsterdam.

Hammond, E.T., Andrews, J., Mott, K.A. and Woodrow, I.E. (1998) Regulation of Rubisco activation in antisense plants of tobacco containing reduced levels of Rubisco activase. *Plant J.* **14**: 101–110.

Heber, U., Schreiber, U., Siebke, K. and Dietz, K.-J. (1990) Relationship between light-driven electron transport, carbon reduction and carbon oxidation in photosynthesis. In *Perspectives in Biochemical and Genetic Regulation of Photosynthesis* (ed. I. Zelitch). Liss, New York, pp. 17–37.

Jacquot, J.-P., Lancelin, J.-M. and Meyer, Y. (1997) Thioredoxins: structure and function in plant cells. *New Phytol.* **136**: 543–570.

Jensen, R.G. (1980) Biochemistry of the chloroplast. In *The Biochemistry of Plants*, Vol. 1, *The Plant Cell* (ed. N.E. Tolbert). Academic Press, New York, pp. 274–313.

Keys, A.J. and Parry, M.A.J. (1990) Ribulose bisphosphate carboxylase/oxygenase and carbonic anhydrase. In *Methods in Plant Biochemistry*, Vol. 3, *Enzymes of Primary Metabolism* (ed. P. J. Lea). Academic Press, London, pp. 1–14.

Martin, W. and Schnarrenberger, C. (1997) The evolution of the Calvin cycle from prokaryotic to eukaryotic chromosomes: a case study of functional redundancy in ancient pathways through endosymbiosis. *Curr. Genet.* **32**: 1–18.

Martin, W., Scheibe, R. and Schnarrenberger, C. (2000) The Calvin cycle and its regulation. In *Photosynthesis; Physiology and Metabolism* (eds R.C. Leegood, T.D. Sharkey and S. von Caemmerer). Kluwer Academic, Dordrecht, pp. 9–51.

Meyer, Y., Verdoucq, L. and Vignois F. (1999) Plant thioredoxins and glutaredoxins: identity and putative roles. *Trends Plant Sci.* **4**: 388–394.

Mott, K.A. and Woodrow, I.E. (2000) Modeling the role of rubisco activase in limiting nonsteady-state photosynthesis. *J. Exp. Bot.* **51** (GMP Special Issue): 399–406.

Parry, M.A.J., Loveland, J.E. and Andralojc, P. J. (1999) Regulation of Rubisco. In: *Plant Carbohydrate Biochemistry* (eds J.A. Bryant, M.M. Burrell and N.J. Kruger). BIOS Scientific Publishers, Oxford.

Peterson, R.B. (1990) The RuBP carboxylase/oxygenase model and photorespiration in C3 leaves. In *Perspectives in Biochemical and Genetic Regulation of Photosynthesis* (ed. I. Zelitch). Liss, New York, pp. 285–299.

Poolman, M.G., Fell, D. A. and Thomas, S. (2000) Modelling photosynthesis and its control. *J. Exp. Bot.* **51**(Special Issue): 319–328.

Portis, A.R. (1995) The regulation of rubisco by rubisco activase. *J. Exp. Bot.* **46**: 1285–1291.

Roy, H. and Andrews, T.J. (2000) Rubisco: Assembly and Mechanism. In *Photosynthesis: Physiology and Metabolism* (eds R.C. Leegood, T.D. Sharkey, and S. von Caemmerer). Kluwer Academic, Dordrecht, pp. 53–83.

Ruelland, E. and Miginiac-Maslow, M. (1999) Regulation of chloroplast enzyme activities by thioredoxin: activation or relief from inhibition? *Trends Plant Sci.* **4**: 136–141.

Salvucci, M. and Ogren, W. (1996) The mechanism of Rubisco activase: insight from studies in the properties and structure of the enzyme. *Photosynth. Res.* **47**: 1–11.

Sanchez de Jimenez, E., Medrano, L. and Martinez-Barajas, E. (1995) Rubisco activase: a possible new member of the molecular chaperone family. *Biochemistry* **34**: 2826–2831.

Schulye, E-D. (1994) *Flux Control in Biological Systems.* Academic Press, San Diego.

Singh, B.K. (ed) (1998) Plant amino acids. In *Biochemistry and Biotechnology.* Marcel Dekker, New York, p. 648.

Schürmann, P. and Jacquot, J-P. (2000) Plant thioredoxin systems revisited. *Annu. Rev. Plant Physiol. Plant Mol. Biol.* **51**: 371–400.

Woodrow, I.E. and Mott, K.A. (1993) Modelling C3 photosynthesis: a sensitivity analysis of the photosynthetic carbon-reduction cycle. *Planta* **193**: 421–432.

Woodrow IE, Kelly ME, Mott KA. (1996) Limitation of the rate of ribulose bisphosphate carboxylase activation by carbamylation and the ribulose bisphosphate carboxylase activase activity: development and test of a mechanistic model. *Aust. J. Plant Physiol.* **23**: 141–149.

Metabolism of photosynthetic products

8.1 Introduction

All the carbon for plant metabolism is provided ultimately by the PCR cycle, so it is difficult to decide the limits for discussion. Purists may argue that events in the reaction center are 'photosynthesis', others that the PCR cycle is the limit; events that consume PCR cycle products are then 'secondary metabolism' and not photosynthesis. However, 'secondary metabolism' usually means consumption of sugars and other products of photosynthesis. The argument is semantic, although for convenience limits must be drawn. Processes are discussed here because they consume products of the light reactions or of the PCR cycle. Greater detail is provided by the reviews in Bryant *et al.* (1999) and Leegood *et al.* (2000).

Photosynthetic processes within chloroplasts are affected by many aspects of metabolism outside in the cytosol, peroxisomes and mitochondria (*Figure 8.1*). These may be short-term effects, involving rapid adjustments to changing conditions in particular compartments. Such conditions include demand for energy and carbon skeletons for synthetic processes caused by the sudden availability of a nutrient. Also, the long-term effects of changing growth and development must play a role. The way that photosynthetic rate is balanced against the demand for assimilates (e.g. for respiration and growth) is being actively analyzed and understanding is growing rapidly. Concepts of feedback regulation of assimilation, based on accumulation of assimilates and 'source–sink' interactions, have long been discussed. The mechanisms by which assimilate production and consumption are related are now better appreciated (Zamski and Schaffer, 1996). There is still need for detailed analysis of the mechanisms and quantification of material and energy fluxes, together with analysis of the effects of 'signals' from other parts of metabolism, before integration of 'source' and 'sink' processes is fully explained.

In steady-state photosynthesis, carbon assimilated in excess of that needed to regenerate RuBP is used within the chloroplast for synthesis of starch, lipids, proteins *et cetera*, or is exported as triosephosphate (dihydroxyacetone phosphate, DHAP). This is the major flux of carbon out of the chloroplast, and the DHAP is consumed in sucrose synthesis in the cytosol. Sucrose is the primary sugar and is transported, via the phloem, throughout the plant. Sucrose is the link between photosynthetic assimilate production in the

185

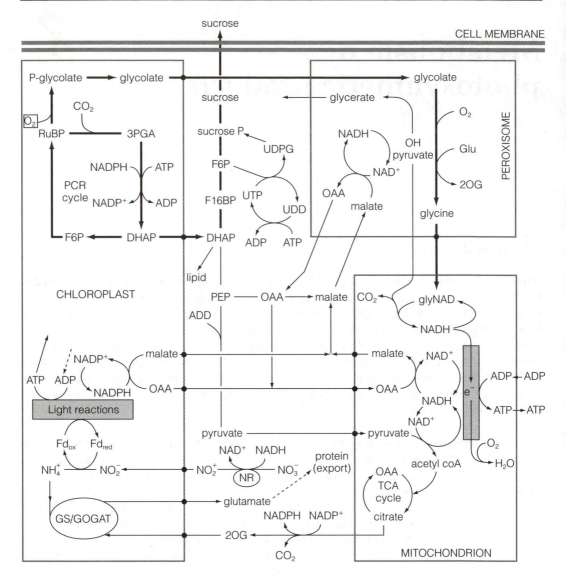

Figure 8.1. Principal fluxes of carbon and their relation to nitrogen assimilation in C3 plant leaf cells in the light

chloroplast and metabolism distant in space and time from it. Carbon assimilated in the PCR cycle is the source of carbohydrate for the great range of metabolic processes that the 'chemical factory', the plant, can accomplish. Carbon also leaves the chloroplast as glycolate, derived from the oxygenase reaction of Rubisco; glycolate is metabolized by reactions involving the three organelles mentioned. Photosynthetic assimilate, in the form of sucrose, starch and other saccharides formed from it, is consumed in glycolysis and respiration (*Figure 8.1*) to provide both carbon compounds and energy for metabolism, and ultimately growth. Many types of sugars (e.g. trioses such as erythrose, pentoses, arabinose and xylulose, and hexoses, glucose, fructose and maltose, as well as polymers such as starch and fructans), are made and accumulate in storage and reproductive organs. Sugar alcohols (e.g. mannitol, sorbitol) are synthesized from assimilates from the chloroplast and play important roles as carbohydrates for energy and synthetic processes and, it is now believed, in regulation of cellular homeostasis and energetics. Amino acids are also synthesized inside and outside the chloroplast using organic acids produced in mitochondria, but derived, ultimately, from sugars provided by the chloroplast (*Figure 8.1*).

8.2 Starch metabolism

Starch is a product of CO_2 assimilation in the chloroplast and is classified as 'transitory starch', to distinguish it from 'storage starch' in vegetative (e.g. tubers) and reproductive (e.g. cereal grains) organs. Starch serves as a most important carbohydrate reserve in plants. In the form of rice, wheat, maize, potato and cassava, starch is the primary dietary carbohydrate source for the human population. Accumulation of starch allows the storage of carbon in excess of immediate need in a form that has little or no effect on cell osmotic potential or ionic balance and can be remobilized readily. Photosynthetic transitory starch is rapidly synthesized when assimilate export from the chloroplast is slower than photosynthesis; it is remobilized when demand for assimilates from the cytosol increases and exceeds supply. Transitory starch is largely remobilized in darkness.

Starch is synthesized in the chloroplast stroma during photosynthesis by many species of plants. It forms large granules 0.2–5 (average 2.5) μm in diameter between the thylakoids. Mutants lacking the ability to transport triose phosphate out of the chloroplast accumulate very large quantities of starch. Some types of plants, for example the grasses (Gramineae), form mainly fructans, polymers of fructose. Others, such as sugar beet, accumulate as much sucrose as starch in leaves. Accumulation of starch is very dependent on the conditions of growth; nutrient deficiencies, low temperature and elevated CO_2, which increase assimilation relative to consumption, lead to starch accumulation, but high temperature and large nutrient supply, which stimulate growth, decrease it. If starch accumulation is prevented, for example by mutations or decreasing the triose phosphate–P_i translocator by antisense methods, plants cannot store sufficient carbon for growth in the subsequent dark period, particularly if the photosynthetic period is short. However, if it is long, increased sucrose synthesis compensates for the decreased starch formation. If transitory starch synthesis is inhibited, by modifying the mechanism in the chloroplast, the photosynthetic rate may decrease. In darkness, starch is remobilized and consumed in respiration and growth. Thus, starch is a very important product of CO_2 assimilation with a role in the regulation of photosynthesis, as well as providing a source of carbon for the cell.

8.2.1 Starch structure

Starch consists of two glucose polymers, amylose and amylopectin. Amylose is a relatively unbranched chain of 500–10000 glucose units joined by α1,4 linkages making up about 25% of the starch grain. Amylopectin consists of short chains of glucose units with α1,4 linkages, joined by α1,6 links, giving a highly branched structure; the α1,6 units determine the grain structure by the way that the chains are arranged. The storage starch grain is a semicrystalline structure, made of alternating, concentric rings of crystalline and noncrystalline (amorphous) starch. The layers are about 9 nm thick, the repeat being due to the length of the chains in amylopectin which vary from about 12 to 16 glucose residues. The chain (40 residues), spans two clusters. The structure is produced rather like tree growth rings, but perhaps with the alternating rings of crystalline and amorphous starch laid down on different days. Details about grain structure and formation and the role of the starch synthases and starch-branching enzymes are given by Smith *et al.* (1997).

8.2.2 Starch synthesis

Starch is synthesized from F6P from the PCR cycle (*Figure 8.2*), which is converted by hexose phosphate isomerase to G6P. The reaction has very small free energy change so that it is not a control step. Production of glucose-1-phosphate (G1P) from G6P by phosphoglucomutase is reversible and gives a ratio of G6P to G1P of about 20. Removal of

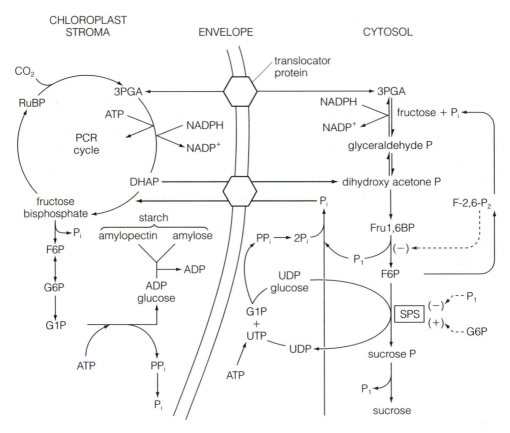

Figure 8.2. Scheme of carbon metabolism in the chloroplast and cytosol in the light, leading to starch and sucrose synthesis (see text for details)

G1P by subsequent reactions encourages synthesis of more G1P. As with many processes in carbohydrate metabolism, nucleotide sugars are involved in the formation of the starch polymer, with ADP glucose synthesized by the enzyme ADP glucose pyrophosphorylase (AGPase; EC 2.7.7.27):

$$\text{ATP + glucose-1-phosphate} \rightarrow \text{ADP-glucose} + PP_i \tag{8.1a}$$

This reaction is irreversible *in vivo* as a pyrophosphatase (which functions under alkaline conditions in the chloroplast) hydrolyzes the PP_i to P_i. A number of mutations in the enzymes of this pathway are well characterized.

ADP glucose is used for starch synthesis, by addition of glucose with α-1,4 bonds to the nonreducing ends of glucan chains:

$$\text{ADP glucose} + (\text{glucose})_n \rightarrow (\text{glucose})_{n+1} + \text{ADP} \tag{8.1b}$$

where $(\text{glucose})_n$ is a preformed polymer (an α-1,4-glucan primer) to which glucose residues are added. The enzyme responsible is starch synthase (ADP glucose: 1,4-α-D-glucan-4-glycosyl transferase), which may be soluble in the stroma and also bound to starch granules, in equal proportion, but their roles are probably similar. In storage organs there are several isoforms of the enzyme, coded by a gene family. Enzyme activity in leaves is barely sufficient to account for rates of starch synthesis. Amylose is converted by a starch-branching enzyme (1-4-α-D-glucan: 1,4-α-D-glucan-6-glycosyl transferase) to amylopectin with α-1,6-branch points.

8.2.3   ADP glucose pyrophosphorylase

Reaction (8.1a) is catalyzed by the plastidic enzyme ADP glucose pyrophosphorylase, which regulates carbon flow. It has rather similar large and small subunits, a molecular mass of 210 kDa, and is encoded by a small gene family. There is considerable variation between species: in potato, there are four different polypeptides, three large and one small subunit, with only two large subunits expressed in leaves but three in the tuber. In maize, three genes encode small subunits and two the large. There is a lysine residue at the active site where two molecules of Mg-ATP bind to the four subunits and then four G1P molecules. The enzyme is controlled allosterically by products of the PCR cycle. The rate is increased 10–20 times by increased concentration of 3PGA and less so by F6P. These accumulate in excess of that needed to regenerate RuBP, and 'signal' the enzyme to proceed with starch synthesis. If 3PGA is depleted, for example, with low rates of photosynthesis or in darkness, the enzyme is inhibited. In addition, ADP glucose pyrophosphorylase is allosterically inhibited by P_i which interacts with 3PGA; consequently a high ratio of $3PGA/P_i$ stimulates starch synthesis. When ATP synthesis is insufficient, 3PGA decreases and P_i accumulates, inhibiting starch synthesis and preventing PCR cycle depletion. If consumption of carbon outside the chloroplast is rapid, then low 3PGA and increased P_i concentrations slow starch synthesis. Sequestering (binding) P_i into metabolites, for example by feeding leaves with mannose, which is phosphorylated by hexokinase but not remetabolized, decreases P_i concentration and stimulates synthesis of starch. A doubling of the P_i concentration from 1 mM in the medium of isolated intact chloroplasts increases the $3PGA/P_i$ ratio in the stroma 10-fold and the rate of starch synthesis 40-fold. Although starch synthesis liberates P_i, the rate of the reaction is inadequate to maintain rapid photosynthesis but allows some CO_2 fixation to continue. Plants deprived of phosphate contain much starch: this is a symptom of nutritional inadequacy caused by lack of P_i preventing growth of the plant which then does not use the excess starch produced. This is called 'sink' limitation.

Remobilization of starch may proceed via G1P (catalyzed by phosphorylase) and G6P, F6P and fructose bisphosphate to DHAP. ATP is required but may be inadequate in darkness. The products of remobilization in darkness, for example hexose phosphates, are transported out of chloroplasts by the appropriate translocator on the chloroplast envelope.

8.3 Sucrose synthesis

Sucrose is the most abundant oligosaccharide in plants, involved in translocation between leaves and other organs and is stored in many tissues, often at large concentrations in specialized organs, for example stem parenchyma of sugar cane and swollen hypocotyls of sugar beet. Sucrose from these sources is important in human nutrition and in international trade. Sucrose is α-D-glucopyranose + β-D-fructofuranose, linked by an α(1–2)glycosidic bond. It is neutral, not ionized, and metabolically inert in the cell (and hence, good for storage), non reducing and very soluble: 1 g dissolves in 0.5 cm^3 of water at 20°C. Sucrose is synthesized (*Figure 8.2*) from fructose-1,6-bisphosphate (Fru1,6BP) by hydrolysis with fructose-1,6-bisphosphatase in an irreversible step, forming F6P:

$$Fru1,6BP + H_2O \rightarrow F6P + P_i \tag{8.2}$$

Fru-6-P reacts with uridine diphosphate (UDP)-glucose, formed from G6P via G1P. The reaction is catalyzed by sucrose phosphate synthase (SPS; systematic name, UDP-D-glucose:D-fructose-6-phosphate-2-glucosyl transferase; EC. 2.4.1.14).

8.3.1 Sucrose phosphate synthase

Transfer of glucose residues from UDP-glucose to F6P by SPS is a reversible reaction:

$$\text{UDPG + F6P} \xrightarrow{\text{Sucrose phosphate synthase}} \text{UDP + sucrose-6'-phosphate + H}^+ \tag{8.3}$$

SPS is important in sucrose synthesis in photosynthetic tissues, and in determining the balance between starch and sucrose production. It is a soluble cytosolic enzyme, probably a dimer of 120–138 kDa subunits (but without intersubunit S–S bonds), with relatively slow activity and complex regulation. The Michaelis constant is about 10 *in vivo*, and 5–65 (depending on pH and Mg^{2+}) *in vitro*. SPS forms less than 1% of the soluble protein in leaves and is relatively unstable on extraction; it may form complexes with other metabolically related enzymes *in vivo*. The N-terminal region of the 120 kDa subunit is very highly conserved, with two regions similar to those in the enzyme sucrose synthase (see section 8.3.2). They probably have similar substrate binding properties. The protein has a molecular mass of 450 kDa, contains SH- groups (required for metabolic regulation rather than catalysis), and is not sensitive to Mg^{2+} or pH. However, it is activated by G6P, which increases the V_{max} and decreases the K_m for F6P. The enzyme is inhibited by P_i but the activity is increased hyperbolically by a large $G6P/P_i$ ratio. Also the products of the reaction, UDP and sucrose P inhibit SPS; there is uncertainty about the control exerted by sucrose with the enzyme from some species apparently being more inhibited than that from others, or it may be unimportant.

SPS regulation.   SPS activity is regulated in a complex way (*Figure 8.3*). In darkness it is largely inactive, strongly inhibited by P_i and phosphorylated. This is catalyzed by SPS kinase, which requires ATP, is stimulated by calcium ions and is inhibited by glucose-6-phosphate. SPS is activated by light (and was earlier classified into different types, depending on modulation by light, but this is not now accepted), but not by a mechanism of the thioredoxin type. In the light, the P_i concentration falls, and glucose-6-phosphate rises and the enzyme is dephosphorylated by SPS protein phosphatase, increasing activity. The effectors play an important role in determining the affinity for the substrates F6P and UDPG and in regulating activity in relation to other characteristics and states of the photosynthetic system. There is a strong sigmoidal dependence of activity on the effectors: both photosynthetic and nonphotosynthetic enzymes are similar but there are differences between species. SPS kinase is a multisubunit complex (45 and 150 kDa, the

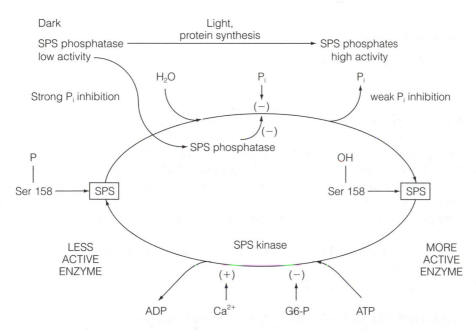

Figure 8.3. Sucrose phosphate synthase (SPS) regulation by protein phosphorylation of serine residue 158 involving SPS kinase activity and calcium ions and glucose-6-phosphate. There is also interaction with inorganic phosphate, which is strongly inhibitory to SPS phosphatase in the dark, but only weakly so in the light. Details in the text

latter probably of two 65 kDa units) responsible for phosphorylation with ATP of ser158 in spinach which inactivates the enzyme. Calcium ions may be involved, especially in regulation of maize SPS. An SPS protein phosphatase, a type 2A protein phosphatase, which is inhibited by large P_i concentration, dephosphorylates the inactive SPS, restoring the activity and decreasing the sensitivity to P_i. Both enzymes are of general occurrence but they vary in activity. The kinase and phosphorylase may form complexes with SPS.

Little is known about the allosteric effector binding sites, which contain sulfhydryl cysteine residues, 10 in total but only one active in regulation and not catalytic. The rate of sucrose synthesis is regulated in the medium to long term by the amount of SPS protein ('coarse control'), probably the principal way, and in the short term by the concentrations of P_i and glucose-6-phosphate ('fine control') but not by pH. A scheme of regulation is given in *Figure 8.3*. Details of regulation are discussed by Huber and Huber (1996). Coarse control is obtained by changing the amount of protein and thereby potential activity over periods of minutes to hours, related to the supply and demand for assimilates and to the light–dark transitions; there is marked endogenous rhythm independent of photosynthesis. The amount of enzyme activity observed is usually just sufficient to account for the flux of C to sucrose, so the enzyme may be important in the long-term regulation of carbon metabolism. The significance of these characteristics is that the flow of C through the enzyme is increased as the substrate F6P increases (G6P is in equilibrium with it) and as the available P_i decreases, plus the activation by light which may be expected to stimulate the overall carbon flux.

8.3.2 Sucrose phosphatase

Sucrose phosphate is dephosphorylated by a specific phosphorylase, sucrose phosphatase (EC 3.1.3.24):

$$\text{sucrose-6}'\text{-phosphate} + H_2O \xrightarrow{\text{Sucrose phosphatase}} \text{sucrose} + P_i \qquad (8.4)$$

Sucrose phosphatase rapidly converts sucrose P to sucrose; the activity in the leaves is usually 10-fold greater than the activity of the synthase. This probably keeps the sucrose P concentration low, enabling SPS to function efficiently, so carbon entering the pathway cannot 'flow back'. Free UDP, sucrose and other sugars stimulate the enzyme. Both sucrose phosphate synthase and the phosphatase may be regulated by Mg^{2+}.

Sucrose is also made from UDPG and fructose by sucrose synthase (EC 2.4.1.13):

$$\text{UDPG} + \text{fructose} \rightarrow \text{sucrose} + P_i \qquad (8.5)$$

However, this enzyme is not believed to be important in leaves.

Sucrose synthesis proceeds in the cytoplasm, for there is delay in ^{14}C labeling after exposure to $^{14}CO_2$ and separation of chloroplast from the cytosol shows UDPG pyrophosphorylase (required for UDPG synthesis) to be cytoplasmic. Carbon for sucrose synthesis comes from DHAP exported from the chloroplast in a strict 1:1 exchange with P_i on the phosphate translocator. Sucrose synthesis consumes energy; one UTP is required in the UDP glucose pyrophosphorylase reaction for every four molecules of DHAP consumed.

8.4 Regulation of sucrose synthesis

Sucrose synthesis is regulated at three main enzymatic control points in the pathway: cytosolic FBPase, SPS and sucrose phosphatase, which has been considered already (*Figures 8.2 and 8.3*). These enzymes catalyze reactions which are far from being in equilibrium, as calculated from the free energy changes derived from estimates of the concentrations of metabolites in the cytosol of leaf cells. The FBPase has a ΔG of -19 kJ mol^{-1}, sucrose phosphatase -11 kJ mol^{-1} and SPS -8 kJ mol^{-1} (close to equilibrium); the concentrations of FBP, UDPG and sucrose P are, respectively, about 0.1, 2 and 0.2 mM in the cytosol, but that of sucrose is much larger (40–50 mM) so the reactions must be far from equilibrium to achieve rapid rates of sucrose production and this makes the reactions important sites for regulation. The sucrose phosphatase reaction ensures that the reaction proceeds towards sucrose synthesis. Sucrose in the cytosol is largely exported to the phloem within a few minutes of synthesis, as shown by measuring the distribution of radio-tracer after $^{14}CO_2$ feeding. Concentrations of sucrose in leaves may increase if export is decreased, for example by cooling growing organs or by deficient nutrient (such as N and P) supply. Some sucrose may be transported into the vacuole but is broken down to hexoses.

Cytoplasmic FBPase, which differs from the chloroplast enzyme, is the first regulatory enzyme in the pathway, and very important. The protein is of 130 kDa molecular mass with a very low K_m for FBP (2–4 µM), requires Mg^{2+} and is not pH-dependent; it shows very strong sigmoidal saturation kinetics with FBP, particularly with AMP present. FBPase is very strongly inhibited (100-fold) by micromolar concentrations of the compound fructose-2,6-bisphosphate (F2,6BP); this is not to be confused with FBP, fructose-1,6-bisphosphate.

8.5 Non-sucrose carbohydrates and sugar alcohols

Sucrose synthesis is the main use of carbon assimilated in the chloroplast, with a relatively small flux to starch. Many types of carbohydrates, in addition to sucrose and starch, are formed and accumulated in plants, with some restricted in distribution to particular families and genera (e.g. fructans, mannose). Fructans are water soluble polymers (varying in size from 5 to 50 kDa and in types of glycoside linkages) of fructose and glucose, stored in vacuoles. They are synthesized from sucrose, without the need for fructose primers, by fructosyl transferases with 1-kestose as the only product from sucrose. They are important storage carbohydrates in some 10% of higher plant species, for example grasses and cereals (Gramineae) and Jerusalem artichoke (*Helianthus tuberosum*, Compositeae), and are related to cold adaptation and over-wintering in grasses. A number of sugar alcohols are made by plants, for example glucitol in apricot (*Prunus armeniaca*) and mannitol in celery (*Apium graveolens*). Such compounds are made by metabolic processes that are closely related to primary carbohydrate production. Mannitol synthesis, for example, occurs in the chloroplast and uses DHAP and NADPH formed there. These compounds are important for drought and salt tolerance, and may have a role in regulating leaf photosynthetic energy balance and in removing free radicals.

Differences in amounts of sucrose synthesized relative to other carbohydrates depends greatly on the species, for example spinach makes much starch in leaves, whereas wheat makes relatively little, and in C4 plants, maize for example, sucrose is probably synthesized in the mesophyll cells and in the bundle sheath at night. Fructans accumulate when sucrose concentrations exceed a particular threshold and are dependent on temperature. Sucrose may be stored in stems, roots and fruits by some species and others may form large amounts of starch. The reasons for this are not understood and further discussion is outside the scope of this text. Fine control balances sucrose production against demands of growth, respiration and synthesis of starch and other oligosaccharides (e.g. fructans) in the short term. Coarse control may adjust the capacity for sucrose synthesis against long-term changes in plant development (e.g. reproduction) or environment (e.g. temperature). Storage of sucrose, starch and fructans allows the capacity of photosynthesis to be fully exploited without feedback inhibition of the rate of photosynthesis and permits growth to proceed over long periods without carbon limitation.

8.5.1 Fructose 2,6 bisphosphate

F2,6BP is a product of carbohydrate metabolism related to the sucrose pathway (*Figure 8.2*), and important in its regulation (Stitt *et al.*, 1987). F2,6BP is in the cytosol where it is synthesized from F6P and P_i by an enzyme, F6P,2 kinase. The concentration of F2,6BP is also dependent on an enzyme, F2,6BP phosphatase, which degrades the molecule to F6P and P_i. This phosphatase is inhibited by F6P and P_i, in contrast to the kinase. The control mechanism suggested is inhibition of FBPase by F2,6BP when the amounts of F6P and P_i increase, for example when flow of C to sucrose slows due to lack of demand. The F6P,2 kinase activity increases, F2,6BP is synthesized and slows the FBPase; consequently, the conversion of FBP to F6P slows. At the same time, the F2,6BPase is inhibited by the effectors, thus maintaining a large F2,6BP concentration. Conversely, when the F6P and P_i concentrations are low but DHAP and 3PGA concentrations are high, the kinase is inhibited and the F2,6BPase is stimulated, thus decreasing the concentration of F2,6BP. This allows FBPase activity to increase and stimulates the flux of C to sucrose, providing a very rapid switching mechanism which can also link PCR cycle activity and glycolysis. At night high F2,6BP concentrations probably inhibit F1,6BPase activity and prevent glucose export. Current assessment is that the role of F2,6BP is not as great as was thought earlier (Sitt *et al.*, 1987). Although there is a correlation between increasing F2,6BP concentrations and increasing starch/sucrose ratio in spinach (Scott *et al.*, 1995), it is not as strong in grasses. Possibly, F2,6BP has a greater role in regulating carbon flow to respiration.

8.5.2 Integration of regulation

With the characteristics of the enzymes in mind, it is possible to understand how sucrose synthesis is regulated. With the onset of rapid photosynthesis, the concentration of

DHAP and FBP in the cytosol increases, thus increasing the concentration of G6P, which activates SPS, and decreasing P_i which removes the inhibition. There may be stimulation of FBPase and decreased synthesis of F2,6BP, but this aspect of control is probably minor in comparison with activation of SPS in the light by dephosphorylation of a serine residue, which inactivates the enzyme in darkness. With increasing concentrations of substrates and activated SPS, the C flux to sucrose increases. However, if demand for sucrose falls, increasing sucrose, sucrose P and decreasing P_i inhibit SPS. Also, the increase in F6P may activate F6P,2 kinase and inhibit F2,6BP phosphatase, resulting in an increase of F2,6BP. This slows the FBPase, resulting in a build-up of triosephosphates and fall in P_i. The system then regulates at a different rate of carbon flux. If the blockage of sucrose use is severe, for example by low temperature or deficient N or P_i, which inhibit growth, then accumulation of triosephosphate stimulates the carbon flux to starch, which accumulates in the chloroplast or storage organs. Conversely, slow rates of photosynthesis but large demand for sucrose in growth, respiration, *et cetera* produce plants with very small carbohydrate contents. Translocation of sucrose out of the cell is an important aspect of how the whole plant regulates carbon assimilation, but is beyond this discussion of photosynthetic production. Phloem loading from the mesophyll and unloading in the sink organs are crucial steps (see van Bel, 1993), as, if export is slowed, sucrose and intermediates accumulate. Then P_i flux into the chloroplast is slowed and inhibition of photosynthesis occurs as the storage capacity of cells in the leaf is reached. This occurs as a consequence of inadequate sink capacity caused, for example, by N or P_i deficiency, which slows growth of the sink organs more than it affects photosynthesis, or of cool temperatures around the roots.

8.6 Biosynthesis of chloroplast lipids

One-third of the chloroplast dry mass is lipid, mainly synthesized during the development of the leaf, although new synthesis and replacement of lipids continues in mature leaves. Synthesis of fatty acids (*Figure 8.4*) such as palmitic, oleic and linoleic acids consumes PCR cycle products, reductant as ferredoxin and NADPH, and ATP. The light reactions, both PSII and PSI, are essential. Synthesis occurs in the stroma where all the required enzymes are found, but involves mitochondria and cytosol. Dihydroxyacetone phosphate moves from the stroma to the cytosol. Pyruvate is formed from the triosephosphate and enters the mitochondria where acetyl-CoA and acetate are produced. Acetate returns to the chloroplasts where acetyl-CoA is again made by a synthetase; it joins to an acyl carrier protein (ACP), a low molecular weight co-factor protein for some 12 enzymes of lipid metabolism, found only in the stroma. ACP is very similar in different organisms (e.g. 40% sequence homology between spinach and *E. coli*).

Malonyl-CoA is also produced in the stroma by carboxylation of acetyl-CoA with bicarbonate by the enzyme acetyl-CoA carboxylase (transcarboxylase) which requires Mg^{2+} (*Figure 8.3*). ATP activates the enzyme complex, which is probably a high molecular mass multifunctional enzyme rather than three separate enzymes: a biotin carrier protein (BCCP), biotin carboxylase and BCCP:acetyl-CoA transcarboxylase, which initiates the reaction. BCCP is bound to the thylakoid lamellae, although the reason for such close proximity to the light reactions is not known. Malonyl-CoA is transferred to ACP giving

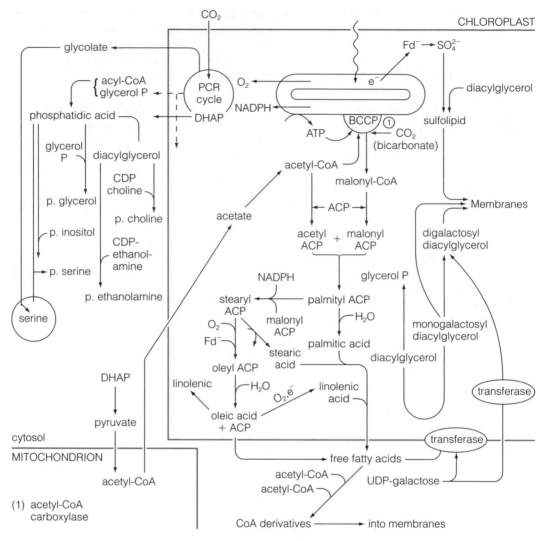

Figure 8.4. Fatty acid synthesis in leaf cells requires products of the light reactions and the PCR cycle and cooperation between organelles (see text). p, phosphatidyl

malonyl-ACP. Condensation of one acetyl-ACP and seven malonyl-ACP followed by reduction with NADPH and NADH gives palmityl-ACP, which is hydrolyzed to palmitic acid or is condensed with more malonyl-ACP and reduced with NADPH to stearyl-ACP. This increases the chain length of the fatty acids from C 16:0 to C 18:0. Stearyl-ACP is reduced (by ferredoxin in the presence of O_2) to oleyl-ACP from which oleic acid is released by hydrolysis. Further modification of the fatty acids occurs in the chloroplast. Palmitoyl- and oleoyl-ACPs may be hydrolyzed on the envelope and enter the cytosol for lipid synthesis. Fatty acids may not be formed directly from PCR cycle products and turnover of membrane lipids is slow, as $^{14}CO_2$ label is found in them several hours after exposure.

8.6.1 Chloroplast lipids

Lipids of chloroplast are complex and synthesized in several parts of the cell; Douce and Joyard (1996) provide details. Glycolipids, of which mono- (MGDG) and digalactosyl dia-cylglyceride (DGDG) are the major chloroplast lipids (indeed MGDG is probably the most abundant lipid in nature), and are synthesized in the chloroplast. MGDG is synthesized by a transferase enzyme bound to the chloroplast envelope which attaches a galactose moiety from UDP-galactose to diacylglycerol, whilst DGDG synthesis is by a soluble, stromal enzyme catalyzing transfer of two galactosyl residues from UDP-galactose to a monoacyl-glycerol. Sulfolipid synthesis (e.g. sulfoquinovosyl diglycerol) is linked to SO_4^{2-} reduction in the chloroplast. Sulfoquinovosyl diglycerol is an ionic lipid, specific to photosynthetic membranes in all organisms (except many photosynthetic bacteria where it is absent or in small amounts), with a unique head group (sulfoquinovose), made of glucose with the 6-hydroxyl substituted by sulfonate. This confers strongly amphipathic features, which are perhaps associated with the unique fluid nature of photosynthetic membranes, but the function is not clear. The role of this S-lipid is intriguing: it is not essential for oxygenic photosynthesis, but there is evidence that it interacts with phospholipids and may substitute for them under phosphate deficiency. Synthesis is probably from UDP-glucose, with incorporation of active sulfate to make UDP-sulfoquinovose, which is then transferred to the *sn*-3 position on diacylglycerol. The enzymes responsible (epimerases, glycosyltransferases and nucleotide hydratases) are encoded by *sqd* A, *sqd* B and *sqd* C genes in higher plants and conserved in many organisms. The *sqd* D gene codes for UDP-sulfoquinovose:diacylglycerol sulfoquinovosyltransferase.

Phosphatidyl glycerol is a major glycerophosphatide of leaves, constituting over 20% of the total lipid of thylakoids and envelope membranes. Synthesis is from glycerol-3-phosphate, produced by phosphorylation of glycerol, acylated by acyl-CoA to phosphatidic acid. Phosphatidic acid is dephosphorylated by phosphatidate phosphatase to form diacylglycerol which either reacts with acyl-CoA (catalyzed by diacylglycerol acyltransferase) to produce triacylglycerol or with CDP-choline or CDP-ethanolamine (with choline and ethanolamine derived from glycine and serine, possibly from the glycolate pathway) to give phosphatidylcholine or phosphatidylethanolamine. Phosphatidylinositol is formed from glycerol phosphate in the mitochondria, where phosphatidylglycerol is formed. The head groups of phospholipids are polar, with the most abundant, phosphatidylcholine and phosphatidylethanolamine, being basic, phosphatidic acid and phosphatidylserine being acidic, and the remainder (phosphatidylinositol, phosphatidylglycerol and the bis-phosphatidylglycerols) neutral. These membrane glycerophosphatides are important components with physical and chemical characteristics allowing subtle interactions with proteins and pigments, which in part determine their movement, and diffusion of substrates and products between the cell compartments.

This brief summary of fatty acid and lipid synthesis serves to show the complex interaction between the light reactions, carbon assimilation and cooperation between cell organelles. Lipid synthesis is a photosynthetic process requiring much energy: 17 and 8 mol of ATP and NADPH, respectively, are needed for a C18 fatty acid. Regulation of lipid synthesis in relation to other chloroplast functions is not well understood.

8.7 Glycolate pathway and photorespiration

Three cell organelles are involved in the glycolate pathway, chloroplasts, peroxisomes and mitochondria (*Figure 8.5*), and more than 15 enzymes and translocator proteins (details are given in Douce and Heldt, 2000). The Rubisco oxygenase reaction (Section

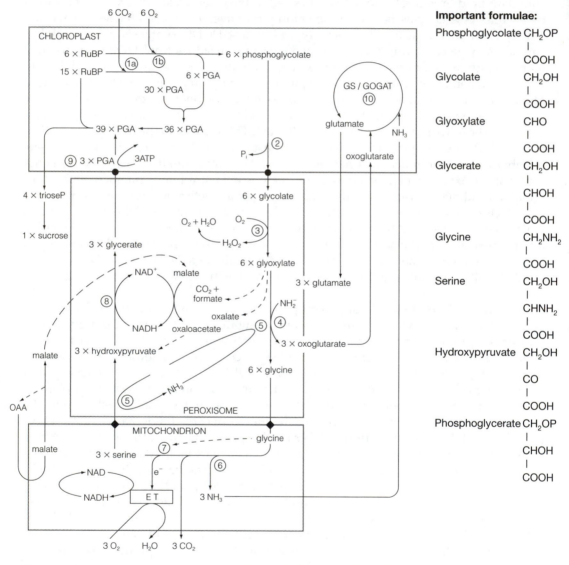

Figure 8.5. Carbon, nitrogen and energy flows in the glycolate pathway and production of CO_2 when photorespiration is one-quarter of the rate of net photosynthesis. Numbers refer to the reactions as follows. Enzymes of the glycolate pathway: (1a) ribulose bisphosphate carboxylase; (1b) RuBP oxygenase; (2) phosphoglycolate phosphatase; (3) glycolate oxidase; (4) glutamate-glyoxylate aminotransferase; (5) serine-glyoxylate aminotransferase; (6) glycine decarboxylase; (7) serine hydroxymethyl transferase (8) hydroxypyruvate reductase; (9) glycerate kinase; (10) glutamine synthetase, GOGAT

7.2.1) forms phosphoglycolate, which enters the chloroplast stroma. P-glycolate does not normally accumulate, but if it does (e.g. when its further metabolism is inhibited photosynthesis rapidly stops). Also, mutants with lesions in the pathway of glycolate metabolism cannot survive in an atmosphere with a low CO_2/O_2 ratio, but they can if it is high enough to prevent the Rubisco oxygenase reaction. P-glycolate is dephosphorylated (*Figure 8.4*) by a specific stromal phosphatase, a dimer with 32 kDa subunits of protein and optimum activity under the alkaline conditions of illuminated chloroplasts. It has high affinity for P-glycolate (K_m less than 100 µM), which does not accumulate. Mutation of this enzyme prevents metabolism and inhibits photosynthesis. Phosphate is used in the stroma and the glycolate produced moves into the cytosol, via an active glycerate/glycolate transporter (in counter-exchange or individually by H^+ symport or OH^- antiport mechanisms) in the chloroplast envelope; from there it permeates into the peroxisome via pores in the single membrane. In peroxisomes glycolate is metabolized by the 'C2 pathway' in which some 11 enzymes recycle part of the carbon back to the chloroplast. In peroxisomes, glycolate is oxidized to glyoxylate by glycolate oxidase, which is at high concentration and induced by light and glycolate. The enzyme is a tetramer of 40 kDa subunits, is imported (as are all peroxisomal proteins) from the cytosol, and has a K_m of about 0.3 mM for glycolate. The reaction reduces the enzymes flavin mononucleotide prosthetic group, is almost irreversible and consumes O_2, producing H_2O_2. As the peroxisome is a detoxifying organelle with large catalase activity, H_2O_2 is rapidly destroyed; plants lacking catalase are killed in the normal atmosphere where photorespiration accounts for up to 40% of the total net carbon flux in photosynthesis. Transgenic plants with only 10% of wild-type catalase suffer damage during photosynthesis in air. Catalase has four 55 kDa subunits, containing heme. Glycolate is transaminated to glycine by glutamate-glyoxylate aminotransferase (GGAT) with glutamate as amino-group donor and by serine-glyoxylate aminotransferase (SGAT), each providing half of the needed amino groups, because of the stoichiometry of the reaction converting serine to hydroxypyruvate (see *Figure 8.5*). Both aminotransferases contain pyridoxal phosphate cofactor bound to the enzyme. GGAT is probably a dimer, of total mass 98 kDa, and the reaction, which uses several amino acids as substrates and may link directly to alanine metabolism of the cell, is reversible. SGAT is a dimer of 46 kDa subunits. It uses serine as amino donor and glyoxalate as acceptor much more efficiently than others, and the reaction is irreversible. Glycolate can also be converted to formic acid, and to oxalic acid, which accumulates in many plants as insoluble crystals of calcium oxalate. Glycine probably diffuses from the peroxisome and enters the mitochondrion; it is unclear whether this is via a glycine–serine counter-exchange transport system (most likely) or by diffusion. One molecule of glycine is decarboxylated and the 1-C fragment and another glycine are condensed to serine [by glycine decarboxylase (GDC) and serine hydroxymethyltransferase (SHMT), respectively, which form a complex] and CO_2 is released and escapes from the mitochondria as photorespiration. Also, NH_3 is released and is re-assimilated (Section 8.7.3). The GDC and SHMT reactions are particularly complex, and are considered in detail in Section 8.7.1. The serine formed leaves the mitochondrial matrix (mechanism either via exchange with glycine or diffusion); the export rate is important for the GDC–SHMT reaction. Serine is converted to hydroxypyruvate by the serine-glyoxylate amino transferase in an almost irreversible reaction. Hydroxypyruvate reductase, a very active peroxisomal dimeric protein of *ca.* 44 kDa subunits, converts hydroxypyruvate to glycerate, using NADH (K_m 6 µM). The NADH is formed in the peroxisomes by a specific malate dehydrogenase. Glycerate leaves the peroxisomes and is transported into the stroma of the chloroplast and phosphorylated to 3-PGA by glycerate kinase, a monomer of 40 kDa with K_m for glycerate of 0.2 mM, requiring Mg-ATP as P-donor, with K_m of 0.25 mM. Regulation is unclear; light probably stimulates it but there is no effect of pH, Mg^{2+} or adenylate energy charge (Section 6.2). The reaction is almost irreversible and is of large capacity.

8.7.1 Glycine decarboxylase

Glycine decarboxylase is a complex of four proteins with different, but closely linked functions (*Figure 8.6*). The H-protein (14 kDa monomer containing lipoamide) has a lipoamide arm, with a terminal S–S bond, in oxidized form, H_{metox}, which reacts with the P-subunit (a homodimer of 105 kDa, with pyridoxal phosphate) and a glycine, releasing CO_2 and attaching a methylamine (CH_2NH_2) to the last S-atom, giving a SH- and a $S-CH_2NH_2$ group. This alters the configuration of H, with hydrophobic residues on the surface of the protein, surrounding the active site, and protects the methylamine group. Thus H, shielded from nucleophilic attack, reacts with another subunit of GDC, T, which catalyzes the reaction with tetrahydrofolate (H_4FGlu_n) to give methylene tetrahydrofolate ($CH_2H_4FGlu_n$). The T-protein is a monomer of 41 kDa. Ammonia is released in this reaction and the photorespiratory nitrogen cycle fluxes are most important as they are

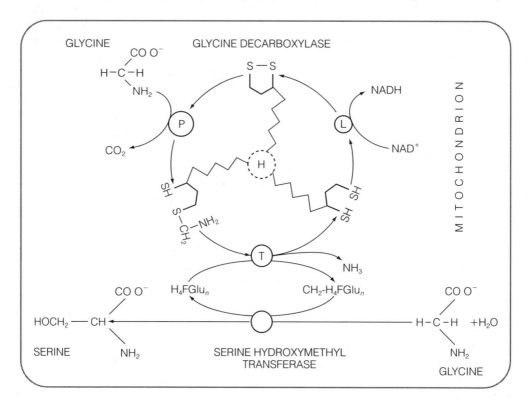

Figure 8.6. The glycine decarboxylase and serine hydroxymethyl transferase reactions of photorespiration; see text for details

several times greater than the net NO_3^- reduction in leaves. The $CH_2H_4FGlu_n$ produced reacts with the other glycine, catalyzed by serine hydroxymethyltransferase, giving serine. Removal of the NH_2CH_2 group in the course of the reaction results in H_{red}, with terminal SH-group. The L-subunit of GDC (a homodimer of 59 kDa total mass, containing an active cystine and flavin-adenine dinucleotide) then oxidizes the molecule to H_{ox} by transferring H to NAD^+, producing NADH. GDH activity is only regulated by the effect of the $NADH/NAD^+$ ratio on the L-protein. Increasing the ratio greatly stimulates activity, so NADH must be rapidly removed if the GDH is to function. Glycine decarboxylase is inhibited by serine, which competes for the reaction site with glycine. The K_i – the Michaelis constant of inhibition – for serine is 4 mM compared to the K_m for glycine binding to the P-protein of 6 mM. NADH also competes with NAD^+ for the active site on the L-protein: the K_i is 15 μM for NADH, compared to the K_m of 75 μM for NAD^+. Co-factors required for the reaction are lipoate, pyridoxal phosphate and tetrahydrofolate. The amounts of these increase when leaves grown in darkness are placed in the light; the activity of GDC also increases 20–fold and that of SMHT 4-fold. These proteins may constitute 40% of the soluble mitochondrial protein, showing the importance of photorespiratory fluxes in C3 leaves. Mitochondria also contain a large concentration (0.2 mM) of folate, over 100-fold that of chloroplasts, and a large proportion of the total folate in leaves. Folate is synthesized from pterin compounds (in the mitochondria, probably exclusively), and are present mainly as polyglutamyl derivatives. The glutamate tails of the folate may aid its location and binding at specific sites on the SHMT and GDC-T protein, ensuring a rapid, efficient reaction by 'channeling' the reactants to the active sites. These complex reactions simplify to:

$$2 \text{ glycine} + H_2O + NAD^+ \rightarrow \text{serine} + CO_2 + NH_3 + 2 \text{ NADH} \tag{8.6}$$

8.7.2   Serine hydroxymethyltransferase

The $CH_2H_4FGlu_n$ produced by the T-protein is recycled to H_4FGlu_n and the CH_2 fragment added to the second glycine by SHMT, which is a homotetramer of about 220 kDa total mass, each subunit with pyridoxal phosphate. The reaction is much faster in the serine to glycine direction than the reverse, so the equilibrium lies far from equilibrium to catalyze the glycine to serine reaction by a very large $CH_2H_4FGlu_n/H_4FGlu_n$ ratio. This requires that there is no removal of $CH_2H_4FGlu_n$ for other reactions. When the glycolate pathway ceases under a high CO_2/O_2 ratio, there is no adverse consequence for metabolism, showing that the cycle is probably self-contained. However, it has been suggested that methyl groups and serine might be supplied by the glycolate pathway for general metabolism.

8.7.3   Photorespiratory N metabolism

The NH_3 (or NH_4^+) released enters the chloroplast, but it is not known if by diffusion or by a specific channel or translocator. There it is assimilated by the same system as NH_3 from nitrate reduction, namely by chloroplastic glutamine synthetase (GS), which catalyzes synthesis of glutamine from glutamate using ATP, and glutamate synthase (GOGAT), which reacts glutamine with 2-oxoglutarate and reduced ferredoxin, giving two molecules of glutamate. These enzymes are only in the chloroplast. Mutants with decreased activity of GS accumulate NH_3 as the photorespiratory flux in air is large and the NH_3 released in the mitochondria cannot be used. NADH from the reaction enters the mitochondrial electron transport chain and, via oxidative phosphorylation, three molecules of ATP are synthesized per two molecules of glycine decarboxylated. One ATP is needed to form glutamine in the cytosol and two ATP enter metabolism for sucrose synthesis, *et cetera*. Alternatively, the electrons may be used in the reduction of oxaloacetate to malate, thereby establishing a shuttle exchange of reducing power with the cytosol. Serine from the mitochondrial reaction is transferred into the peroxisomes, where it is deaminated by serine-glyoxylate aminotransferase.

8.7.4 Photorespiratory flux and regulation

The flux of carbon through the pathway is large, about five times that of the TCA cycle, and there is a large capacity of GDC and SHMT; together they form 50% of the total protein in the mitochondrial matrix. This complex, multistage process returns 75% of the C exported from the chloroplast as glycolate back to the chloroplast as 3PGA; the rest is lost in photorespiration. Flux of carbon through the pathway consumes NADPH (or NADH) from the light reactions but generates ATP either by direct donation of electrons to the mitochondrial electron transport chain, or from malate–oxaloacetate shuttles. The glycolate pathway is an important way of regulating the energy and reductant balances of leaves. Photorespiration is important as an oxidative process consuming PCR cycle products and energy. Stoichiometry of the pathway and of glycine decarboxylation is such that for every two molecules [i.e. four carbons (4C)] of glycolate entering the peroxisomes one CO_2 (one carbon atom), or 25% of the carbon is lost when two molecules of glycine (4C) are converted to serine (3C). The carbon flux to glycine depends on the ratio of RuBP oxygenase to RuBP carboxylase reactions (Chapter 7), i.e. on the O_2/CO_2 ratio. With CO_2 shortage, for example when stomata are closed because of water stress, the flux of carbon through the glycolate pathway increases relative to that through the PCR cycle (*Figure 8.1*), and a greater proportion of newly assimilated CO_2 is lost by photorespiration. A simple model of the system (see *Figure 12.6* and equation 7.6) uses the characteristics of Rubisco in relation to O_2 and CO_2 concentration. When gross photosynthesis and photorespiration are equal (in leaves at the CO_2 compensation concentration), the oxygenase to carboxylase ratio, α, is 2 and the flux into the glycolate pathway is four times that in photosynthesis. Photorespiration and glycolate pathway metabolism are a drain on photosynthetic CO_2 fixation, and use extra NAD(P)H and ATP, despite ATP synthesis in the mitochondria. Net carbon fixation decreases because photorespired CO_2 offsets gross assimilation. When O_2 is removed photorespiration stops and gross assimilation increases as the ATP and NADPH consumed by photorespiration become available to synthesize extra RuBP. The amount of NADPH and ATP required for synthesis of a molecule of sucrose depends on the pathway by which the precursors are formed; in an O_2-free atmosphere, 37 ATP and 24 NADPH are required in total for all reactions. However, in air, consumption is 58 and 45 molecules, respectively. Sucrose synthesis with 50% photorespiration demands 72 ATP and 59 NADPH. Competition for PCR cycle carbon between RuBP and sucrose synthesis increases as photorespiration increases and, as a consequence, carbon flux to sucrose is slowed. At very large α, a flux of carbon from storage into the PCR cycle would be necessary to keep the cycle running.

Biochemically, the glycolate pathway, involving three cellular organelles, may be a method by which C3 plants have partially overcome the effects of a large O_2 and small CO_2 concentration on Rubisco. The pathway scavenges some of the carbon and dissipates reducing power in the cytosol, but generates essential ATP; Rubisco oxygenase activity is 'inevitable' due to the high O_2/CO_2 ratio in the atmosphere, so that production of phosphoglycolate cannot be prevented. The oxygenation would be the best point for regulation of photorespiration, as once phosphoglycolate is produced it must be used as productively as possible to avoid inhibition of photosynthesis and to increase efficiency. Rubisco evolved when CO_2 pressure was very large and O_2 very small and its catalytic mechanism has remained relatively unchanged (see Section 7.2.2), despite the apparent disadvantages of its slow reaction rate (turnover of sites $2 \, s^{-1}$) and oxygenase activity. As Rubisco has a central role in assimilation, its structure has been strongly conserved; the multiple mutations in amino acid residues thought necessary to increase efficiency must all happen together and this would have a very low probability of occurring; single mutations might be lethal, as lack of photosynthesis would destroy the whole organism and so tend to slow selection of an improved enzyme reaction. Other parts of the photosynthetic system seem to be more 'flexible' and have evolved around Rubisco, for example in C4 metabolism (Chapter 9).

8.7.5 Rationale and regulation of photorespiration

Regulation of the fluxes of carbon, nitrogen and energy associated with the photorespiratory pathway is not understood in depth. For example, removal of amino acids (serine) for protein synthesis, or of CH_3 units for lipid metabolism, requires that the stoichiometry of the cycle must be altered and this could have adverse effects on the process. The pathway is regulated only to ensure the maximum flux to glycerate; it simply uses glycolate and there is no feedback regulation as expected if the dominant, inevitable, aspect of the flux is production of potentially damaging P-glycolate by Rubisco. Alterations of the photorespiratory flux by changing 'downstream' control processes will then have possibly damaging effects. Changes in photorespiration rate relative to net or gross CO_2 assimilation in C3 plants, and the very good simulation of the processes by models based on Rubisco characteristics (Chapter 12), suggests that there is little control exerted within the cycle. Hence, reduction in photorespiration by altering regulation appears not to be feasible. Attempts have been made to find chemicals which block the glycolate pathway and prevent waste of assimilated carbon by photorespiration. Sodium bisulfate reacts with glycolate and forms the α-hydroxy sulfonate, disodium sulfoglycolate, which inhibits glycolate oxidase. Other α-hydroxysulfonates with the formula R-CHOH-SO_3Na, such as HPMS (α-hydroxy-2-pyridine methane sulfonic acid), are effective inhibitors, although they may undergo reactions to produce glycolate bisulfate, which is the inhibitor. Isonicotinyl hydrazide (INH) causes accumulation of glycolate by blocking the glycine to serine conversion in the mitochondria. However, inhibitors or mutations that block glycolate metabolism after RuBP oxygenase stop carbon flow in the pathway. This inhibits photosynthesis because carbon accumulates and is not available for resynthesis of RuBP. Thus, increasing CO_2 assimilation cannot be achieved by blocking the glycolate pathway. It is necessary to stop the oxygenase activity either chemically or genetically by modifying the enzyme if photosynthesis is to be increased. This goal is yet to be achieved despite much active work.

The glycolate pathway links the metabolism of nitrogen inside and outside the chloroplast. For each CO_2 released in photorespiration one NH_3 is also produced. With photorespiration 25% of gross photosynthesis, the NH_3 assimilation may be about 5 μmol NH_3 m^{-2} leaf s^{-1}. If re-assimilation of ammonia is blocked by methionine sulfoxime (MSO), which inhibits glutamine synthetase, then ammonia accumulates in tissues and photosynthesis stops. Nitrogen turnover during photorespiration is many times greater than the net nitrogen reduction in the chloroplast. Photorespiration thus involves combined carbon and nitrogen cycles which link the energy and metabolites in chloroplasts, mitochondria and cytosol. A major aspect of the glycolate pathway in leaf biochemistry and physiology is that it consumes excess reduced pyridine nucleotide, 'burning off' reductant and allowing a faster turnover of NADH in the peroxisomes and NADPH in the chloroplast. This is regarded as crucial under stress conditions, e.g. when CO_2 supply is limiting. Also, the decarboxylation of glycine may provide a method of generating ATP in the mitochondria in the light if synthesis in the chloroplast is inadequate, for example for sucrose synthesis; the NADH from photorespiration can be consumed in the mitochondrial electron transport chain, but quantitative aspects of the fluxes are still very unclear. Malate can be produced by reduction of oxaloacetic acid in the mitochondria and used in the peroxisomes for reduction of hydroxypyruvate. As chloroplast NADPH can also be used for synthesis of malate from OAA imported from the mitochondria, and malate is freely exported to the cytosol by transporter–shuttle systems, it is clear that very extensive interactions occur between all cell compartments. Such systems then become regulatory devices to balance cellular functions, particularly the most crucial redox-state. Consumption of excess reductant will be important when photosynthesis is severely restricted, when the electron transport chain is fully reduced and chlorophyll absorbs excess energy forming excited states and damaging products (e.g. superoxide, H_2O_2). Photorespiration decreases the energy burden, particularly on PSII which is sensitive to photoinhibition and may function together with the carotenoids in quenching chlorophyll excited states and singlet O_2 (see Chapter 3). The loss of carbon by photorespiration of C3 plants is clearly inefficient in terms of carbon gain. It is perhaps surprising that such a large proportion of the world's flora is C3 and that mechanisms like C4 photosynthesis (Chapter 9), which minimize photorespiration, have not dominated. C3 plants are predominantly from well-watered, often low-light environments where the additional energy costs of C4 metabolism outweigh the benefits. Perhaps carbon loss may be relatively unimportant ecologically, compared to the 'safety valve' offered under temporarily adverse conditions. However, even photorespiration cannot protect C3 plants from the effects of long exposure to strong illumination, in hot and dry conditions in the current low atmospheric CO_2 concentrations.

8.7.6 Glycolate metabolism and C4 plants

C4 plants maintain a very large CO_2 and low O_2 concentration at the RuBP carboxylase active site so that RuBP oxygenase activity is small and little phosphoglycolate is formed (Chapter 9 and 12). Leaves of C4 plants assimilating $^{14}CO_2$ in air produce little ^{14}C-labeled glycine and form serine by non-glycolate routes. This, together with virtual absence of photorespiration measured as release of CO_2 in the light, suggests no glycolate metabolism. However, C4 plants have enzymes for the synthesis and metabolism of glycolate in small quantities. Bundle sheath cells can decarboxylate glycine at large rates. Conditions such as water stress stimulate the formation of glycine and the glycolate pathway functions in C4 plants but it is much smaller than in C3 species under comparable conditions. Additionally, any photorespiratory CO_2 is efficiently removed by the PEP carboxylase reaction, hence the very small photorespiratory flux. C3–C4 intermediates have mechanisms for re-assimilating photorespiratory CO_2, suggesting that this is an advantage in the hot conditions in which they grow. Details are given in Chapter 9.

8.8 Chloroplast respiration

In higher plants, dark respiratory and photorespiratory CO_2 is released from mitochondria. However, a form of respiration occurs in chloroplasts of unicellular green algae and higher plants: the plastoquinone pool is reduced in darkness by a dehydrogenase and oxidized by an oxidase using oxygen as terminal acceptor, contributing to generation of the pH gradient. Thylakoids function in photosynthesis and respiration, using part of the same mechanism (e.g. cytochrome *b/f*) with competition for electrons controlling PSII activity. Chlororespiration occurs in chloroplasts of higher plants, where 11 gene products with high homology to NADH dehydrogenases found in mitochondria occur; seven are chloroplast encoded and four nuclear encoded (three of these are expressed in stromal thylakoids). A pathway may exist for respiratory oxidation of starch and reducing equivalents, providing chloroplasts with reductant and substrates and regulating the redox-state of plastoquinone under particular physiological conditions.

8.9 Interaction of dark respiration and photosynthesis

Transition between light and dark occurs daily for plants and rapid fluctuations between bright and dim light are common in many habitats due to clouds or sunflecks in vegetation. As the light energy incident on the leaf changes, so do the fluxes of energy and materials in photosynthetic cells and the ratio of assimilation to 'dark respiration'. How do photosynthesis and respiration interact and what are the controls operating during changes in photon flux and with darkness? Regulation of dark respiration is needed to balance growth with assimilation and prevent depletion of metabolites and 'futile cycles'. This is a complex area of metabolic interaction and poorly understood; only some aspects of the problem are considered here.

8.9.1 Dark respiration

Dark respiration is the consumption of carbohydrates and other compounds to produce energy and carbon substrates (e.g. organic acids) for cellular synthetic processes. The process consumes O_2 and produces CO_2 and H_2O; starch and sugars are degraded to pyruvate by glycolysis, generating ATP. The pyruvate is converted to acetyl-CoA by pyruvate dehydrogenase, situated in the mitochondria. There, acetyl-CoA enters the tricarboxylic acid cycle (TCA cycle or Krebs or citric acid cycle) in the mitochondria, where it is metabolized to organic acids (e.g. citrate, fumarate, 2-oxoglutarate), releasing CO_2 and forming NADH. NADH is oxidized by the mitochondrial electron transport chain with O_2 as terminal electron acceptor, and electron transport is coupled to phosphorylation of ADP to ATP.

Details of mitochondrial functions in ATP synthesis and use of reductant are given by Vedel *et al.* (1999) and Atkin *et al.* (2000). Mitochondrial inner membrane complex I of the electron transport chain oxidizes NADH and also NADPH and electron transport is coupled to ATP synthesis. NAD(P)H dehydrogenases oxidize NADH or NADPH from the intermembrane space, which is in direct communication with the cytosol. Thus, reductant from the cytosol may be oxidized. As reductant produced in the chloroplast interacts with that in the cytosol via transport and shuttle mechanisms (Section 7.6), so reductant balance of the organelles and the cell as a whole are regulated interdependently. Mitochondria may play a larger role in regulation of reductant and ATP in the photosynthetic cell than previously recognized, for example when CO_2 assimilation is restricted and photosynthetic electron transport is maintained so that reductants accumulate. Then mitochondrial consumption of reductant and synthesis of ATP may be important.

Alternative oxidase.   Plant mitochondria have a cyanide-insensitive respiratory pathway, called the alternative pathway, in addition to the 'normal' pathway of electron transport to oxygen, which is sensitive to cyanide. This involves an alternative oxidase on the inner membrane which uses electrons from the mitochondrial ubiquinone pool (when it is very reduced), and reduces oxygen to water in a single four-electron reaction. The production of ATP related to electron transport is small and probably variable, allowing flexibility in regulation. It removes reductant, and thereby probably diminishes the synthesis of singlet oxygen in PSI, but without the need for ATP synthesis. Under CO_2 deficiency, for example, ATP/ADP is high so consumption of reductant without ATP synthesis would be valuable. The alternative oxidase can be induced by environmental conditions, and in particular stages of development, offering 'coarse control' (i.e. dependent on gene expression and protein accumulation) of energy metabolism. The mechanism of regulation of the alternative oxidase by 'fine control' (i.e. allosteric regulation) is complex, depending for example on the concentration of pyruvate and NADPH, both stimulating the enzyme, providing a possible feedforward regulation to stimulate the oxidase if respiratory and metabolic substrates accumulate. The different mitochondrial respiratory systems provide a very flexible mechanism for regulating electron flow and reductant and adenylate status, not only of the mitochondria but also of the cytosol and chloroplasts.

TCA cycle.   Photorespiration is three to eight times greater than TCA cycle respiration (measured in darkness) in C3 plants. TCA cycle respiration proceeds in the light but it may be partially inhibited when photosynthesis and photorespiration are very active. As NADH produced in the glycolate pathway and TCA cycle respiration both use the mitochondrial electron transport chain, respiration could proceed during photosynthesis as the chain *per se* is not inhibited by light. Indeed, there is no evidence that respiration cannot take place during photosynthesis; however, there are many points at which the TCA may be regulated.

TCA cycle respiration is saturated at low O_2 concentration (1–2 kPa) but photorespiration increases with increasing O_2. Measurements of respiration in the light at different

O_2 concentrations extrapolated to zero O_2 suggest that 'dark' respiration is inhibited, but measurements of CO_2 evolution with increasing light at low CO_2 concentration, which prevents net photosynthesis, suggests that dark respiration continues. However, at very small photosynthetic carbon flux in the cell, respiration may be stimulated, particularly if the ATP/ADP ratio is low. Incorporation of metabolites into TCA cycle intermediates in the light has been measured to clarify the role of dark respiration. Long-term pulse chase studies suggest that ^{14}C enters TCA cycle intermediates in the light. $^{14}CO_2$ from photosynthesis enters some organic acids and alanine, which is formed via pyruvate, if assimilation is rapid. However, it does not quickly (30 min) enter other organic acids or all amino acids. Thus, glutamate becomes radioactive only in some experiments. However, transfer of $^{14}CO_2$-labeled leaves from light to darkness causes an influx of ^{14}C into the TCA cycle, as expected if the flux is inhibited in the light. Leaves fed with ^{14}C-acetyl-CoA, TCA acids, amino acids or pyruvate (from glycolysis) form labeled TCA cycle acids, glutamate and aspartate, consistent with an active TCA cycle. The distribution of $^{14}CO_2$ into assimilates shows continuation of the cycle. The metabolic controls of the TCA cycle are possibly bypassed when large concentrations of substrate are added to photosynthesizing tissues, allowing respiration to proceed even if normally blocked or much reduced. Inhibition of photosynthesis by DCMU showed that respiration occurs in the light; however, preventing assimilation could stimulate respiration because photosynthetic carbon and energy fluxes are inhibited.

A simple interpretation is that the TCA cycle continues in the light with a slow rate of CO_2 production when photosynthesis and photorespiration are large, but the rate increases at lower rates of photosynthesis. The requirements for continued TCA activity and mitochondrial electron transport and ATP synthesis seem well established. The TCA cycle provides carbon skeletons for amino acid synthesis, particularly 2-oxoglutarate for glutamate, and this will be required in the light. Mitochondria supply cells in darkness with ATP and NADH, and probably also in the light for sucrose synthesis, for example. Probably, in the light photophosphorylation and photosynthetic electron transport provide most of the ATP and NADPH required in the cytosol, but the TCA cycle contributes when required.

8.10   Adenylates in cells

Control of ATP concentrations and ATP/ADP ratios in cells and cellular compartments during transitions between dark and light and under other conditions has been examined by rapidly separating chloroplasts, mitochondria and cytosol from leaves or, somewhat easier, from protoplasts. Protoplasts are prepared by incubating leaf slices in an enzyme preparation which digests the cellulose walls. The released protoplasts are separated by centrifugation in density gradients of sucrose and sorbitol under controlled conditions, then broken by osmotic shock and the organelles are separated on density gradients. The purity of cell fractions is checked by measuring marker enzyme complements, allowing corrections for cross-contamination. As adenylate pools turnover very quickly, the reactions must be stopped and organelles separated rapidly to allow adenylates to be measured. Such studies have provided detailed analyses of adenylates in tissues.

In darkness leaves have some 30% less ATP, ADP and AMP than in the light but ATP/ADP ratio and AEC (Chapter 6) are only slightly smaller. The ATP/ADP ratio is larger in the cytosol than in the mitochondria and chloroplast stroma; oxidative phosphorylation in the mitochondria and export of ATP maintains a very high ATP/ADP ratio and AEC in the cytosol. Shuttles of triosephosphate and 3-phosphoglycerate between chloroplast stroma and cytosol regulate the chloroplast's adenylate concentration without direct transfer of ATP, although ATP may be imported by chloroplasts via the adenylate translocator. Chloroplasts in the dark require ATP to maintain ion balance and concentrations of photosynthetic intermediates ready for the onset of photosynthesis; they are also needed for metabolic processes such as protein synthesis. Protein synthesis continues in darkness in the cytosol, so that high ATP/ADP and AEC are necessary.

On illumination, photophosphorylation in the chloroplast stroma increases rapidly (within 30 s) so that the ATP concentration increases and the ATP/ADP ratio also, but AMP decreases, resulting in a large AEC. However, the sum of ATP, ADP and AMP remains constant, showing that little direct adenylate exchange occurs between chloroplast and cytosol. The cytosolic ATP/ADP ratio and AEC increase greatly and those in the mitochondria decrease, suggesting that the cytosol exercises control over respiration. However, after a few minutes illumination, the cytosolic ATP/ADP ratio and AEC fall but to values greater than in darkness and the mitochondrial values rise but to less than those observed in darkness. In the chloroplast stroma, ATP/ADP ratio is typically 2.4 and it remains at about this, even with changing conditions, although large perturbations do cause transients. The cytosolic ratios are greater (3–4) than the stromal. Even in the light, the chloroplast AEC is lower than the cytosolic, so the photosynthetic carbon cycle and other stromal activities (e.g. protein synthesis) presumably operate at lower AEC than those in the cytosol. The reason for, and consequences, of this are not understood.

Small differences between rates of ATP synthesis and consumption in different cell compartments would have rapid and very large effects on ATP/ADP ratios. The stability observed therefore indicates a very integrated and, as is well known, complex homeostatic, adenylate system which minimizes change. Rapid fluctuations in adenylates would have the most damaging effects on long-term stability of the whole system. However, because the different cell compartments are in rapid communication (e.g. via the adenylate transporter) and there are multiple ways of generating ATP, the system itself is very stable. Even P_i must be balanced in the cell compartments to accompany adenylate changes. The fluxes involved in the adenylate metabolism are not well quantified, so it is not possible to assess the contributions of different pathways of synthesis and consumption. Stoichiometry of ATP production per e^- and H^+ is crucial (see Noctor and Foyer, 2000). Probably, ATP supply resulting from linear electron transport is inadequate to meet cell demands and a Q cycle plus facultative cyclic and pseudocyclic phosphorylation probably operate, but the rates may be low (Chapter 5). A number of metabolic processes operate to exchange adenylates between cell compartments and affect the ratios of ATP/ADP, for example the malate–oxaloacetate, glutamate–oxoglutarate and triosephosphate-3 phosphoglycerate shuttles all affect ATP production and ATP to reductant synthesis and consumption. Photorespiration is also a major flux, with NADH produced oxidized in part by the mitochondria, thus generating abundant ATP and increasing the ATP/ADP ratio in the cytosol. Demand for ATP is, of course, greatly dependent on metabolism. Nitrate reduction and synthesis of amino acids requires 7.5 times less ATP per electron than CO_2, thus increasing the proportion of nitrate assimilated will substantially alter the ATP to reductant balance.

In the light, a high cytosolic ATP/ADP ratio and high AEC are generally thought, with little evidence, to depress mitochondrial oxidative phosphorylation and glycolysis. Regulation of the rate of ATP synthesis in mitochondria is not well understood. Mitochondrial adenylate would seem to be controlled by cytoplasmic AEC which remains high at all times, possibly because there is no AMP in the cytosol and no adenylate kinase. Regulation of ATP synthesis in mitochondria is probably flexible, depending on reductant availability, ATP production in other cell compartments, respiratory conditions *et cetera*. The rate of ATP synthesis in the mitochondrion is linked flexibly to the proton motive force, because of interaction between ATP synthase, the ADP/ATP translocator, and adenylate kinase in the inter-membrane space. When ADP concentration is small (upto about 100 µM), ATP is made largely by oxidative phosphorylation. At larger ADP concentrations more ATP is made by the adenylate kinase reaction, especially when magnesium is present (MgADP is the true substrate of ATP synthase). Reductant generated in the glycolate pathway and glycine oxidation may be used by the mitochondrial electron transport chain in the light. However, electrons from glycine oxidation may be transferred into the cytosol by malate and aspartate shuttles and not enter mitochondrial electron transport, so it may not be essential for mitochondria to function in the light. However, the weight of evidence is that respiration continues and that ATP synthesis by mitochondria is important for sucrose synthesis (and other ATP demanding processes) in the cytosol. On darkening cells there is a decrease in the ATP/ADP ratio and EC values in the chloroplast stroma, and concomitantly a transient decrease in the cytosol and a transient increase in mitochondrial ATP/ADP ratio, before they return to values seen after long periods in darkness.

8.11 Interaction between dark- and photorespiration, photosynthesis and adenylates

A working model of the interaction between dark respiration, photorespiration, photosynthesis and the adenylate systems is the following. The TCA cycle operates in darkness and transfer of adenylate energy keeps the cytosol ATP/ADP ratio and AEC high and transport of ATP into the chloroplast also maintains minimal ATP/ADP ratio and AEC in the stroma. In light, chloroplasts produce ATP, increasing the chloroplast ATP/ADP ratio and AEC and transferring energy to the cytosol, which experiences greatly increased ATP/ADP ratio and AEC; these may decrease mitochondrial activity, allowing the ATP/ADP ratio and AEC to re-equilibrate, but providing the necessary organic acids for synthetic reactions. With a small photon flux, adenylates from the chloroplast might not be sufficient for sucrose and other synthetic reactions, and mitochondria provide ATP to the cytosol, leading to a high rate of dark respiration relative to CO_2 fixation. With active photorespiration, the consumption of reductant in the mitochondria coupled to ATP synthesis will increase the ATP/ADP ratio there and raise the ratio in the cytosol. Although photorespiration is a drain on assimilation and decreases production, it provides a method of balancing cell processes against ATP supply and reductant. Under conditions where ATP synthesis may be inhibited (e.g. with water stress) respiration and photorespiration may increase and maintain cellular adenylate concentrations. If the demand for energy and substrates between parts of the cell becomes unbalanced, then the dynamic equilibria between cell compartments provides for that most important of cellular characteristics, stability.

8.12 Chlorophyll synthesis

Synthesis of chlorophyll *a* and *b* requires light in higher plants, which become chlorotic if grown in darkness; green plants also lose chl when placed in darkness because the chlorophyll is broken down as the chloroplasts degenerate. Chlorophyll synthesis is essential for the formation of the mature chloroplast and thylakoid system in higher plants, as the coordinated construction of the system during chloroplast development requires that the subunits, proteins, cofactors and pigments, are available. All enzymes of the biosynthetic pathway of chl are contained in the chloroplast stroma or are membrane bound. The pathway (*Figure 8.7*) is closely linked by a common precursor to protein synthesis, so that the stoichiometry of pigment and protein components is correct and thereby so is the structure of the mature system. Formation of δ-aminolaevulinic acid (ALA), a nonprotein amino acid, is a major step in the process of synthesis not only of chl but also heme and other tetrapyrroles. This step is also light-regulated with the activity of the protochlorophyllide reductase complex stimulated. Different points of entry of carbon into the tetrapyrrole synthetic pathway (5-C pathway) are known, with succinyl-CoA and glycine, 2-oxoglutaric acid or with glutamate providing the C-chain of δ-aminolaevulinate. The Shemin pathway in mammals and fungi catalyzes condensation of succinyl-CoA with glycine to give ALA. However, in most bacteria, cyanobacteria, green algae and higher plants, glutamate is the source (supporting a cyanobacterial origin for chloroplasts). Glutamate must first be activated at the α-carboxyl by ligating to tRNA^{GLU}. The reaction is catalyzed by glutamate tRNA ligase, which requires ATP and Mg^{2+}. The single tRNA^{GLU} gene (*trn*E), which is highly conserved, is coded in chloroplast DNA in all higher plants, and may also be responsible for activating glutamate during protein synthesis. Thus, it coordinates the two pathways in relation to the supply of glutamate and, thence, to nitrogen supply and energy. The nucleotide sequences of tRNA^{GLU} are known for a number of plants, and 5-methylaminomethyl-2-thioridine, characteristic of bacterial tRNAs, is probably essential for recognition by the tRNA^{GLU} synthetase (ligase). This enzyme is a dimer of *ca* 51–56 kDa mass and occurs in two forms, one chloroplastic and the other cytosolic; both are nuclear encoded and translated on chloroplast ribosomes, the chloroplastic form being imported. Despite the similarity between cytoplasmic and chloroplastic forms of the tRNA^{GLU,} they are not interchangeable. The key position of tRNA^{GLU} in chl and protein synthesis suggests an important regulation to balance the processes. Competition between the two in the chloroplast is regulated by an elongation factor (EF-Tu) for protein synthesis and binding of the tRNA to glutamyl-transfer RNA reductase determines ALA synthesis. The concentration of tRNA^{GLU} in the chloroplast is considered crucial to synthesis.

The gutamyl-tRNA^{GLU} is reduced by NADPH (but its specificity is not well assessed) and a specific glutamyl-transfer RNA reductase, located in the stroma; it may be a regulatory

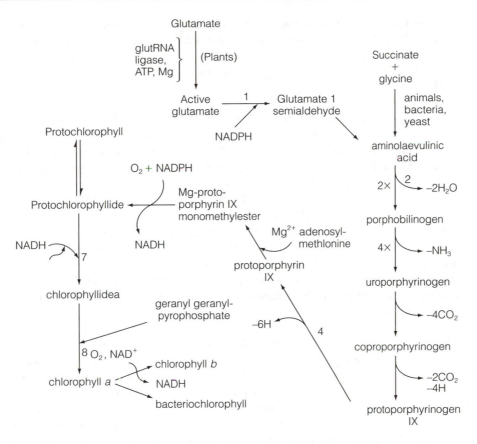

Figure 8.7. Outline of the biosynthesis of chlorophylls and related molecules

site. The purified reductase protein is an oligomer of 270 kDa (monomer 54 kDa) in barley and *Arabidopsis*: enzymes are specific to the substrate. The glutamate-1-semialdehyde produced is converted to the δ-aminolaevulinate by glutamate-1-semialdehyde-2,1–aminomutase, an abundant, stable stromal enzyme of about 45 kDa. Its sequence is known from a number of higher plants and is highly conserved. The protein is encoded by a nuclear gene in barley (GSA1) and other plants, and by GSA1 and GSA2 in *Arabidopsis*. The translation of the mRNA for the enzyme may be light-regulated.

For chl synthesis, the rate of ALA production is the rate-limiting step in higher plants. Transcriptional and translational regulation are important in relation to environmental signals and feedback from precursors of the pathway. Two molecules of δ-aminolaevulinate are condensed by the corresponding dehydratase enzyme, giving porphobilinogen. The enzyme is a hexamer of 250–300 kDa molecular mass. Four molecules of porphobilinogen are polymerized by porphobilinogen deaminase (also called urogen I synthase), giving the open ring tetrapyrrole hydroxymethylbilane. Hydroxymethylbilane is directly converted to uroporphyrinogen III by the enzyme uroporphyrinogen III synthase, in the stroma. Further steps in chlorophyll synthesis involve the conversion of uroporphyrinogen III to coproporphyrinogen III (enzyme unknown in plants but assumed to be a urogen III decarboxylase) and, hence, to protoporphyrinogen IX. This intermediate is oxidized by protogen oxidase to protoporphyrin IX which, when Mg^{2+} is inserted into the tetrapyrrole, gives Mg-protoporphyrin IX. The latter step is unique to chlorophyll synthesis and is catalyzed by magnesium chelatase, a nuclear gene product, which requires Mg^{2+} and ATP.

Further steps in synthesis of chlorophyll are the addition of the propionic acid side chain to ring C of Mg-protoporphyrin IX in a reaction requiring S-adenosylmethionine bound to the enzyme S-adenosylmethionine:Mg-protoporphyrin O-methyltransferase. Protochlorophyllide is formed when the Mg-protoporphyrin IX monomethyl ester side chain is cyclized. This requires two enzymes, a soluble and a membrane-bound cyclase, and requires oxygen and NADH or NADPH. A vinyl reductase reduces the vinyl group at position 8 to an ethyl using NADPH. Protochlorophyllide is converted to chlorophyllide by NADPH:protochlorophyllide oxidoreductase and requires light and NADPH. This enzyme, a single polypeptide of 36–38 kDa, is a major protein of the prolamellar body in etioplasts (chloroplasts partially developed in darkness). It uses light (acting as a substrate) and NADPH and is associated with a protochlorophyllide holochrome which converts the pigment to the chlorophyllide using NADPH and undergoes a characteristic spectral shift.

Chlorophyll *a* is produced from chlorophyllide *a* by addition of the phytol group, synthesized via the isoprenoid pathway. Mevalonic acid, a C6 compound, provides the starting point of the pathway. After decarboxylation (loss of a CO_2), removal of water and activation by pyrophosphate, a C5 compound is formed (isopentenyldiphosphate or IPP) which contains a double bond. By joining these C5 units into chains the alternating double-bond structures of the isoprenoids are formed, for example the C15 compound farnesyldiphosphate contains three double bonds and the geranylgeranyldiphosphate (GGPP) is a C20 compound with four double bonds. GGPP is synthesized by prenyltransferase in the plastid; it is the precursor of phytol, giving the phytyl chain of chlorophyll. Chlorophyllase or chlorophyll synthase may be responsible for the addition of the chain to the tetrapyrrole moiety. Chlorophyll *b* is probably made by a light-dependent enzyme which oxidizes the methyl group of ring B; chlorophyll *b* is needed for the formation of the chlorophyll *a/b* binding protein in thylakoids.

8.12.1 Regulation of chlorophyll synthesis

Regulation of the synthetic pathway occurs by feedback inhibition by heme, and light and oxygen are also very important. In angiosperms in darkness, synthesis stops at protochlorophyllide, which accumulates; teleologically it is sensible not to make a light-harvesting pigment in darkness. Light stimulates amino acid incorporation into proteins, ALA, *et cetera*. Analysis of promoter regions of genes (e.g. GSA) shows motifs known to be involved in light activation. In *Arabidopsis* only one GSA gene is thought to be light regulated. There is, however, considerable variation in the response of the genes in different plants to light, possibly because of the different light-signal transduction pathways operating with different photoreceptors (e.g. red and blue light, with the red acting via phytochrome; Section 10.5.1). The amount of mRNA of NADPH-protochlorophyllide oxidoreductase and the large subunit of Mg-chelatase are also light regulated. Oxygen is also required for chl synthesis, but the mechanisms are not known. There is close co-ordination in pigment formation, and the tRNA[GLU] and glutamic acid tRNA ligase are also most important in coordinating protein and chlorophyll synthesis. Protochlorophyllide reductase synthesis is regulated by phytochrome and plays a central role in coordinating chloroplast development. Many mutants of chl synthesis are known (it is easy to visually screen populations to select chlorotic mutant plants!) and transgenic plants have been made with antisense RNA to the glutamyl-transfer RNA reductase and other enzymes. Class I mutants, and the transgenics mentioned, have lesions in ALA synthesis which are lethal (effects on protein synthesis), class II have lesions between ALA and protochloro-phyllide, and class III lesions after protochlorophyllide synthesis. Inhibition of specific steps in the pathway is a specific target for development of herbicides which are plant-specific and do not affect other organisms.

During senescence of leaves, the chloroplast structures and the chlorophylls and other pigments are broken down in a regulated way, which in many plants may serve to remo-bilize the materials for synthetic or storage processes elsewhere. Leaf longevity for assimilation is important in crop production and variation in senescence is thus a poten-tially important factor in crop breeding. Mutations in plants may cause premature senes-cence but others induce a 'stay-green' phenotype. Stay-greens may result from late onset of senescence, which then proceeds as normal for that species or variety, or from normal onset but slow rate of senescence. Some stay-greens retain chlorophyll because the enzymes of chlorophyll breakdown are impaired. One type lacks the enzyme which con-verts the C7 formyl group of chlorophyll *b* to the *a*-type methyl structure, allowing the macrocycle to be opened. Another type has no phaeophorbide *a* oxygenase which opens the macrocycle. However, other aspects of senescence proceed normally (see Thomas and Howarth, 2000) so photosynthesis is not maintained. In others, photosynthetic activity continues. Adverse environmental conditions, such as water or light stress, can induce pigment loss and decrease photosynthesis; the mechanisms include oxidative damage by reactive oxygen species

8.13 Isoprenoid synthesis

Isoprenoids are compounds with the basic C5 unit (isopentenyl diphosphate, IPP). They are a vast group of very important biological molecules of great functional and structural diversity (*Figure 8.6*). All are formed from IPP by addition of isoprene units to make chains of different lengths, with modification to give cyclic structures, addition of oxygen and so on. Isoprenoids include isoprene, large quantities of which are released in bright light and hot conditions from leaves of some species (especially trees), phytol in chloro-phylls, plastoquinones (redox carriers in the photosynthetic electron transport chain), steroids and carotenoids. Many important regulatory and signalling compounds in plants are products of the pathway, for example the sesquiterpene farnesol, the diterpene gibberellins and the tetraterpene derivative abscisic acid are all plant growth regulators. The tetraterpene carotenoids and the oxygenated derivatives, the xanthophylls, are an important family of compounds formed early in the evolution of plants. For many years, their synthesis was thought to be by a common pathway in animals, fungi and plants, with IPP made by the acetate/mevalonate pathway. More recently old and new evidence (see Lichtenthaler, 1999) supports synthesis from an alternative route of synthesis in chloroplasts, from 1-deoxy-D-xylulose-5-phosphate (DOXP).

8.13.1 DOXP pathway

An alternative route of synthesis, the DOXP pathway, has recently been described. It occurs in eubacteria, cyanobacteria, algae and higher plants, where it is located in chloroplasts. The DOXP pathway uses glyceraldehyde-3-phosphate and pyruvate as starting materials; pyruvate reacts with thiamine pyrophosphate and this complex with glyceraldehyde-3-phosphate to give DOXP, followed by intramolecular rearrangement of DOXP producing IPP. This mechanism explains, better than the mevalonate pathway, the radiotracer labeling pattern of isoprene, carotenoids, phytol and plastoquinone and mono- and diterpenoids from plants given labeled glucose or CO_2. However, sterols and ubiquinone are from the mevalonate path. There is possibly cooperation from the two pathways, which reflect different origins in evolution. Perhaps the DOXP pathway will clarify some biochemical and ecological problems, e.g. why isoprene is released from the leaves of many plants in the light, constituting a major source of atmospheric organic matter, why bright light and heat stimulate the emission, and details of the mechanisms. Changing energy and reductant, or substrate synthesis, may play a part and the loss may be a means of regulating conditions within the chloroplast. Further examination of the DOXP path may shed light on metabolic processes which have long been regarded as well established.

Abscisic acid (ABA) is a plant growth regulator with many effects, for example increasing concentrations decrease growth, alter gene expression and induce stomatal closure. Therefore, it is of considerable importance for photosynthesis and its adaptation to environmental conditions. ABA synthesis is probably by enzymatic cleavage (9-*cis*) of the xanthophylls violaxanthin or neoxanthin via xanthoxin, and is greatly stimulated by water deficits, cold and other stress conditions. Where, ABA is made, in the cell, the mechanisms which produce it and the conditions that are required are not well understood.

8.13.2 Synthesis of carotenoids

For synthesis of carotenoids by the mevalonate pathway, carbon from acetyl-CoA passes via mevalonic acid and geranyl pyrophosphate into end products such as phytoene and phytol. In the process, the isoprene unit is joined into chains of increasing length. From phytoene, successive didehydrogenations (catalyzed by a desaturase) produce lycopene, and then cyclization forms carotenoids with the α- and β-ionone rings at the ends of the molecules. The bonds may be single or double; β-carotene is a linear structure of alternate single and double bonds and ionone rings, which provides the required molecular energy levels to capture light, accept triplet excitation energy from chlorophyll and quench singlet oxygen. Oxygenation of carotenoids produces xanthophylls, such as violaxanthin, antheraxanthin and zeaxanthin; these form the xanthophyll cycle (Chapter 3) which is a mechanism to dissipate excitation energy in the light-harvesting system of thylakoids. Many herbicides block the desaturation steps and thereby carotenoid synthesis. In light, treated plants cannot dissipate excess energy and die. Carotenoids probably functioned as light-harvesting pigments in early photosynthesis and later, after development of the chlorophylls, became more important in regulating the energy state of the chemical reaction centers. When oxygen accumulated in the atmosphere, carotenoids assumed a protective role. The synthesis of carotenoids in bacteria may be prevented by quite low concentrations of oxygen.

References and further reading

Atrin, O.K., Millar, A.H., Gardeström, P. and Day, D.A. (2000) Photosynthesis, carbohydrate metabolism and respiration in leaves of higher plants. In *Photosynthesis: Physiology and Metabolism* (eds R.C. Leegood, T.D. Sharkey and S. von Caemner). Kluwer Academic Publishers, Dordrecht, pp. 153–175.

Bartley, G.E., Coomber, S.A., Bartholomew, D.M. and Scolnitz, P.A. (1991) Genes and enzymes for carotenoid biosynthesis. In *The Photosynthetic Apparatus, Molecular Biology and Operation* (eds L. Bogorad and I.K. Vasil). Academic Press, San Diego, CA, pp. 331–346.

Benning, C. (1998) Biosynthesis and function of the sulfolipid sulfoquinovosyl diacylglycerol. *Annu. Rev. Plant Physiol. Plant Mol. Biol.* **49**: 53–75.

Bonnemain, J.L., Delrot, S., Lucas, W.J. and Dainty, J. (1991) *Recent Advances in Phloem Transport and Assimilate Compartmentation.* Quest Editions, Presses Académiques, Nantes.

Bourguignon, J., Rébeillé, F. and Douce, R. (1998) Serine and glycine metabolism in higher plants. In *Plant Amino Acids* (ed. B. Singh). Marcel Dekker, New York, pp. 111–146.

Bryant, J.A., Burrell, M.M. and Kruger, N.J. (eds) (1999) *Plant Carbohydrate Biochemistry.* BIOS Scientific Publishers, Oxford.

Buchanan-Wollaston, V. (1997) The molecular biology of leaf senescence. *J. Exp. Bot.* **48**: 181–199.

Büchel, C. and Garab, G. (1997) Respiratory regulation of electron transport in chloroplasts: chlororespiration. In *Handbook of Photosynthesis* (ed. M. Pessarakli). Marcel Dekker, New York, pp. 83–93.

Cunningham, F.X. Jr. and Gantt, E. (1998) Genes and enzymes of carotenoid biosynthesis in plants. *Annu. Rev. Plant Physiol. Mol. Biol.* **49**: 557–583.

Day, D.A. and Wiskich, J.T. (1995) Regulation of alternative oxidase activity in higher plants. *J. Bioenerget. Biomembranes* **27**: 379–385.

Douce, R. and Heldt, H.-W. (2000) Photorespiration. In *Photosynthesis: Physiology and Metabolism* (eds R.C. Leegood, T.D. Sharkey, and S. von Caemmerer). Kluwer Academic, Dordrecht, pp. 115–136.

Douce, R. and Joyard, J. (1996) Biosynthesis of thylakoid membrane lipids. In *Oxygenic*

Photosynthesis: the Light Reactions (eds D.R. Ort, and C.F. Yocum). Kluwer Academic, Dordrecht, pp. 69–101.

Douce, R., Bourguignon, J., Macherel, D. and Neuberg, M. (1994) The glycine decarboxylase system in higher plant mitochondria – structure, function and biogenesis. *Trans. Biochem. Soc.* **22**: 184–188.

Heber, U. *et al.* (1996) Photorespiration is essential for the protection of the photosynthetic apparatus of C3 plants against photoinactivation under sunlight. *Botanica Acta* **109**: 307–315.

Heldt, H.W., Flügge, U.-I., Borchert, S., Brückner, G. and Ohnishi, J.-I. (1990) Phosphate translocators in plastids. In *Perspectives in Biochemical and Genetic Regulation of Photosynthesis* (ed. I. Zelitch). Liss, New York, pp. 39–54.

Huber, S.C. and Huber, J.L. (1996). Role and regulation of sucrose-phosphate synthase in higher plants. *Annu. Rev. Plant Physiol. Mol. Biol.* **47**: 431–444.

Huber, S.C., Kaiser, W.M., Toroser, D., Athwal, G.S., Winter, H. and Huber, J.L. (1999) Regulation of sucrose metabolism by protein phosphorylation: stimulation of sucrose synthesis by osmotic stress and 5-aminoimidazole-4-carboxamide riboside. In *Plant Carbohydrate Biochemistry* (eds J.A. Bryant, M.M. Burrell and N.J. Kruger). Bios Scientific Publishers, Oxford.

Jordan, P.M. (1991). The biosynthesis of 5-aminolevulinic acid and its transformation into uroporphyringen III. In *Biosynthesis of Tetrapyrroles* (New Comprehensive Biochemistry, Vol. 19), Elsevier, Amsterdam, pp. 1–66.

Kannangara, C.G. *et al.* (1994) Enzymatic and mechanistic studies on the conversion of glutamate to 5-aminolevulinate. In *The Biosynthesis of Tetrapyrrol Pigments* (Ciba Foundation Symposium no. 180) (eds D.J. Chadwick and K. Ackrill). John Wiley, Chichester, pp. 3–25.

Keys, A.J. and Parry, M.A.J. (1990) Ribulose bisphosphate carboxylase/oxygenase and carbonic anhydrase. In *Methods in Plant Biochemistry*, Vol. 3, *Enzymes of Primary Metabolism* (ed. P.J. Lea). Academic Press, London, pp. 1–14.

Kozaki, A. and Takeba, G. (1996) Photorespiration protects C3 plants from photooxidation. *Nature* **384**: 557–560.

Kumar, A., Schaub, U, Söll, D and Ujwal, M.L. (1996). Glutamyl-transfer RNS: at the crossroad between chlorophyll and protein biosynthesis. *Trends Plant Sci.* **1**: 371–375.

Leegood, R.C., Sharkey, T.D. and von Caemmerer, S. (eds) (2000) *Photosynthesis: Physiology and Metabolism*. Kluwer Academic, Dordrecht.

Lichtenthaler, H.K. (1999) The 1-deoxy-D-xylulose-5-phosphate pathway of isoprenoid biosynthesis in plants. *Annu. Rev. Plant Physiol. Plant Mol. Biol.* **50**: 47–65.

Lichtenthaler, H.K., Schwender, J., Disch, A., and Rohmer, M. (1997) Biosynthesis of isoprenoids in higher plant chloroplasts proceeds via a mevalonate-independent pathway. *FEBS Lett.* **400**: 271–274.

Matile, P. , Hörtensteiner, S. and Thomas, H. (1999) Chlorophyll degradation. *Annu. Rev. Plant Physiol. Plant Mol. Biol.* **50**: 67–95.

Möhlmann, T., Tjaden, J. and Neuhaus, H.E. (1999) Starch synthesis in plant storage tissues: precursor dependency and function of the plastidic ATP/ADP transporter. In *Plant Carbohydrate Biochemistry* (eds J.A. Bryant, M.M. Burrell, and N.J. Kruger). Bios Scientific Publishers, Oxford.

Nam, H.G. (1997) The molecular genetic analysis of leaf senescence. *Curr. Opin. Biotechnol.* **8**: 200–207.

Noctor, G. and Foyer, C.H. (1998) A re-evaluation of the ATP: NADPH budget during C3 photosynthesis: a contribution from nitrate assimilation and its associated respiratory activity? *J. Exp. Bot.* **49**: 1895–1908.

Noctor, G. and Foyer, C.H. (2000) Homeostasis of adenylate status during photosynthesis in a fluctuating environment. *J. Exp. Bot. (GMP Special Issue)* **51**: 319–328.

Oliver, D.J. (1998) Photorespiration and the C$_2$ cycle. In *Photosynthesis: a Comprehensive Treatise* (ed. A.S. Raghavendra). Cambridge University Press, Cambridge, pp. 173–182.

Patrick, J.W. (1997) Phloem unloading: sieve element unloading and post-sieve element transport. *Annu. Rev. Plant Physiol. Plant Mol. Biol.* **48**: 191–222.

Quick, P. W. and Schaffer, A.A. (1996) Sucrose metabolism in sinks and sources. In *Photoassimilate Distribution in Plants and Crops* (eds E. Zamski and A.A. Schaffer). Marcel Dekker, New York, pp. 115–156.

Rao, I.M., Arulanantham, A.R. and Terry, N. (1990). Diurnal changes in adenylates and nicotinamide nucleotides in sugar beet leaves. *Photosynth. Res.* **23**: 205–212.

Rébeillé, F. and Douce, R. (1999). The folate status of plant mitochondria. In *Plant Carbohydrate Biochemistry* (eds J.A. Bryant, M.M. Burrell and N.J. Kruger). Bios Scientific Publishers, Oxford.

Rüdiger, W. and Schoch, S. (1988) Chlorophylls. In: *Plant Pigments* (ed. T.W. Goodwin). Academic Press, London, pp. 1–59.

Schoefs, B. and Bertrand, M. (1997) Chlorophyll biosynthesis. In *Handbook of Photosynthesis* (ed. M. Pessarakli). Marcel Dekker, New York.

Scott, P. Lange, A.J., Pilkis, S.J. and Kruger, N.J. (1995) Carbon metabolism in leaves of transgenic tobacco (*Nicotiana tabacum* L.) containing elevated fructose-2, 6-bisphosphate levels. *Plant J.* **7**:461–469.

Singsaas, E.L., Lerdau, M., Winter, K. and Sharkey, T. D. (1997) Isoprene increases thermotolerance of isoprene-emitting species. *Plant Physiol.* **115**: 1413–1420.

Smith, A.M., Denyer, K. and Martin, C. (1997) The synthesis of the starch granule. *Annu. Rev. Plant Physiol. Plant Mol. Biol.* **48**: 67–87.

Stitt, M., Huber, S. and Kerr, P. (1987) Control of photosynthetic sucrose formation. In *The Biochemistry of Plants,* Vol. 10 (eds M.D. Hatch and N.K. Boardman). Academic Press, London, pp. 328–409.

Thomas, H. and Howarth, C.J. (2000) Five ways to stay green. *J. Exp. Bot.* **51** (GMP Special Issue): 329–337.

van Bel, A.J.E (1993) Strategies of phloem loading. *Annu. Rev. Plant Physiol. Plant Mol. Biol.* **44**: 253–282.

Vanlerberghe, G.C. and McIntosh, L. (1997). Alternative oxidase: from gene to function. *Annu. Rev. Plant Physiol. Plant Mol. Biol.* **48**: 703–734.

Vedel, F., Lalanne, E., Sabar, M., Chétrit, P. and De Paepe, R. (1999) The mitochondrial respiratory chain and ATP synthase complexes: Composition, structure and mutational studies. *Plant Physiol. Biochem. 37*: 629–643.

Von Wettstein, D., Gough, S. and Kannangara, C.G. (1995) Chlorophyll biosynthesis. *Plant Cell* **7**: 1039–1057.

Whitehouse, D.G. and Moore, A.L. (1995) Regulation of oxidative phosphorylation in plant mitochondria. In: *The Molecular Biology of Plant Mitochondria* (eds C.S. Levings II and I.K. Vasil). Kluwer, Dordrecht, pp. 313–344.

Wildermuth, M.C. and Fall, R. (1998). Biochemical characterization of stromal and thylakoid-bound isoforms of isoprene synthase in willow leaves. *Plant Physiol.* **116**: 1111–1123.

Young, A.J., Phillip, D. and Savill, J. (1997) Carotenoids in higher plant photosynthesis. In: *Handbook of Photosynthesis* (ed. M. Pessarakli). Marcel Dekker, New York, pp. 575–596.

Zamski, E. and Schaffer, A.A. (eds) (1996) *Photoassimilate in Plants and Crops.* Marcel Dekker, New York.

C4 photosynthesis and crassulacean acid metabolism

9.1 Introduction

The PCR cycle is the only process giving a net increase in the chemical energy of the biosphere and occurs in all photosynthetic organisms which assimilate carbon dioxide. This basic mechanism of carbon assimilation, considered in Chapter 7, is the core process of photosynthetic CO_2 assimilation in plants with C4 and CAM variants. The basic mechanism and its regulation are very similar in all photosynthetic organisms, although as quantitative understanding of the processes improves, great subtlety is apparent in the way that different groups of plants produce assimilates in very different environments. In the PCR cycle the first stable product of the reaction between RuBP and CO_2 is 3PGA, a three-carbon compound, so this mode of carbon assimilation is called C3 photosynthesis (*Figure 9.la*) and the very large number of species with this mechanism are called C3 plants. However, in many families of higher plants, two additional metabolic systems have evolved for accumulating CO_2 and passing it to the PCR cycle. This increases the efficiency of photosynthesis in adverse environments, particularly where heat, drought and salinity cause an imbalance between the requirements for CO_2 assimilation and those for maintenance of other plant functions. These systems are C4 (see *Figure 9.1b*) and crassulacean acid metabolism (CAM for short; see *Figure 9.1c*).

9.1.1 C4 mechanism – outline

The C4 mechanism involves the primary fixation of CO_2 into the three-carbon precursor phosphoenol pyruvic acid (PEP) to produce oxaloacetic acid, a four-carbon carboxylic acid as the first product, hence the name C4 plants. Oxaloacetate is rapidly metabolized to other four carbon acids and amino acids. The system involves separation of functions between cells in two major types of leaf tissue, the mesophyll and bundle sheath (Chapter 4). The acids are formed in mesophyll cells and transferred to bundle sheath cells where they are decarboxylated, providing a large concentration of CO_2 so that Rubisco (which occurs only in the bundle sheath) is effectively CO_2 saturated. The CO_2 is assimilated by the PCR cycle; this is the true energy-capturing part of the processes, as synthesis of PEP cannot be achieved without it. There is considerable variation in anatomy and metabolism between different species of C4 plants, which are grouped into three main types, depending on the characteristic enzymes. The NADP-ME types of C4 plants

(a) C3 photosynthesis (3-carbon primary product)

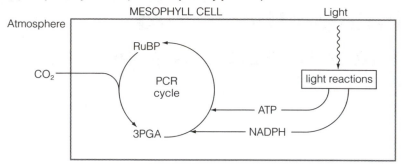

(b) C4 photosynthesis (4-carbon primary product)

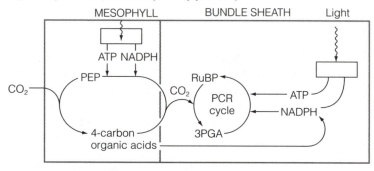

(c) CAM photosynthesis (4-carbon organic acids produced in darkness)

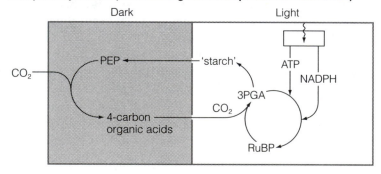

Figure 9.1. Variation in photosynthetic mechanisms in higher plants: (a) C3 photosynthesis by PCR, and synthesis of dicarboxylic acids in (b) C4 photosynthesis and (c) CAM cycle

include many tropical grasses, predominantly from hot, moist conditions, and the NAD-ME types of hot very dry conditions such as arid grasslands in North America and Australia. C4 metabolism is an ecological adaptation to intense light under low atmospheric CO_2 concentration, which also allows the photosynthetic system to operate efficiently within a leaf with smaller stomatal conductance (Chapter 12) than is possible with C3 metabolism in the low-CO_2 concentration of the current atmosphere. The small stomatal conductance results in relatively slower rates of transpiration in C4s compared to C3s in comparable conditions. C4 photosynthesis is not a single mechanism, rather it has evolved independently many times (31 on some estimates) in 18 different families of very different types, and is found in some 8000 species.

9.1.2 Crassulacean acid metabolism – outline

Crassulacean acid metabolism (*Figure 9.1c*) also involves the initial formation of four-carbon organic acids from the reaction of CO_2 and PEP, but this synthesis proceeds in darkness using energy accumulated by the operation of the PCR cycle during previous illumination. The acids formed in darkness are decarboxylated in the subsequent light period and the CO_2 is assimilated by the PCR cycle with light energy. CAM plant leaves are not structurally differentiated into tissues with different photosynthetic biochemistry, but CO_2 accumulation and PCR cycle assimilation are separated in time. CAM occurs in many succulent plants (with strongly developed water-storage 'hydrenchyma') including, but not confined to, the family Crassulaceae, hence the name, and is related to water conservation in dry environments. CAM is relatively uniform in mechanism but has probably evolved many times: it occurs in 33 families and about 16 000 species, with habitats ranging from submerged aquatic to epiphytic as well as deserts. The main ecological value is that CAM provides a method of conserving water yet maintains CO_2 assimilation. The changes associated with C4 and CAM photosynthesis are ecologically important and comparison of mechanisms of C3, C4 and CAM plants in relation to the environments in which they live has provided much understanding of the factors which determine plant adaptation, survival and productivity.

9.2 C4 photosynthesis

C4 plants generally have higher rates of CO_2 assimilation than C3 plants; their photosynthesis does not saturate in bright light, continues at very low concentrations of CO_2 and is insensitive to O_2. Associated with these features are more rapid growth rates, higher dry matter production and larger economic yields than C3 plants (Chapter 13). C4 assimilation occurs in diverse families of angiosperms (but not in gymnosperms), particularly the monocotyledons and, within this group, tropical panicoid grasses (Poaceae) such as sugar cane (*Saccharum officinarum*), maize (*Zea mays*) and sorghum (*Sorghum vulgare*) which are of great economic importance. C4 metabolism does not occur in trees, the reasons for which are not known. However, C4 mechanisms of CO_2 assimilation also occur widely in dicotyledons, for example, in the Chenopodiaceae and Compositae. Even within a genus, some species have C4 metabolism whilst others have C3; for example in *Atriplex* (Chenopodiaceae), *A. sabulosa* is C4 and *A. hastata* is C3. There are a number of genera and families which include species which are intermediate between the 'true' C3 and C4 types; they are discussed in Section 9.3.

9.2.1 C4 plant anatomy

Within the C4 plants there is considerable variation in anatomy (discussed in detail for maize in Chapter 4) but the bundle sheath, a ring of large, closely packed cells around the vascular tissue, is distinctive. The compact structure of the bundle sheath offers a low surface area to the intercellular spaces and in many C4 types (e.g. C4 sedges), a mestome sheath of schlerenchymatous cells separates the bundle sheath and mesophyll, but in maize it is not represented. The arrangement of the chloroplasts in bundle sheath cells is related to the type of C4 metabolism. Large chloroplasts with many thylakoids are arranged (*Table 9.1*) around the outer wall of the cells (centrifugal) in some that form aspartate (e.g. *Panicum maximum)* and malate (e.g. *Zea mays)*. Chloroplasts are on the walls nearest to the vascular tissue (centripetal) in NAD-ME types (e.g. *Amaranthus edulis*) which produce aspartate, and contain many more large mitochondria close to the vascular tissue than other C4 plants. Bundle sheath cells are larger than mesophyll cells and have a much smaller surface-to-volume ratio and dense walls, with a double layer of suberin which has a very low permeability to gases. Bundle sheath cells are probably one to two orders of magnitude less permeable to CO_2 than C3 cells. This is particularly important in allowing very large CO_2 concentrations, resulting from the decarboxylation of the C4 cycle, in cells where Rubisco is located and it restricts diffusion of O_2 into them. As O_2 concentration in the atmosphere is so large (210 kPa) it is very difficult to decrease the O_2 metabolically and also its diffusion into cells is difficult to restrict; the question of bundle sheath 'leakiness' is discussed in Section 12.3.

Other anatomical features are clearly of significance but are not universally present. For example, the peripheral reticulum (Section 4.4.2) increases the internal membrane surface area for transport of assimilates via plasmodesmata, but the extent differs in the different types of C4s. There are also differences in the thylakoid structure. In some malate producers (e.g. maize) only stromal thylakoids are present, whereas some aspartate-producing C4 plants (e.g. *Sporobolis aeroides)* have pronounced grana. In both malate- and aspartate-forming types the mesophyll chloroplasts are granal. Such differences reflect differences in the distribution of PSII and other components (Section 4.4).

9.2.2 C4 metabolism

The four main steps in C4 photosynthesis (*see Figure 9.2a*) are: (1) carboxylation of PEP produces organic acids in the mesophyll cells without a net gain in energy; (2) organic acids are transported to the bundle sheath cells; (3) organic acids are decarboxylated there, producing CO_2 , which is assimilated by Rubisco and the PCR cycle with a net gain in energy; and (4) compounds return to the mesophyll to regenerate more PEP. The steps are common to all C4 types. However, the decarboxylation enzymes and the compounds transported between mesophyll and bundle sheath differ. These conclusions are based on the proportions of $^{14}CO_2$ in assimilate products and the time course of labeling and on the distribution of enzymes between cell organelles and tissue (*Table 9.1*); an account of the discovery and analysis of the C4 syndrome is given by Hatch (1999). Gas exchange and biochemical studies showed that some plants had (in comparison with others now known to be C3s) large rates of photosynthesis. Also, they were relatively insensitive to CO_2 and O_2 concentration and, when provided with $^{14}CO_2$, the primary radioactive products were oxoacetic, malic and aspartic acids. Only after a delay did ^{14}C accumulate in PCR cycle compounds (e.g. 3PGA), suggesting flow of carbon from organic acids to the PCR cycle. Studies of gas exchange with radiotracers allowed extrapolation of the ^{14}C content (measured at steady state, when the fluxes of carbon between pools are constant) back to zero time, showing that more than 95% of ^{14}C entered organic acids and less than

5% entered 3PGA directly. Only in very large atmospheric CO_2 concentration did Rubisco use atmospheric CO_2 directly. Thus, in the C4 leaf, Rubisco is effectively separated from the atmosphere both physically and by an intermediate biochemical step. Realization that there was a different type of photosynthesis from that described by Calvin (see Chapter 7) led to detailed examination of the anatomy, biochemical processes and the distribution of the enzymes responsible for the characteristics of the plants. Rubisco is now known to be in the bundle sheath, with activity and content one-third of that in C3 leaves; it has specific activity (mol substrate assimilated/unit of protein, range 2300–4000), twice that of C3 plants. However, the affinity for CO_2 (K_m 28–34 µM) is small compared to C3 plants (13–30 µM). As an important aside, the small amount of Rubisco required for high rates of CO_2 assimilation increases the nitrogen use efficiency (slope of the curve relating photosynthetic rate to nitrogen content of the leaf); averaged over a range of C4s it is 50% greater than in C3s at *ca* 25°C and 100% greater at *ca* 35°C. This shows the beneficial effects of C4 metabolism at higher temperatures. The decreased N content also means that less N is needed for C4s than C3s per unit of production. In a North American prairie, with C3 and C4 species, C4s may dominate in N-deficient conditions and have large N-use efficiency. However, if some additional N is available C3s totally dominate. Thus, the biochemical, physiological and ecological consequences of the C4 mechanism are established. Returning to the enzyme kinetics, there is an interesting and presumably important variation in the K_m for CO_2 between the PEP-CK and the NAD- and NADP-ME types (35, range 28–41, and 53, range 41–63, respectively). The larger V_{max} and K_m reflect the adaptation of the enzyme to the very large CO_2 concentration in the bundle sheath that allow rapid reaction. However, this occurs despite the similarity of the specificity factor for C3 and C4 plants. The specificity factor (mol CO_2 assimilated/mol O_2 assimilated, see Section 7.2.2) or C4 Rubisco is *ca* 70 (range 55–85) compared to 85 (range 82–90 mol mol^{-1}) in C3 plants (note that there is some uncertainty about the exact magnitude of the specificity factor due to the difficulty of calculating the CO_2 concentration in solution). Many other enzymes, common to C3 and C4 plants, have similar features but with important variations.

Differences in metabolism between mesophyll and bundle sheath.

The marked differences in metabolism between mesophyll and bundle sheath cells have been shown by a range of techniques. These include measurement of enzyme distribution between the cell types after separating the tissues by differential grinding or rolling of leaves to disrupt the 'softer' mesophyll cells and release their contents, leaving the 'harder' bundle sheath cells. Cells or their contents are then separated by centrifugation. Also, enzymes are used to disrupt the cell walls of the tissues at different rates. The cells of mesophyll and bundle sheath are distinguished by different enzymes, chlorophyll *a/b* ratios, *et cetera* (see *Table 9.1*), so corrections for contamination are possible. Chloroplasts and other organelles from the cells can be obtained by normal methods of cell disruption and separation. Distributions of enzymes in tissues and cell organelles clearly show the spatial separation between the carboxylation and decarboxylation processes. Enzymes of initial CO_2 fixation and amino acid synthesis are in mesophyll cells; those involved in the decarboxylation of amino acids and carboxylation by the PCR cycle activity are in the bundle sheath, not in the mesophyll where the only PCR cycle enzymes are for reduction of 3PGA to DHAP. Enzymes and metabolites of systems regulating the active oxygen species and associated compounds (see Chapter 5) in leaves of maize are very differently

Table 9.1. Characteristics of different types of C4 plants, and examples of species with this form of metabolism, including distribution of chloroplasts and enzymes in bundle sheath and mesophyll cells, compared with C3 species

Type of photosynthesis	NADP-ME	PEP-CK	NAD-ME	C3
Examples	*Zea mays* *Sorghum sudanense* *Saccharum officinarum*	*Panicum maximum* *Chloris gayana* *Sporobolis fimbriatu* *Urochloa panicoidess*	*Atriplex spongiosa* *Portulaca oleracea* *Amaranthus edulis* *Eleusine indica*	*Triticum aestivum* *Glycine max* *Pisum sativum*
Bundle sheath	Yes	Yes	Yes	No
Cell size (µm)	113 × 18 (longer inner sheath)	35 × 35	40 × 24	
No. chloroplasts/cell	42		39	
bs chloroplast position	Centrifugal	Centrifugal	Centripetal and centrifugal	
Suberized layer	Yes	Yes	No	
Grana	Very reduced	Yes	Yes	
Mitochondria	Few	Few?	Many	
Mestome Sheath	No	No	Yes	No
Mesophyll				
Cell size (µm)	56 × 16		38 × 8	60 × 20
No. chloroplasts/cell	31		10	90
Grana	Yes	Yes	Yes	Yes
Major organic acid	Malate	Aspartate	Aspartate	—
Decarboxylation	bs chloropl	bs cytosol	bs cytosol	
Enzyme activity				
Carbonic anhydrase	High meso	High meso	High meso	High
NADP malate dehydrogenase	Very high meso chloropl	Low meso and some bs	Medium meso and some bs	Very low
NADP malic enzyme	None mc/very high bs chloropl	No mc/low bs	Very low meso and low bs	Very low
Aspartate aminotransferase	Some meso chloropl low bs	Very high cytosol meso and high bs	Very high meso and some bs cytosol and mitochondria	Very low
PEP carboxykinase	Almost no meso and bs	No mc/high bs cytosol	Almost no meso and bs cytosol	Very low

NAD malic enzyme	Very low meso, some bs	Low meso and some bs	Low meso high bs mitochondria	Very low
NAD malate dehydrogenase	None	Low	High bs mitochondria	Very low
Alanine aminotransferase	Low meso and bs	High bs and some meso cytosol	High bs and meso cytosol	Very low
PEP carboxylase	High meso cytosol/low bs	Very high meso cytosol/low bs	Very high meso cytosol/low bs	Very low
Pyruvate, P_i dikinase	Very high meso chloropl/low bs	High meso chloropl/almost no bs	High meso chloropl/almost no bs	None
3PGA kinase	High meso and very high bs chloropl	High meso chloropl some bs	High meso and medium bs chloropl	High chloropl
Rubisco	Almost none mc/high bs chloropl	Very little mc/ bs chloropl	Almost no mc/ bs chloropl	High chloropl
PCR cycle enzymes	bs chloropl	bs chloropl	bs chloropl	Chloropl
Photorespiratory enzymes	Some meso high bs	Some meso high bs	Some meso high bs	High
Light reactions	PSI, II meso PSI bs	PSI, II meso PSI, II bs	PSI, II meso PSI, II bs	PSI, II —

bs, bundle sheath; meso, mesophyll; chloropl, chloroplast.

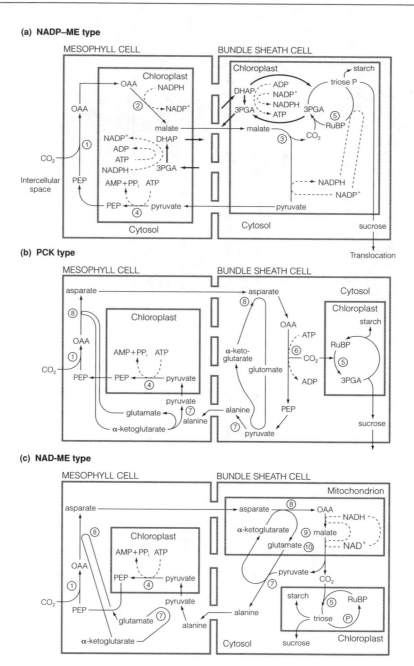

Figure 9.2. Photosynthetic metabolism of C4 plants, compounds transferred between mesophyll and bundle sheath decarboxylation with: (a) NADP requiring malic enzyme or 'NADP-ME' type; (b) aspartate-forming and PEP type of C4 metabolism; (c) aspartate-forming and NAD requiring malic enzyme 'NADME'-type of C4 metabolic enzymes listed below: (1) PEP carboxylase; (2) NADP malate dehydrogenase; (3) NADP malic enzyme; (4) pyruvate, P; dikinase; (5) RuBP carboxylase/oxygenase; (6) PEP carboxykinase; (7) alanine aminotransferase; (8) aspartate aminotransferase; (9) NAD malate dehydrogenase; (10) NAD malic enzyme

distributed in the two tissues. Glutathione reductase, dehydroascorbate reductase and H_2O_2 are in the mesophyll and ascorbate, ascorbate peroxidase and SOD are in the bundle sheath. Catalase, monohydroascorbate reductase and reduced glutathione are equally distributed in both tissues. The distribution relates to the site of production of reductant and probably there is cycling of reduced and oxidized forms between tissues.

Enzymes of C4 metabolism. PEP carboxylase (PEPc) (EC 4.1.1.31). In all C4 plants, the initial CO_2 fixation (a carboxylation but not to be confused with the energy accruing carboxylation of RuBP by Rubisco in the PCR cycle) occurs in the mesophyll cell cytosol and is catalyzed by PEPc. PEPc uses bicarbonate ions, not CO_2 in the reaction with PEP, forming oxaloacetic acid (OAA):

$$PEP + HCO_3^- + H^+ \longrightarrow OAA + P_i \tag{9.1}$$

The reaction is irreversible with a large negative free energy change (-31.8 kJ mol^{-1}) and is therefore a control step in C4 assimilation (see Section 7.4 for discussion of limiting reactions). The activity of PEPc is about 13–27 µmol min^{-1} mg^{-1} chlorophyll in all C4s. The reaction rate is very fast, the purified enzyme having a molar activity of almost 10 000 mol substrate min^{-1} mol^{-1} enzyme at pH 7 and 22°C. PEPc is a tetramer of identical subunits of 100 kDa and requires Mg^{2+} (forming a bidentate metal complex coordinated to the enolate oxygen, the carbonyl oxygen and the phosphate oxygen). PEP binds before HCO_3^-, then phosphate transfers from PEP, forming the enolate form of pyruvate and carboxyphosphate, which undergoes rearrangement to give P_i and CO_2 bound to the enzyme. Then the enolate form of pyruvate (stabilized by the metal) reacts with the CO_2, giving oxaloacetate. The active site structure is not known from X-ray crystallography, but the reaction mechanism is established (Chollet *et al.*, 1996). PEPc is activated by light 10–20-fold, via a phosphorylation at a serine residue at the N-terminal end of the protein, catalyzed by a protein kinase specific to the enzyme (*Figure 9.3*). The phosphorylated enzyme is more active and sensitive to glucose-6-phosphate and triosephosphates which enhance catalysis, and less sensitive to those (such as malate and aspartate) which inhibit catalysis. Effectors have greater effect in the low pH of the cytosol (pH 7.3) than at the pH optimum of about 8. PEP has a poor affinity (large K_m) of 1–2 mM for PEP, which is in very high concentration in actively

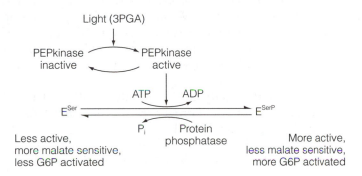

Figure 9.3. Scheme of phosphoenol pyruvate carboxylase regulation by phosphorylation involving PEP kinase and protein phosphorylase

photosynthesizing mesophyll cells. The K_m for bicarbonate is about 2–20 µM, which is about that in equilibrium with the internal CO_2 concentration of C4 leaves in air, about 0.5–1 Pa CO_2. Hence, this reaction is able to remove bicarbonate ions from water in equilibrium with an atmosphere containing an extremely small concentration of CO_2. PEPc has no oxygenase activity, in contrast to Rubisco, to offset CO_2 fixation. Thus, the enzyme is very efficient in assimilating CO_2. The PEP reaction acts as an effective 'CO$_2$ pump', supplying CO_2 to the PCR cycle, greatly slowing RuBP oxygenation, and is efficient at high temperatures, to which many C4 plants are adapted. This gives very small compensation concentration, that is the CO_2 concentration of a leaf when enclosed in an airtight chamber in the light. C4 plants remove almost all CO_2 (5–10 µmol mol^{-1}) in contrast to C3 plants (50–60 µmol mol^{-1}). PEP is allosterically controlled by many photosynthetic assimilates (i.e. they change the rate of reaction without entering into it). If metabolism of the primary products is inhibited, for example by slow PCR cycle turnover, malate accumulates and PEP consumption slows, thus decreasing the drain of carbon from the PCR cycle. With rapid CO_2 assimilation and an increase in glucose-6-phosphate, an important metabolite in starch synthesis, the reaction becomes more rapid thus stimulating the rate of assimilation and decreasing the drain of carbon to storage.

The many PEPc isoforms are encoded by small multigene families, for example three nuclear encoded in *Sorghum* and five in *Zea*. Some are specific to particular tissues. Light induction occurs in C4 plants by light responsive elements unlike those of Rubisco. However, there is little information about the genetic regulation controlling transcription of different PEPc genes in different plants. PEPc and its regulation appears very flexible, for example in *Mesembryanthemum crystallinum* two isoforms exist, and expression of one is greatly increased as this facultative CAM plant switches from C3 to CAM. Analysis of PEPc suggests that CAM evolution preceded that of C4. All PEPc genes derived from a single common ancestor, and possibly gene duplication and changes could have preceded divergence of the different CAM and C4 groups. As PEPc removes HCO_3^-, CO_2 from the atmosphere is rapidly converted to HCO_3^- by carbonic anhydrase, even in low CO_2, in which C4 plants can photosynthesize but C3 plants cannot.

Carbonic anhydrase (carbonate hydratase, EC 4.2.1.1, CA) occurs in the mesophyll cell cytosol. It catalyzes the interconversion of CO_2 and bicarbonate ions, and increases the rate of reaction. It is not found in the bundle sheath where CO_2 is required for Rubisco. Carbonic anhydrase has activities of 30–90 µmol min^{-1} mg^{-1} chlorophyll in C4s, although it is generally two to three times greater in NAD-ME types (80–90 µmol min^{-1} mg^{-1} chlorophyll) than PCK types (30 µmol min^{-1} mg^{-1} chlorophyll). It is an enzyme with a very rapid catalytic rate and large specific activity.

Oxaloacetate is not translocated directly to the bundle sheath, but is first reduced to malate by NADP-dependent malate dehydrogenase (EC 1.1.1.82, NADP-MDH) in the mesophyll, a reaction with $\Delta G^{\circ\prime}$ of +30 kJ mol^{-1}:

$$\text{Oxaloacetate} + \text{NADPH} + \text{H}^+ \longrightarrow \text{malate} + \text{NADP}^+ \tag{9.2}$$

NADP-MDH from maize is a homodimer of 43 kDa subunits with a molecular activity of 60 000 at pH 8.5 and 25 °C. It maintains the reaction towards malate synthesis, has a high affinity (and is specific) for NADPH (24 µM) and OAA (56 µM), NADP$^+$ (73 µM) and malate (32 mM). The enzyme activity is also altered by the NADPH/NADP$^+$ ratio in the mesophyll in the light. The enzyme is light activated by the ferredoxin-thioredoxin *m* system (or by reduced substances *in vitro*), by reduction of a disulfide bridge (Section 7.3). Formation of malate is characteristic of maize and sugar cane, which have NADP malate dehydrogenase; malate is transported to the bundle sheath (*Figure 9.2c*).

In the bundle sheath, malate is decarboxylated in chloroplasts by NADP-specific malic enzyme (EC 1.1.1.40, NADP-ME), hence, this is the 'NADP-ME' type of C4 assimilation. The reaction has a $\Delta G^{\circ\prime}$ of -1.3 kJ mol^{-1}:

$$\text{malate} + \text{NADP}^+ \longrightarrow \text{pyruvate} + CO_2 + \text{NADPH} + H^+ \tag{9.3}$$

The reaction of reversible decarboxylation is 10-fold greater than carboxylation at pH 8.0 and 0.6 mM CO_2. NADP-ME from sugar cane is very active at pH 8.0 and 30°C. It is a homotetramer with 62 kDa subunits at pH 8.0, and active in the light, but is a homo-dimer at pH 7.0 and is rather inactive. It may be activated by light via the ferredoxin–thioredoxin system. There is requirement for Mg^{2+} for the very complex allosteric regulation with acetyl-CoA, FBP and Mn^{2+} ions.

Another decarboxylating enzyme of importance is PEP carboxykinase (EC 4.1.1.49, PEPCK), which reacts OAA with ATP in the presence of manganese ions to give PEP and ADP ($\Delta G^{\circ\prime} = 0$ kJ mol^{-1}). PEPCK is located in the bundle sheath cytoplasm, and is a hexamer of 68 kDa subunits and requires manganese. The K_m values for OAA, ATP and CO_2 are about 18 and 20 µm and 2 mM, respectively. It is activated in the light by phosphorylation.

Pyruvate from decarboxylation of malate is recycled to form PEP, by a reaction with ATP and P$_i$ catalyzed by pyruvate P$_i$ dikinase (EC 2.7.91, PPDK), in a reaction unique to C4 plants:

$$\text{pyruvate} + \text{ATP} + P_i \longrightarrow \text{PEP} + \text{AMP} + PP_i \tag{9.4}$$

The enzyme from maize leaves is a homodimer of 94 kDa subunits, molecular activity of 2600 at pH 7.5 and 22°C. It has K_m values (all µM) of 250 for pyruvate, 1500 for P$_i$, 15 for ATP, 140 for PEP, 40 for pyrophosphate and less than 10 for AMP. PEP is formed under cellular conditions, despite the reversibility of the reaction, because the pyrophosphate is removed by a pyrophosphatase and adenylate kinase. The enzyme reaction mechanism involves phosphorylation of histidine residues at the active site; in darkness the active enzyme with the phosphorylated histidine is phosphorylated at a threonine residue. Both the phosphorylation and dephosphorylation of the threonine are done by the same regulatory protein.

Although the PEPc- and PPDK-mediated reactions are common to all C4 species, there are differences in the metabolic routes by which pyruvate is formed, both in the enzymes and their location. A major difference is in the transport of aspartate rather than malate from mesophyll to bundle sheath. Aspartate is rapidly formed in mesophyll cells by transamination of oxaloacetate by aspartate aminotransferase, with glutamate as the amino group donor:

$$\text{OAA} + \text{glutamate} \longrightarrow \text{aspartate} + \alpha\text{-ketoglutarate} \tag{9.5}$$

The enzyme aspartate aminotransferase (EC 2.6.1.1) occurs in different isoforms. One is in the cytoplasm of mesophyll cells; it is a homodimer of 42 kDa subunits with molecular activity of 18 200 at its pH optimum of 8.0 and 25°C, and a high affinity for substrates (K_m for aspartate, α-ketoglutarate, glutamate and OAA of about 2, 0.2, 15 and 0.06 mM, respectively). However, it is inhibited by malate. In bundle sheath cells it deaminates aspartate. Aspartate is transported to the bundle sheath by a dicarboxylate shuttle. Alanine aminotransferase (EC 2.6.1.2) reacts alanine with 2-oxoglutarate, giving pyruvate and glutamate, in a reversible reaction. Several isoforms of the enzyme exist in plants, for example three in *Panicum miliaceum* (NAD-ME type).

$$\text{Alanine} + \text{2-oxoglutarate} \longrightarrow \text{pyruvate} + \text{glutamate} \tag{9.6}$$

There is distinction between C4 plants in the enzyme decarboxylating organic acids in the bundle sheath (*Table 9.1*). In malate formers, as already mentioned, decarboxylation is by NADP-requiring malic enzyme (ME) in the chloroplast of bundle sheath cells (NADP-ME type). In one group of aspartate formers, the PCK type, aspartate is probably deaminated in the bundle sheath cell cytosol by aspartate aminotransferase, giving OAA. This is phosphorylated by PEPCK to PEP, releasing CO_2 which is fixed in the PCR cycle (*Figure 9.2b*) and PEP is converted to pyruvate, which is transaminated to alanine (by a specific enzyme with glutamate as amino group donor). Alanine returns to the mesophyll cell where it is deaminated to pyruvate and phosphorylated to PEP.

The NAD-ME type of C4s converts aspartate to oxaloacetate which is reduced to malate by the bundle sheath cell mitochondria and decarboxylated by an NAD-specific ME (EC 1.1.1.39). It also is active in PEP-CK types:

$$\text{malate} + \text{NAD}^+ \longrightarrow \text{pyruvate} + CO_2 + \text{NADH} \tag{9.7}$$

There are a number of different isoforms of the enzyme; in *Elusine coracana* it is a homo-octomer of 63 kDa subunits which requires Mn^{2+}. The enzyme is strongly activated by acetyl CoA, and fructose 1,6-bisphosphate. The K_m for malate and NAD^+ of the enzyme activated with fructose 1,6-bisphosphate is 2 and 0.6 mM, respectively.

The NAD-ME type contrasts with the NADP-ME type, located in the chloroplast, discussed earlier. Pyruvate is converted to alanine by the specific aminotransferase and returned to the mesophyll cell cytosol for conversion to the acceptor.

9.2.3 Transport of metabolites

Movement of organic and amino acids between the mesophyll and bundle sheath cells is probably by diffusion through the plasmodesmata, which occupy about 3% of the surface of the cells and cross the densely suberized layer. This requires theoretical concentration gradients of metabolites of the order of between 10 mM, for the centrifugally arranged bundle sheath chloroplasts with short diffusion path in *Zea*, and 30 mM for centripetal chloroplasts with long diffusion pathway, e.g. in *Amaranthus*. However, experimental estimates of diffusion constants suggested that these gradients were overestimated 5–15-fold. Measurements of gradients are difficult because of the large contents of metabolites in non-photosynthetic tissues and in vacuoles. Gradients vary considerably (e.g. 10 mM for triose P but 43 mM for malate and 1 mM for pyruvate). Rates of diffusion of small molecular weight compounds is of the order of 2–5 mmol min^{-1} (mg chlorophyll) per mmol of gradient of concentration between the isolated bundle sheath and medium. However, there are intracellular transporters to carry the large fluxes of metabolites in C4 photosynthesis. C4 acids are carried by specific translocators; that carrying OAA from cytoplasm to chloroplast in NADP-ME types has a high affinity for OAA (K_m 45 mM) but very low affinity for malate. A C4 dicarboxylic acid transporter (similar to that in C3 plants and with a K_m of 0.5 mM for malate) carries malate from mesophyll chloroplasts to the cytosol. Pyruvate uptake occurs by a specific transporter on the mesophyll cell membrane. PEP is carried by a P_i translocator which also carries 3-PGA and triose phosphate. The capacity of C4 transporters is generally very much greater (up to 50-fold for PEP) than in C3s. The pyruvate transporter is substantially stimulated by light and the ΔpH. Bundle sheath chloroplasts have transporters that are similar to those of C3s: for phosphate, and glycolate, malate and pyruvate depending on the type of metabolism. Clearly, the differences between C3 and C4 metabolism and between the C4 types involve not only the mechanisms of CO_2 accumulation and fixation, but all aspects of the transport of metabolites. Therefore, importing the C4 syndrome into C3 plants by genetic manipulation requires very considerable alterations to many aspects of anatomy and biochemistry.

9.2.4 Energetics of C4 photosynthesis

C4 plants require an extra two molecules of ATP for fixation of one CO_2, viz. five ATP and two NADPH, compared with C3 plants because of the conversion of pyruvate to PEP by the action of pyruvate, P_i dikinase and adenylate kinase in mesophyll chloroplasts. The NADP-ME type requires two ATP and one NADPH in the mesophyll cells, giving five ATP and two NADPH per CO_2 assimilated, and this is the same for the NAD-ME type. Requirements of the PCK type are more difficult to assess, as the proportion of the PEP carboxykinase and NAD-malic enzyme are not known. Assuming that three ATP are produced by consumption of one NADH in the mitochondria, only 0.5 ATP is needed to make PEP per CO_2 fixed. This would mean that 3.5 ATP and 2.25 NADPH per CO_2 would be needed. However, these values may be too small and 3.6 ATP and 2.3 NADPH are likely.

Photochemical competence of C4 photosynthesis depends on many factors, including the leakiness of CO_2 from the bundle sheath, the costs of the regeneration of PEP and decarboxylation steps, plus, of course, the reduction of 3-PGA to triose in the PCR cycle. A consistent demand for 5.7 ATP and 2 NADPH per CO_2 fixed is obtained with different methods, including estimation of the flux of oxygen and CO_2, which are very close to 1:1 in many conditions. There is a very small rate of respiration: about 4% of O_2 uptake in NADP-ME types (which produce ATP by cyclic photophosphorylation, i.e. without O_2

release, and with very low photorespiration), 9% for NAD-ME types (which may have greater pseudocyclic photophosphorylation) and 22% in PEP-CK types (which produce ATP by mitochondrial respiration). These are small compared to C3 plants. The minimum quantum requirement of maize is about 17 quanta/CO_2 fixed (a maximum quantum yield of 0.059) and calculations based on the biochemistry of the processes are close to this. Maximum quantum yields of C3 and C4 plants are similar at 25°C at 36 Pa CO_2, as the costs of photorespiration in C3 and of the concentrating and carbon transport mechanisms in C4 balance. Increasing temperatures above this, at the same CO_2 favors C4 metabolism as does decreasing CO_2. The quantum yield of C3 plants in elevated CO_2 (100 Pa) is greater than that of C4s in 36 Pa because photorespiration is decreased or eliminated. However, C4s in very large CO_2 concentration (10%) have the same quantum yield per O_2 evolved as C3s in high CO_2, about 0.1.

Photochemical competence differs between forms of C4 plants (*Table 9.1*). Bundle sheath chloroplasts of NADP-ME types have little PSII and may synthesize ATP by cyclic photophosphorylation. This greatly decreases oxygen accumulation in bundle sheaths. NADPH is supplied in NADPME types from the mesophyll by a malate shuttle; for each malate decarboxylated 1 NADPH is released. If NADPH supply is insufficient for the PCR cycle, reductant might be provided by a shuttle of DHAP from the mesophyll chloroplast into the bundle sheath. DHAP is oxidized to 3PGA, giving NADPH and ATP.

$$DHAP + ADP + NADP^+ \longrightarrow 3PGA + ATP + NADPH \qquad (9.8)$$

The 3PGA from the bundle sheath is recycled back to the mesophyll to provide the substrate. Bundle sheath cells of NAD-ME types have some noncyclic electron flow to $NADP^+$; those of PCK types have less PSII activity and more cyclic photophosphorylation. Mesophyll chloroplasts of NAD-ME type have larger capacity for cyclic ATP synthesis than the NADP-ME types, which like the PCK types have both PSI and II. Aspartate formers probably have a normal balance of PSI and II activity in the bundle sheath chloroplast and appear not to require transport of reductant. The advantages of the modifications to the photochemical apparatus and of the complex shuttles of reductant and ATP are explained by advantages to particular forms under certain conditions, allowing survival. However, they may not compete directly so one type does not predominate. C4 plants are able to absorb incident radiation at high intensity and use it for CO_2 assimilation, but they are less efficient at low intensity. Light harvesting and distribution to the different photosystems may be more efficient if the functions are separated. ATP production by cyclic photophosphorylation (minimizing the production of O_2 and reductant) may overcome a limitation to the rate of PCR cycle activity, which is necessary for the efficient CO_2 accumulating system of PEP carboxylase to be fully exploited. Decreased PSII in bundle sheath chloroplasts decreases O_2 production, so that Rubisco, which has somewhat higher K_m for CO_2 in C4 compared to C3 plants, functions at low O_2 and high CO_2. PEPc is very efficient at assimilating CO_2 from very low concentrations, allowing CO_2 to accumulation in the bundle sheath. When organic acids from 10 mesophyll cells supply one bundle sheath cell, decarboxylation gives a 10-fold increase in CO_2 concentration. In air this may give a partial pressure of 150–300 Pa of CO_2; modeling (see Chapter 12) suggests a 15-fold increase. The concentration depends not only on the CO_2 inside the leaf and thus on the ambient CO_2, but on release of CO_2 inside the cells, and leakage from them, and therefore on the flux of C4 acids in the plasmodesmata from mesophyll to bundle sheath. Also, the solubility of CO_2 (principally affected by temperature) and particularly the diffusion (leakage) of CO_2 across the cell membrane and walls is important but difficult to estimate: measurements from isotope exchange, ^{14}C labeling, and O_2 sensitivity give 8–50% of the rate of C4 cycle assimilation. A 10–20% rate of leakage may be reasonable. Despite this, the CO_2 concentration can reach 30–75 μM, three to eight times larger than CO_2 in the chloroplast of C3 plants (see Kanai and Edwards, 1996). With very large external CO_2 concentration (10%, 10 kPa) the C4 mechanism may be bypassed and Rubisco fixes CO_2 atmospheric directly, thus avoiding the costs of the C4 cycle. PEPc and the mesophyll reactions are very efficient, providing Rubisco with CO_2 and ensuring that it is fully effective. When Rubisco activity is decreased by antisense methods, the bundle sheath CO_2 increases and PEPc is down-regulated, in part, showing that the C4 and C3 cycles are not tightly correlated.

Oxygen production in the bundle sheath chloroplast is small, at least in the NADP-ME type, but estimates of O_2 concentrations suggest that they may be considerably greater than ambient. However, as the CO_2/O_2 ratio is still very high, Rubisco oxygenase activity is much slower than in C3s. Thus, C4 photosynthesis is insensitive to O_2 and photorespiration is very small, probably less than 5% of the rate of CO_2 assimilation. Thus, C4 plants have smaller light and CO_2 compensation concentration than C3 plants, larger carboxylation efficiency and higher rates of photosynthesis. This is achieved with a smaller amount of Rubisco per unit leaf area. If C4 and C3 plants (e.g. maize and bean) are illuminated together in the chamber, the C4 plants remove CO_2 from the atmosphere more efficiently than the C3 species, which deplete their carbohydrate reserves and eventually die. This method has been used to try and identify 'C4-like' C3 plants, but without success. Thus, efficient photosynthesis and C4 metabolism which eliminates or minimizes photorespiration are linked. However, some phosphoglycolate is formed and there is carbon flux through the glycolate pathway. Under conditions such as water stress, where CO_2 supply is limited by stomatal closure, an increase in glycine and serine formation indicates that more carbon enters the glycolate pathway but photorespiration, measured as CO_2 production and from the difference in assimilation between low (1 kPa) and high (21 kPa) O_2, does not increase. As PEPc is efficient, any CO_2 produced is re-assimilated and only under extreme conditions does CO_2 escape from maize leaves in the light. C4 plants have a greater photosynthetic rate than C3 plants at smaller stomatal conductance, so that they have a larger water use efficiency (WUE = photosynthesis/transpiration) and smaller total water loss in drought-prone environments. This slows the onset of water stress and may decrease its severity under conditions of intermittent rainfall.

9.2.5 Ecology and evolution of C4 photosynthetic types

What is the ecological rationale of C4 metabolism? It imposes an additional energy burden compared to C3 photosynthesis. Also, there are large fluxes of assimilates across cell membranes, but probably transport between cells is by simple diffusion through plasmodesmata, and does not involve translocator or transport mechanisms.

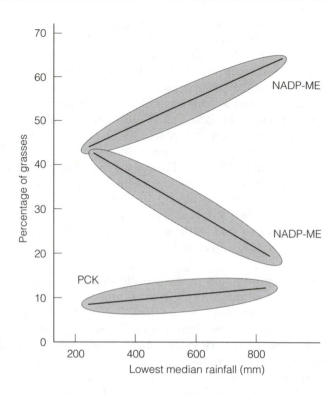

Figure 9.4. Illustration of the distribution of different metabolic types of C4 grasses in relation to the lowest median rainfall in Australia

C4 photosynthesis is superior to C3 because it allows efficient CO_2 assimilation; even in dilute CO_2 stomatal conductance is smaller in C4 than C3 plants, helping to conserve water, but the resulting low internal CO_2 concentration (see Chapters 11 and 12) has little effect on photosynthesis as PEPc is so efficient. Differences such as those in photosystems and chloroplast structure in C4 types probably relate to the requirements for balanced energy supply and reductant and the importance of the CO_2/O_2 ratio in their different environments. C4 plants predominate in bright sunlight in hot, often dry through to desert, conditions, where they form the greater proportion of species and produce most of the biomass. Hence they are very well represented in the tropics, where temperatures exceed about 16°C for the mean maximum warmest month and 6–12°C for the mean minimum warmest month, between the 20° North–South latitudes. C4s decrease markedly in species abundance and as a proportion of the biomass at higher latitudes where temperatures are cool. Grasses are the dominant C4 types and in areas such as the prairies of North America their representation is low in the cooler, wetter periods and increases greatly as the temperatures rise and drought increases. In particular, the NAD-ME types dominate in the more extreme hot, dry conditions (*Figure 9.4*), but the reasons for this are not understood. However, C4 species are also found in temperate and in shaded conditions. The C4 temperate grass *Spartina townsendii* is found in salt marshes, where C4 metabolism may be an advantage under salt and osmotic stress, and where low temperatures are tolerable and competition from C3 plants better adapted to cool, moist conditions is decreased by salinity.

Evolution of the C4 syndrome is a recent event in the history of plants. It only occurs in angiosperms, mainly in the monocotyledons, and predominantly in annual, herbaceous grasses, is also represented in a wide range of dicotyledonous families, but is not found in trees. There is no evidence (e.g. from carbon isotope ratios and anatomy) of C4-photosynthesis more than some 12.5 million years before the present but by 7 million years ago, $\delta^{13}C$ values of -10 to $-14\%o$ and a fossil C4 grass show that C4 metabolism was well established. The atmospheric CO_2 concentration in most of earth's history was considerably greater than that required to saturate Rubisco, despite the increase in oxygen concentration over the last 2 billion years. Only since photosynthesis fixed massive amounts of carbon, leading to the formation of coal, oil and gas in the Carboniferous period (360–280 million years ago) and removed large quantities of CO_2 from the atmosphere and increased the O_2 concentration did atmospheric CO_2 decrease substantially. However, the decrease in CO_2 to such small values that Rubisco oxygenase became important with the large O_2 concentration, is relatively recent, probably only in the last 20 million years, during the Pleistocene. Thus, the trigger for evolution of C4 photosynthesis appears to have been selection pressure from carbon availability under conditions where photorespiration becomes a large part of photosynthesis, for example in hot conditions and where water stress is prevalent, and oxygenation exceeds 20–30% of carboxylation. The 3–4-billion years of Rubisco evolution under very large CO_2 concentrations, much of it with low O_2, has optimized the enzyme so that a rapid change in direction, such as 'redesigning' the active site of Rubisco, required to overcome oxygenation of RuBP and thus remove photorespiration at source, is extremely difficult. Probably many changes would also be necessary in the regulation of other parts of the PCR cycle. It may not be possible to achieve such major changes in advanced organisms without jeopardizing their survival. However, the adaptation of existing structures and enzymes to form a more efficient system for CO_2 assimilation appears, from its development in many families relatively rapidly (10–20 million years or, assuming annual plants, generations), to be relatively 'easy' in evolutionary terms. Changes to gene expression have been of the greatest importance (see Monson, 1996), so that GDC activity is restricted to mitochondria in the bundle sheath. Altered PSII activity reduces O_2 production in the bundle sheath in some C4s, to increase activity of PEPc and carbonic anhydrase in certain tissues. However, some changes are insufficient to achieve the full C4 efficiency (e.g. C3–C4 intermediates with altered mitochondrial and GDC distributions, although more efficient in recycling CO_2 from photorespiration are not nearly as efficient as true C4 types). For the full syndrome anatomical changes have been very important, as they provide an enclosure for the large increase of CO_2 around Rubisco which is so important for greatly decreasing Rubisco oxygenation. The efficiency of C4 species has been achieved by extensive modification of the 'normal' C3 plant. Anatomy is differentiated at subcellular and cellular level and the enzyme complement is modified quantitatively (e.g. more PEP). Chloroplast structure and light reactions are modified to optimize ATP and NADPH supply. Attempts have been made to transfer C4 features by crossing C3 and C4 species of *Atriplex* [e.g. *A. triangularis* (C3) and *A. rosea* (C4)]. This gave fertile Fl hybrids, which were intermediate in morphology and anatomy, with a weakly developed bundle sheath. The enzymes were C4, with PEPc and Rubisco, but the enzymes were not compartmented as in the C4 parent, for example Rubisco was in all cells and not restricted to the bundle sheath. Also, C4 acids were synthesized but not effectively metabolized. This led to inefficient photosynthesis and growth.

9.3 C3–C4 intermediates

The C3–C4 intermediates are an interesting and important group of plants for understanding the evolution of the C4 mechanisms and of photosynthesis in general. They have characteristics of both metabolic types, with lower CO_2 compensation point and decreased photorespiration. They are regarded as in the process of evolution of the full C4 syndrome. In monocotyledonous and dicotyledonous plants, some 24 species, in eight genera, in six families are C3–C4 intermediates; all but two of these genera have true C3 and C4 types. In general these species are short-lived weedy herbs of dry, hot habitats. The genera *Panicum* (Poaceae) and *Moricandia* (e.g. *M. arvensis*, Brassicaceae) and species of *Flaveria* (Asteraceae) have very diverse features: two species (*F. pringlei* and *F. cronquistii*) are C3, four (*F. australasica, F. trinervia, F. bidentis* and *F. palmeri*) are C4 and five are C3–C4 (e.g. *F. brownii, F. floridana,* and *F. linearis*). *F. brownii* has some features more like C4 than the other two. Many intermediates have large PEPc CO_2 fixation and carbon isotope discrimination, but have Rubisco in mesophyll cells (although perhaps inactive). Other intermediates are very C3-like, with no PEPc activity, a C3 plant range of carbon isotope discrimination but with low CO_2 compensation point and decreased photorespiration. Probably a 'Kranz-type' anatomy of a distinct layer of cells around the minor, but not major, veins allows photorespiratory CO_2, which is evolved in the bundle sheath, to be absorbed by the surrounding chloroplasts. Kranz anatomy is better developed in C3–C4 species of *Moricandia* than in C3 species but not in C3–C4 *Flaveria* species. Generally mitochondria and peroxisomes are more concentrated in the bundle sheath, which is much more developed in intermediates of different genera than in the corresponding C3 species. P-glycolate formed in the mesophyll chloroplasts (*Figure 9.5*) is metabolized by the glycolate pathway to glycine that diffuses out of peroxisomes but is not consumed in the mesophyll mitochondria, which do not possess the P subunit of glycine decarboxylase (GDC). Rather it passes to the numerous mitochondria, on the inner wall of the bundle sheath, which have a large activity of GDC, and is metabolized to serine. The CO_2 released is re-assimilated by the bundle sheath chloroplasts. Serine is converted to hydroxypyruvate and thence to 3PGA. Consequently, the CO_2 compensation point and apparent photorespiration are decreased and photosynthesis enhanced.

Evolution of the intermediate species is, of course, complex and involves selection based on morphology and reproduction, not only on the characteristics of the photosynthetic system. However, it is possible that the morphological separation occurred after separation of the C4 and C3–C4 lines, as analysis of the genes for the H-protein of GDC and of the morphological criteria which separate the groups indicates (see Sage and Monson, 1996). Probably, C4 photosynthesis has evolved twice in different lines of *Flaveria* and it is likely that the C3–C4 intermediates have given rise to C4 types, as expected if C3 photosynthesis was the starting point for C4. The main changes in photosynthetic processes in C3–C4 intermediates is in the localization of the mitochondria and peroxisomes in a bundle sheath-like structure where CO_2 evolution from GDC in the mitochondria (arranged on the inner tangential walls of the bundle sheath) is concentrated. Also, the GDC in the mesophyll mitochondria of the C3–C4 intermediates is lacking the P-subunit and therefore inactive. So the metabolism of glycolate is restricted to the inner part of the vascular system. The diffusion of glycine to the bundle sheath and release of CO_2 there, coupled with location of the chloroplasts around

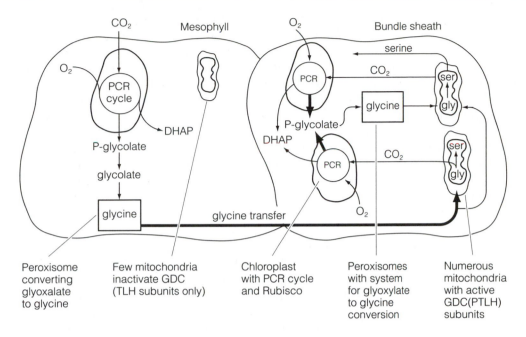

Figure 9.5. Schematic of carbon assimilation and photorespiration in the C3–C4 intermediate *Moricandia arvensis*

the mitochondria, increases the CO_2 concentration at the chloroplast, maximizing assimilation of CO_2. Thus, photorespiration appears to be decreased due to efficient recycling of CO_2 and the increase photosynthesis is due to the efficiency of assimilation by Rubisco. This increases the rate of assimilation by 30% compared to the C3 types at equivalent CO_2 concentration, decreases the compensation point from some 5 to 1–2.5 Pa and increases the quantum yield of photosynthesis. However, CO_2 assimilation is O_2-dependent, there is a post-illumination burst, and discrimination against the heavy isotope of carbon ($\delta^{13}C$) for C3–C4 intermediates is −22 to −31‰, the same as that of C3 plants, as is PEPc activity. Regeneration of PEP therefore is a factor limiting development of more C4-like characteristics. The C3–C4 intermediates are probably the first stage in development of C4 photosynthesis. In *Moricandia* there may have been a parallel development of anatomical specialization and development of a CO_2 pump, providing the next stage in C4 development. Water use efficiency is regarded of major importance for the success of C4 plants compared to C3 in hot, dry conditions: C3–C4 intermediates of *Flaveria* are similar to C3 species but the C4-like *F. brownii* had improved and the C4 *F. trinerva* had substantially improved water use efficiency, showing the advantage provided by the development of C4 characteristics.

The genome of C3 predecessors has been, presumably, modified as a result of evolutionary selection through heat, drought and low CO_2 partial pressure. Genetic processes responsible for the development of C4 characteristics and the role of selection pressure from the environment are not well understood, but whilst the main function of many genes has remained very similar to that in C3 plants, extensive modification to the regulatory sequences and to promoter regions has occurred. Changes to enhancing and silencing mechanisms, in the use of 5′ and 3′ untranslated regions to achieve very varied transcriptional and post-transcriptional regulation and evolution of novel transcription factors have all contributed to the rapid evolution of C4s. There is substantial cell and tissue specificity in the expression of particular proteins in types of cells, for example genes encoding for nitrate and nitrite reductase, ferredoxin-dependent glutamate synthase and glutamine synthase are preferentially expressed in mesophyll cells, whereas Rubisco genes are expressed in the bundle sheath, not mesophyll, particularly in the light. Light has substantial effects on the formation of the C4 syndrome, and is perceived by the red, blue and UV receptors. Genes for C4 metabolism are expressed in leaves but not stems or roots and are frequently associated with wide changes in *cis*-acting regulatory elements, but *trans*-acting factors are also involved. In leaves, there is coupling between vein development and the formation of the mesophyll and bundle sheath cells, again under light-regulation. In some species of higher plants, there is a transition between C3 and C4 metabolism depending on the environment. A sedge, *Eleccharis vivipara*, is C3 when submerged but when growing in dry conditions, its C4 genes are expressed resulting in fully developed krantz anatomy, enzymes and photosynthesis; the change can be induced by abscisic acid, which is a plant growth regulator induced by stress conditions.

C3–C4 intermediates provide valuable material for understanding evolution of photosynthetic characteristics, particularly the changes in genes and their expression associated with the altered anatomy and protein amounts and characteristics, which give rise to changes in response to environmental conditions. If C4 photosynthesis has arisen more than once (as the different metabolic types suggest) then present day C3 species may include individuals with more C4-like characteristics, which might be more tolerant of hot, dry or saline conditions and therefore be more productive. Considerable effort has been made to select C4 (or at least C3–C4) types of C3 agricultural crops for conditions but no high efficiency or tolerant plants have been obtained.

9.4   Crassulacean acid metabolism

Crassulacean acid metabolism is an adaptation to a shortage of water, with diurnal stomatal closure and generally extreme succulence. There are few CAM plants grown as food crops (in contrast to C4s). One is pineapple (*Ananas comosus*), which is very widely grown and accumulates very large biomass, up to 35 t ha^{-1} year^{-1}. Many CAM plants grow in areas which are hot (but often cold at night), under large photon flux. Epiphytic CAM species tend to grow in the upper canopy of forests, however some CAM species grow in deep shade (but may exploit sunflecks). It is a major physiological adaptive feature of plants probably developed from C3 plants before the C4 syndrome, and is mutiphylletic, arising in three groups, the Orchidaceae, Crassulaceae and Caryophyllales (including a common lineage for the Cactaceae and Mesembryanthaceae) in about 7% of all species. CAM plants have a PEPc CO_2 concentrating mechanism and the Rubisco carboxylating system in the PCR cycle. Both activities occur in the same mesophyll cells but are separated in time. CAM plants assimilate CO_2 and synthesize organic acids, mainly malate, during darkness when stomata open, offering a low resistance to CO_2 diffusion into leaves, but assimilate little CO_2 in the light when stomata are generally closed

(*Figure 9.6*). The internal structure of CAM succulents is characterized by very dense mesophyll with little air space (8% of total volume), which offers a large limitation to CO_2 flux (small internal conductance). In darkness, ambient CO_2 concentrations may be large and partially compensate for small conductance; leaves are also cooler and the vapor pressure of the air is much greater than in the light so the gradient of vapor pressure between the mesophyll cell surfaces in the leaf and the air is small, greatly decreasing the transpirational water flux. CAM imposes limitations on metabolism because of the need to separate the two phases of CO_2 accumulation and decarboxylation and PCR cycle activity. Also, the size of cells, and the large vacuoles, may affect metabolism compared to C3 plants.

9.4.1 CAM photosynthetic mechanism

CAM is a CO_2-concentrating mechanism primarily, which greatly increases CO_2 at the active site of Rubisco in some species. However, there is also substantial direct assimilation of CO_2 by Rubisco, so generalizations must be made with care. As in C4 plants the mechanism is based on carboxylation of the β-position of PEP, by PEPc, forming oxaloacetate, probably in the cytosol (*Figure 9.6*). PEPc is down-regulated when the tissue is illuminated during Phase II (*Figure 9.7*), although up to 50% of total daily CO_2 assimilation occurs in this phase. PEPc is regulated by reversible phosphorylation, which alters sensitivity to malate. In darkness the phosphorylated, malate-insensitive form phase is functional, and is dephosphorylated (perhaps rather slowly over hours) in the light during phase II. Probably PEPc remains active for much of the light period.

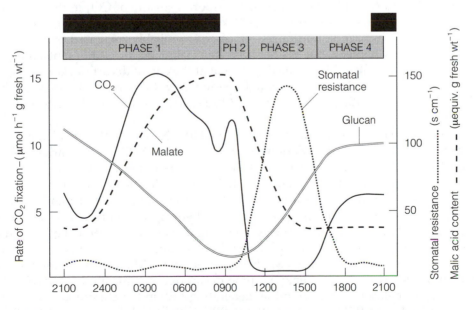

Figure 9.6. Diurnal course of assimilation, stomatal resistance and malate and glucan accumulation in a generalized CAM plant, showing the nocturnal fixation of CO_2 and accumulation of malate at the expense of stored glucan; well-watered plant with stomatal opening in darkness shown by black bars.

Oxaloacetate from the PEPc reaction is reduced by NADH to malate by cytoplasmic (and also mitochondrial) NAD-malate dehydrogenase. Malic acid is transported as hydrogen malate across the tonoplast and accumulates (150 µ equivalents per gram fresh weight) in the vacuole; there may be other pools in leaf cells. The pH of the vacuole becomes very acidic, although metabolism is protected by compartmentation. PEP is derived from storage carbohydrate, probably glucans (starch and a dextran, a glucose polymer), mobilized to produce 3PGA and PEP; ATP is consumed and synthesized in these reactions so that the energy consumption is balanced. NADH from oxidation of glyceraldehyde phosphate reduces oxaloacetate; as with ATP, storage carbohydrate supplies the energy for CO_2 fixation in the dark. Early in the dark period malate accumulation and CO_2 fixation are slow, but they increase in the middle of the dark period before decreasing. This pattern results from stimulation of CO_2 fixation by a higher PEP concentration which overcomes the inhibition of PEPc by malate. Phosphofructokinase which converts fructose-6-phosphate to fructose bisphosphate is 100-fold less sensitive to PEP than that of C3 plants.

With illumination, the stomata of CAM leaves may remain open for a period and CO_2 fixation by PEPc and Rubisco may occur. Generally this CO_2 fixation from the atmosphere decreases rapidly to a very low rate as the stomata close. Malate is transported out of the storage compartment and is decarboxylated by NADP malic enzyme, and perhaps NAD malic enzyme, which is found in many families with CAM, or by PEP carboxykinase, which is more important in others (e.g. Bromeliaceae and Liliaceae). They are designated NADP malic enzyme and PEP carboxykinase types, respectively. The CO_2 released, probably in the cytosol, enters the chloroplast and is assimilated by the PCR cycle with ATP and NADPH from electron transport. Pressure of CO_2 in the tissue may reach 2500 Pa and oxygen rises from 21 to 26 kPa, but the RuBP carboxylase reaction is favored and photorespiration minimized. Rubisco is probably regulated by the mechanisms discussed in Chapter 8, but has not been well examined for CAM plants. It is probably increasingly activated during the day, so that at the end of decarboxylation it would be able to use external CO_2. In the early light period, low Rubisco activity may be a way of preventing conflict with slowly deactivated PEPc, which is much more efficient.

Malate decarboxylation is controlled by a low $NADP^+/NADPH$ ratio, as $NADP^+$ is the acceptor for reductant. Decarboxylation could occur with $NADP^+$ in the dark but ATP is required for decarboxylation and is available (in the required amount) only in light. Pyruvate is phosphorylated to PEP and recycled to triosephosphate and thence to carbohydrates (mainly glucan), which are stored. PEP carboxykinase CAM plants generate oxaloacetate from malate and NADH by malate dehydrogenase in the cytosol (or possibly chloroplast). In the later part of the light period, malate may become depleted, intercellular CO_2 concentration falls and photosynthesis (within the tissue) is decreased. In darkness the cycle starts again. The two distinct types of CAM metabolism are the malic enzyme and PEPCK. The malic enzyme type decarboxylates malate by NADP-ME or NAD-ME in the cytosol or mitochondria, generating PEP in the chloroplast from pyruvate by PPDK. In the PEPCK type, without PPDK, the OAA is converted to PEP by PEPCK in the cytosol. There is considerable variation in the storage carbohydrates (e.g. sucrose and other soluble sugars, starch, fructan or galactomannan). Some eight combinations of decarboxylation enzymes and storage reserves have been identified. Also the

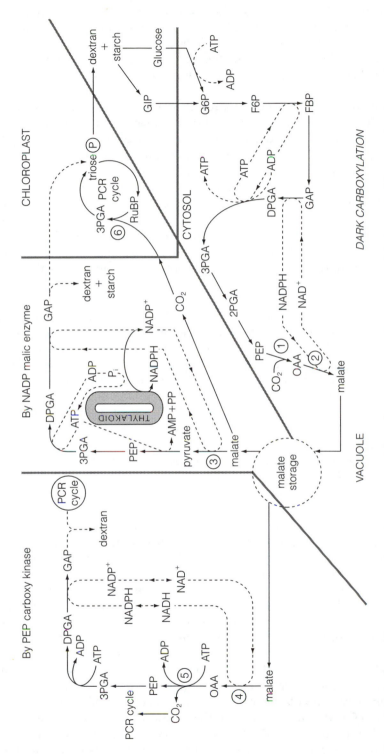

Figure 9.7. Scheme of crassulacean acid metabolism (CAM) showing the reactions responsible for dark CO_2 assimilation with formation of organic acid, storage and subsequent decarboxylation in the light. Numbers refer to the following enzymes: (1) PEP carboxylase; (2) malate dehydrogenase; (3) NADP malic enzyme; (4) NAD malic enzyme; (5) PEP carboxykinase; (6) RuBP carboxylase/oxygenase

intensity of the different phases depends on species and environment. Early in the light period there is little CO_2 fixation because PEP carboxylase is inhibited by malate, but as this is decarboxylated the PEPc fixes more CO_2 in the light. After long periods in the light and extensive decarboxylation, CO_2 from the atmosphere is assimilated directly by Rubisco and the PCR cycle, giving rise to storage glucans.

Photorespiration occurs in CAM plants as a result of a low CO_2/O_2 ratio, when decarboxylation has depleted the storage materials and under severe stress. When stomata do not open there is CAM idling, with decreased but active acid synthesis and CO_2 assimilation, which may provide an important protection against photoinhibition. The coincidence of decarboxylation with periods of very intense photon flux may enable this energetically unfavorable process to operate effectively. During decarboxylation, the small stomata and internal conductances slow loss of CO_2, but as electron transport continues, so O_2 concentration rises, perhaps as high as 42 kPa (42%). This may increase photorespiration and the risk of generating active oxygen. Also, the Mehler-ascorbate reaction may increase and contribute to decreasing over-reduction of the electron transport system. Unsurprisingly therefore, large xanthophyll cycle activity is also observed in many CAM plants.

The evolutionary pressures and molecular changes that accompanied the development of CAM are now being actively examined. CAM evolved earlier than C4 and is more widespread, but the possible differences in the mechanisms of adaptation in the two groups are only now being analyzed. CAM are very varied with great adaptation to particular environmental conditions. Facultative CAM plants offer an interesting model system for understanding regulation of gene expression under different environmental conditions. One, *Mesembryanthemum crystallinum*, the ice plant, is C3 under cool, well-watered conditions, but increasing salinity and seasonal temperature changes with decreasing water supply induce CAM metabolism. For example, dehydration and osmotic stress as well as sodium and chloride ions stimulate (by transcriptional activation which controls mRNA accumulation) the expression of *Ppc*1, a gene encoding a CAM-specific isoform of PEPc which is encoded by a small gene family.

The metabolic plasticity and capacity to cope with large changes in fluxes of metabolites in CAM plants is achieved by expression of enzymes, including those of primary metabolism but also for secondary carbohydrate metabolism and transport, particularly isoforms of enzymes and carrier proteins specific to CAM. Possibly these genes are the result of gene duplication, recombination and changes in the promoters and regulatory mechanisms of ancestral C3-type PEPc. Synthesis of mRNAs occurs very quickly in response to salinity (also to drought or other stresses). Control of gene expression is often at transcription, controlled by interacting *cis*-acting sequences and *trans*-acting factors. Transcription rates are very dependent on species, stress and the protein; in *M. crystallinum* rates increase six-fold on exposure to sodium chloride. Protein synthesis is likewise variable, and is regulated after transcription and translation. There is considerable complexity in regulation of protein synthesis, as well as in activation and metabolic control, in CAM plants to achieve the temporal sequences. Indeed, there is pronounced circadian rhythm in CAM metabolism which may persist under constant conditions for a long period. The basis for this is not understood, and probably involves events in signaling within the photosynthetic system and outside it. As with C4 plants, but to a less obvious extent, proteins in CAM plants may be preferentially expressed in different cells and tissues, for example photosynthetic enzymes are expressed in the photosynthetic mesophyll not in the water-holding parenchyma, which is such a marked feature of many CAM plants.

CAM permits CO_2 accumulation in darkness when stomata are open, without substantial loss of water, as the plants are predominantly from arid habitats with hot, dry and very bright days, which would cause much water loss if stomata were open. Also, CAM plants occur frequently as epiphytes in forest canopies, where water supply may be erratic and seasonal, and there is little light. Often, growth rates in such environments are slow, and survival in very prolonged drought is a prerequisite. However, CAM plants can grow rapidly and be very productive and dominant in adverse environments. A very high ratio of CO_2 fixation to water loss is achieved under conditions that are lethal to most C3 plants. Energy accumulated in the light is used to generate the substrate for the PEPc reaction at night when low temperatures (characteristic of desert and alpine environments) favor malate accumulation. Large vacuoles enable CAM plants to store both malate and water, which are obviously important as a buffer against desiccation. There is considerable complexity in regulation of the different aspects of metabolism within the same cells at different times and under a range of conditions. The CAM system is characterized by great plasticity, which confers survival ability in extreme environments but with little competition.

References and further reading

Apel, P. (1994) Evolution of the C4 photosynthetic pathway: a physiologist's point of view. *Photosynthetica* **30**(4): 495–505.

Borland, A.M., Maxwell, K. and Griffiths, H. (2000) Ecophysiology of plants with Crassulacean acid metabolism. In *Photosynthesis: Physiology and Metabolism* (eds R.C. Leegood, T.D. Sharkey and S. von Caemmerer). Kluwer Academic, Dordrecht, pp. 583–605.

Ceerling, T.E., Wang, Y. and Quade, J. (1993) Expansion of C4 ecosystems as an indicator of global ecological change in the late Miocene. *Nature* **361**: 344–345.

Chollet, R., Vidal, J. and O'Leary, M.H. (1996). Phosphoenolpyruvate carboxylase: a ubiquitous, highly regulated enzyme in plants. *Annu. Rev. Plant Physiol. Mol. Biol.* **47**: 273–298.

Cushman, J.C. and Bohnert, H.J. (1999) Crassulacean acid metabolism: molecular genetics. *Annu. Rev. Plant Physiol. Plant Mol. Biol.* **50**: 305–332.

Doulis, A.D., Debian, N., Kingston-Smith, A.H. and Foyer, C.H. (1997) Differential localization of antioxidants in maize leaves. *Plant Physiol.* **114**: 1031–1037.

Furbank, R.T. (1998) C4 photosynthesis. In: *Photosynthesis: a Comprehensive Treatise* (ed. A.S. Raghavendra). Cambridge University Press, Cambridge, pp. 123–135.

Hatch, M.D. (1999) C4 photosynthesis: a historical overview. In *C4 Plant Biology* (eds R.F. Sage and R.K. Monson). Academic Press, San Diego, CA, pp. 377–410.

He, D. and Edwards, G.E. (1996) Estimation of diffusive resistance of bundle sheath cells to CO_2 from modeling of C4 photosynthesis. *Photosynth. Res.* **49**: 195–208.

Henderson, S., Hattersley, P. , von Caemmerer, S. and Osmond, C.B. (1994) Are C4 pathway plants threatened by global climatic change? In E*cophysiology of Photosynthesis* (eds E.-D. Schulze and M.M. Caldwell). Ecological Studies 100. Springer, Berlin, pp. 529–549.

Kanai, R. and Edwards, G.E. (1996) The biochemistry of C4 photosynthesis. In *C4 Plant Biology* (eds R.F. Sage and R.K. Monson). Academic Press, San Diego, CA, pp. 49–88.

Leegood, R.C. and Edwards, G.E. (1996) Carbon metabolism and photorespiration: temperature dependence in relation to other environmental factors. In: *Photosynthesis and the Environment* (ed. N.R. Baker). Advances in Photosynthesis Series. Kluwer Academic, Dordrecht, pp. 191–221.

Lüttge, U. (1997) *Physiological Ecology of Tropical Plants.* Springer, Heidelberg, p. 384.

Lüttge, U. (1998) Crassulacean acid metabolism. In *Photosynthesis: a Comprehensive Treatise* (ed. A.S. Raghavendra). Cambridge University Press, Cambridge, pp. 136–149.

Monson, R.K. (1996) The origins of C4 genes and evolutionary pattern in the C4 metabolic phenotype. In *C4 Plant Biology* (eds R.F. Sage and R.K. Monson). Academic Press, San Diego, CA, pp. 377–410.

Monson, R.K. and Rawsthorne, S. (2000) CO_2 assimilation in C_3–C_4 intermediate plants. In *Photosynthesis: Physiology and Metabolism* (eds R.C. Leegood, T.D. Sharkey and S. von Caemmerer). Kluwer Academic, Dordrecht, pp. 533–550.

Nelson, T. and Langdale, J.A. (1992) Developmental genetics of C4 photosynthesis. *Annu. Rev. Plant Physiol. Plant Mol. Biol.* **43**: 25–47.

Orsenigo, M., Patrignani, G. and Rascio, N. (1997) Ecophysiology of C3, C4 and CAM plants. In *Handbook of Photosynthesis* (ed. M. Pessarakli). Marcel Dekker, New York, pp. 1–25.

Osmond, C.B. and Grace, S.C. (1995) Perspectives on photoinhibition and photorespiration in the field: quintessential inefficiencies of the light and dark reactions of photosynthesis? *J. Exp. Bot.* **46**: 1351–1362.

Osmond, C.B., Maxwell, K., Popp, M. and Robinson, S. (1999) On being thick: fathoming apparently futile pathways of photosynthesis and carbohydrate metabolism in succulent CAM plants. In *Plant Carbohydrate Biochemistry* (eds J.A. Bryant, M.M. Burrell and N.J. Kruger). BIOS Scientific Publishers, Oxford, pp. 184–200.

Rawsthorne, S. and Bauwe, H. (1998) C3–C4 intermediate photosynthesis. In *Photosynthesis: a Comprehensive Treatise* (ed. A.S. Raghavendra). Cambridge University Press, Cambridge, pp. 150–162.

Sage, R.F. (1996) Why C4 photosynthesis? In *C4 Plant Biology* (eds R.F. Sage and R.K. Monson). Academic Press, San Diego, CA, pp. 3–16.

Sage, R.F. and Monson, R.K. (eds) (1996) *C4 Plant Biology*. Academic Press, San Diego, CA.

Sage, R.F. and Pearcy, R.W. (2000) The physiological ecology of C4. In *Photosynthesis: Physiology and Metabolism* (eds R.C. Leegood, T.D. Sharke and S. von Caemmerer). Kluwer Academic, Dordrecht, pp. 497–532.

Sheen, J. (1999) C4 gene expression. *Annu. Rev. Plant Physiol. Plant Mol. Biol.* **50**: 187–217.

Smith, B.N. (1997) The origin and evolution of C4 photosynthesis. In *Handbook of photosynthesis* (ed. M. Pessarakli). Marcel Dekker, New York, pp. 977–986.

t'Hart, H. and Eggli, U. (1995) *Evolution and Systematics of the Crassulaceae*. Backhuys Publishers, Leiden, p. 192.

Ting, I.P. (1985) Crassulacean acid metabolism. *Annu. Rev. Plant Physiol.* **36**: 595–622.

von Caemmerer, S., Millgate, A., Farquhar, G.D. and Furbank, R.T. (1997) Reduction of ribulose-1,5-bisphosphate carboxylase/oxygenase by antisense RNA in the C4 plant *Flaveria bidentis* leads to reduced assimilation rates and increased carbon isotope discrimination. *Plant Physiol.* **113**: 469–477.

Winter, K. and J.A.C. Smith (eds) (1996) *Crassulacean Acid Metabolism*. Springer, Berlin.

Molecular biology of the photosynthetic system

10.1 Introduction

The photosynthetic apparatus has been considered in this book at different levels of structure and function. Of central importance to the understanding of how those structures are formed and how they function, is analysis of the genetic and molecular biological processes which ensure the development of such a complex and efficient (in terms of self-replication if not energy use) photosynthetic system. Determination of function by structures derived from expression of genetic information within the organisms is a central tenet of biology, including plants (Feifelder, 1987). It is a considerable intellectual challenge, and of great potential practical importance, to understand how a functional photosynthetic system is achieved. How is the genetic information organized? What does it code for, how is it transcribed and translated into the components of the system? How are the components assembled into the final structures and what are the mechanisms? In a mature photosynthetic system how is regulation achieved by genetically determined mechanisms? Renewal of short-lived or damaged components and regeneration and senescence processes are regulated in photosynthetic tissues; how do genetic and metabolic processes interact to achieve correct integration? Such genetic–molecular–physiological questions are being asked and answered. These questions are general for all organisms, and are dealt with in the discipline of molecular biology rather than as part of a study of photosynthesis, but without such understanding an important facet of the process and an important tool for analyzing higher-level photosynthetic processes is lost.

Plants have unique features of genetic organization, with DNA in the chloroplast as well as the nucleus and mitochondria. The nucleotide base sequences of DNA in the genome contain the information required for synthesis of proteins with structural, catalytic and transport functions, which make the complex structures of photosynthetic organisms, for example light-harvesting complexes, the stromal enzyme Rubisco and the phosphate translocator in chloroplasts. Enzymatic proteins synthesize other components of the photosynthetic system (e.g. lipids and energy-gathering pigments of thylakoid membranes). For active transcription and translation of genetic information into proteins, the components of the protein-synthesizing machinery (e.g. ribosomal proteins, rRNA) are required. Some aspects of the molecular biology of photosynthesis are considered to provide a brief guide to the whole process from the plant's genome through to formation of the chloroplast. Detailed discussion of many of the topics touched upon here is contained in the references. Whilst concentrating on the photosynthetic system, it should not be forgotten that photosynthetic organisms depend on the regulated expression of other, nonphotosynthetic, features of their genome which determine their ability to function in a complex environment. Here, the higher plant chloroplast and nuclear genomes, the links between them and how this genetic information leads to the photosynthetic system, are considered and discussed. The effects of environment, particularly light, on the mechanisms and regulatory processes in gene expression required for development of the photosynthetic system are outlined.

10.2 Cellular organization and genetic information

Photosynthesis occurs in both prokaryotes and eukaryotes. The higher plant chloroplast is derived from a form of oxygen-evolving prokaryotic cyanobacterium, incorporated into a eukaryotic cell in a symbiosis between early prokaryote and eukaryotic organisms in the course of evolution (Reumann and Keegstra, 1999). There are great similarities, but also substantial differences, between chloroplasts and cyanobacteria in their organization. These differences lie not only in photosynthetic structures and metabolism, but in the way their genetic material is organized. The genome of prokaryotes is not contained within a nucleus, and neither is the photosynthetic apparatus separated from the body of the cell, as it is in eukaryotes with their membrane-bound nuclei, chloroplasts and mitochondria which all contain DNA. Nuclear, chloroplastic and mitochondrial DNAs are referred to as nDNA, cpDNA and mtDNA, respectively. The eukaryotic nuclear genome is separated from the protein synthesizing machinery in the cytosol by membranes, but eukaryotic chloroplasts, as in prokaryotes, contain the necessary components for producing some of the types of proteins of which they are made. The others are made in the cytosol, and encoded in the nucleus. Eukaryotes have complex interactions between the organellar genomes and their products. Nuclear genes for proteins found exclusively in chloroplasts, for example, are transcribed in the nucleus and the mRNA species produced are exported to the cytosol and translated there; the proteins are transported into the chloroplast. Transport of chloroplast-coded proteins from the plastid into the cytosol or nucleus has yet to be demonstrated. Chloroplast genomes and the proteins encoded by the genes, together with development of chloroplasts in photosynthetic eukaryotes, will be considered in some detail, but the nuclear genome is only briefly discussed in relation to photosynthesis, although it provides most of the information required for the photosynthetic system and the plant's development. The mitochondrial contribution is ignored. Here the discussion is restricted, for reasons of space and coherence, largely to photosynthetic components.

10.3 Genetics of nuclear and chloroplast information

Many of the characteristics of higher plants are transferred between generations with segregation of nuclear genes and their combination according to Mendel's laws of inheritance. However, certain features of higher plants, such as pigmentation of leaves, are non-Mendelian, as they are derived from the maternal germline and cytoplasmically transmitted, and thus maternally inherited. Part of the plant's genome is chloroplastic and this DNA is transferred from generation to generation via the plastids of the maternal parent. Plastids may not be transferred through the pollen tube to the zygote or maternal cytoplasm in the egg cell, preventing expression of paternal traits. Inheritance of plastid characteristics may also be biparental. Information from both parents is incorporated through the plastids, although to different degrees, depending on the species of plant. In zonal *Pelargonium* (*Pelargonium* × *Hortorum*) biparental control of chloroplast pigmentation (chlorophyll content, green or white plastids) is under nuclear control via regulation of plastid DNA replication. In approximately two-thirds of about 60 higher plant genera examined, some components of the photosynthetic system are maternally inherited; the remaining third have biparental inheritance. There is no simple taxonomic basis for this. Regulation of the mode of inheritance appears to change easily with only a few genes involved. Different forms of inheritance of specific chloroplast components have now been shown, even within an enzyme complex. For example, the one to four different forms of the small subunit (SSU) protein of ribulose bisphosphate carboxylase-oxygenase (Rubisco, see Chapter 7) are nuclear encoded and inherited in a Mendelian fashion. However, the three forms of the chloroplast encoded large subunit (LSU) protein of the same enzyme are maternally inherited in *Nicotiana gossei* x *N. tabacum* crosses. This was demonstrated by using differences in behavior of the polypeptides under isoelectric focussing.

10.4 Organization of the chloroplast genome in higher plants

Genetic continuity between generations via the maternal germ line and the plastids was eventually associated with the physical location of DNA in organelles. Chloroplasts contain DNA, shown by staining with Feulgen and other indicator dyes and by optical microscopy. Also, areas of low electron density in chloroplasts are associated with DNA. Isolation of cpDNA from plastids, free of nuclear DNA contamination, was eventually achieved despite the large, coiled cpDNA molecule, the strands of which sheer during isolation and are also cut by nuclease activity in plants. Information about the chloroplast genome, the molecular biology of photosynthesis and understanding of the way in which the nDNA and cpDNA contribute to the formation of structures in the photosynthetic system have grown dramatically following identification of the separate forms of DNA in 1962. Amounts and characteristics of cpDNA have been analyzed in many organisms, including photosynthetic bacteria and algae. Complete genome analysis (nucleotide sequences) of cpDNA from at least six unicellular eukaryotic algae [e.g. the photosynthetic green alga *Chlorella ellipsoidea* and the red alga *Porphyra purpurea*], a liverwort (*Marchantia polymorpha*) and several vascular plants, including tobacco (*Nicotiana tabacum*), maize (*Zea mays*), rice (*Oryza sativa*), a pine (*Pinus thunbergii*) and some legumes, has been carried out. The structure of cpDNA and details of specific proteins for which it encodes are now well understood. Approximately 120 highly conserved genes occur in all. A group of about 50 genes encode components of plastid gene expression (e.g. ribosomal proteins). Another 40 genes encode components of the photosynthetic apparatus (e.g. of photosystems, ATP synthase). A third group encodes for components which are of poorly known function, but are required for accumulation and proper functioning of the photosynthetic complexes, and some for cell survival (see Rochaix, 1997). Despite the chloroplast's ability to make some of its own components, its competence is very limited; no thylakoid complex is made entirely of proteins encoded in cpDNA – protein products of nDNA are always involved. The large subunit of Rubisco is the only part of any photosynthetic carbon reduction cycle enzyme encoded by cpDNA (Chapter 7). In addition, protein synthesis in the chloroplast involves nDNA as well as cpDNA products. Chloroplasts depend upon the nucleus for a major part of their protein complement, which is of the order of 400 different proteins (both structural and functional). In theory, with an average genome of 150 kbp, each cpDNA could code for 120–130 products (only 20–30% of the total complement). Some 100 polypeptides are made by isolated chloroplasts with 80 stromal, 20 thylakoid and a few envelope components; the remaining polypeptides derive from genes encoded by nuclear DNA. Cytosolic proteins, encoded by nDNA, are integrated into the chloroplast to provide the effective photosynthetic structures. Components encoded by the cpDNA are synthesized in the chloroplast. Methods by which proteins are transported between different parts of the cell have been described and much has been discovered about the transport mechanisms (Robinson and Mant, 1997; Robinson *et al.*, 2000).

Following early success in isolating chloroplasts which could carry out photosynthetic carbon metabolism, the concept of chloroplast autonomy prevailed and attempts were made to maintain chloroplasts *in vitro* by supplying only simple 'nutrients', such as inorganic phosphate and CO_2. However, higher plant chloroplasts are not autonomous. A constant supply of many structural components is needed for their function. They are intimately integrated developmentally as well as metabolically into the life of the cell, despite their complexity and separation from the cytosol. Following symbiosis of the two organisms, integration into a very tightly regulated system has presumably been achieved by the evolution of complex biochemical control mechanisms and appropriate genetic mechanisms, including redistribution of genes from 'chloroplast' to nucleus (Gray, 1996). There are many reasons why chloroplasts are regarded as prokaryotic organelles whereas the nucleus and cytosol are eukaryotic. There are, for example, differences in the transcription of DNA into mRNA by RNA polymerase, which binds to a region of DNA called a gene promoter – a sequence of bases recognized by the α-subunit of polymerase. In prokaryotes there are several known promoter sequences, such as the Pribnow 'box' sequence of bases, variations on TATAAT, acting as transcription or regulatory sequences. Promoter regions of eukaryotic genes have the TATA or Hogness box and enchancer elements, sequences which regulate the intensity of gene expression. In eukaryotes a number of specific transcription factors are needed plus a larger range of RNA polymerases; the primary mRNA transcript is usually much longer than the mRNA on which the protein chain is made and the intervening noncoding base sequences ('introns') are excised and the remaining coding exons are rejoined or spliced by complex enzymatic systems before the protein is made. RNA polymerases of bacteria and chloroplasts are homologous, suggesting common early ancestry. In prokaryotes, genes of related functions are often adjacent and polycistronic transcription is more common than monocistronic, enabling the proteins of a metabolic sequence of enzymes to be made in one event (and presumably in a fixed stoichiometry) from the 'reading frame' of the RNA. Eukaryotic genes, in contrast, are often monocistronic and in different parts of the genome. Their transcription therefore requires more complex regulation with a single environmental or metabolic stimulus acting as a signal coordinating both timing and stoichiometry (see Stern *et al.*, 1997).

Translation of mRNA differs in prokaryotes and eukaryotes. In the former, chain initiation requires N-formylmethionine transfer ribonucleic acid, tRNA^{fmet}, but the latter use methionine tRNA, tRNA^{met}. Differences are observed in the ribosomes of prokaryotes and eukaryotes, which are 70S (2.5×10^6 molecular mass) and 80S (4.3×10^6), respectively, made up of different sizes of large and small subunits, numbers and types of proteins and rRNA molecules. Ribosomes of prokaryotes are of 50S and 30S subunits whereas those of eukaryotes are composed of 40S and 60S subunits. Ribosomal functions are also affected differently in the prokaryotes and eukaryotes by antibiotic inhibitors of protein synthesis, for example prokaryotic ribosomes are inhibited by chloramphenicol but not by the cytoplasmic ribosomal inhibitor cycloheximide. Hence, many aspects of chloroplast gene expression are prokaryotic.

10.4.1 The chloroplast genome

Chloroplasts are partially autonomous genetically, containing DNA which replicates independently of other parts of the genome. They are highly polyploid organelles, containing many identical copies of cpDNA per chloroplast (in wheat up to 300). Molecules of cpDNA are often aggregated and linked to the thylakoid membranes forming 'nucleoids' resembling the organization of the bacterial genome. Most mature chloroplasts have 10–20 nucleoids with 2–24 cpDNA molecules in each. During cell and plastid division, nucleoids are transferred with thylakoids more or less equally into daughter chloroplasts. The location of cpDNA molecules in chloroplasts is species dependent, for example centrally in dicotyledons and peripherally in monocotyledons. As leaves expand (as a consequence of cell growth) cpDNA is synthesized faster than the chloroplasts divide and the number of copies per chloroplast increases and may rise to 12 000 per cell. Later, chloroplasts divide without further cpDNA synthesis, decreasing the cpDNA content per chloroplast. In a cell with, say, 20 chloroplasts, each with 20 nucleoids containing 20 cpDNA genomes, a total of 8000 copies of each single gene exist. This contrasts with a single nucleus per cell and relatively few copies of nuclear genes even in polyploid plants. The cpDNA is a double-stranded, single covalently closed circular DNA molecule (*Figure 10.1*), of density 1.679–1.699, with a molecular mass of 82–96 million Dalton (MDa) and a circumference of 40–50 µm depending on the species of plant. It is tightly coiled in a spiral with a large proportion of adenine and thiamine (61–64%), suggesting a high content of noncoding nucleotide base sequences. DNA of chloroplasts is not associated with histones, as is DNA in nuclei. Differences in size between the largest and smallest cpDNAs are considerable – in algae from 81 to 275 kbp and, in the angiosperms, 111 to 182 kbp in monocotyledonous plants and 117 to 165 kbp in dicotyledons. This is about 10 times greater than the mitochondrial genome of mammals, twice that of yeast but only a twenty-fifth of the *E. coli* genome and one-thousandth the size of the very small nuclear genome, of the angiosperm, *Arabidopsis thaliana*. Recent advances in sequencing the base structure of organisms have been substantial and the genomes of yeast, rice and *Arabidopsis* are now completed or nearing completion.

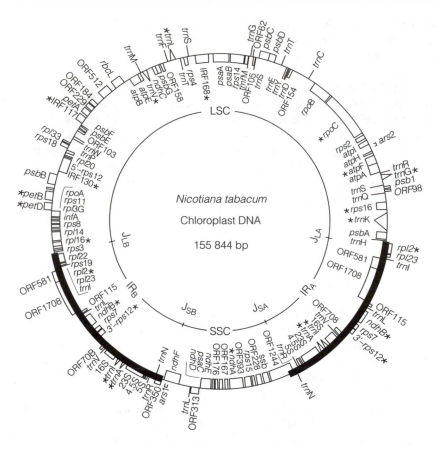

Figure 10.1. The tobacco chloroplast genome is a circular, double-stranded molecule. Genes are indicated, e.g. *rbc*L for the large subunit of Rubisco, *atp* for genes encoding proteins of the coupling factor complex. Genes on the outside of the circle are on the A strand and are transcribed anti-clockwise. Genes inside the circle are encoded on the B strand and transcribed clockwise. The inverted repeats (IRA and IRB) are shown as heavy lines. The positions of the small single copy (SSC) region and large single copy (LSC) region are indicated. Some of the genes shown are identified in *Table 10.1* and discussed in the text. (Reproduced from *Molecular Biology of Photosynthesis* (1998) pp. 1–25, Govindjee *et al.*, Fig. 1, with kind permission of Kluwer Academic Publishers)

10.4.2 Chloroplast DNA gene maps

Gene maps for cpDNA have been constructed by using restriction endonucleases to cut the molecule at particular sites, followed by restriction fragment length polymorphism (RFLP) analysis. These maps are complete for a number of plants, including the bryophyte *Marchantia*, and higher plants such as tobacco and rice. Structures of the higher plant cpDNA molecules differ in details but they are basically very similar, with considerable homology in the position and location of the coding sequences and their products.

The tobacco genome.   A gene map for *Nicotiana tabacum* is shown in *Figure 10.1*. Shinozaki *et al.* (1989) sequenced the entire cpDNA by cloning a set of overlapping restriction fragments of cpDNA which is 155 844 bp in size. One strand of the circular molecule is called the A strand and is transcribed anti-clockwise. The gene for the large subunit of Rubisco, *rbc*L, is on the A strand. The complementary strand (B) is transcribed clockwise. This cpDNA shows a characteristic feature of most higher plant cpDNAs examined to date (legumes are the exception) viz. the presence of two large inverted repeat (IR) sequences of 25 339 bp, separated by a large and a small single copy region (LSC and SSC, respectively) where the genes are only present once. In *N. tabacum* the LSC and SSC are 86 684 and 18 482 bp long, respectively. About 120 gene products have been identified, including RNA polymerases and other necessary enzymes for components of protein synthesis (*Table 10.1*). They include all rRNAs and tRNAs (four and 30 products, respectively, in tobacco), 23 ribosomal proteins (total approximately 60, less the remainder are from cytosolic synthesis) and about 20 (plus *ca* eight putative) proteins. Of the 53 proteins, nine are predicted but not yet identified in the chloroplast. Most of the membrane protein complexes (Chapter 5) have components coded in the cpDNA, e.g. coupling factor, PSI and PSII complexes and cytochrome b_6–f complex, but the LHCP is entirely nuclear encoded. Surprisingly only one soluble stromal polypeptide of the photosynthetic carbon reduction cycle (the LSU of Rubisco) is coded in cpDNA, all others are in nDNA.

Chloroplast genes encoding RNAs and proteins.   Chloroplasts contain the means of expressing the information in cpDNA. There are genes for four different rRNAs in the IRs, that is two copies of each, and they resemble genes in *Anacystis*, a cyanobacterium. The 4.5S rRNA is very similar to the 3′ end of the prokaryotic 23S rRNA. The plant 70S ribosomes contain about 60 proteins, of which 20 have been identified in the tobacco genome (plus three repeats), and some contain introns. Introns do occur in cpDNA but the number and size is variable (more than 100 of 400–1000 bp in size), and they seem to change quickly and are not conserved even where the evolutionary origin of the plants is similar. Introns occur in the ancestors of *Marchantia* and the angiosperms, for example two in *trn*I and *trn*A were acquired 450–500 million years ago in the *Chlorophyceae*, the common ancestor, and suggests strong conservation of cpDNA. Introns of 0.5–2.5 kbp occur in six different tRNA genes (greater than in tRNAs of any other organisms) and in nine protein genes (including *atp*F, *ndh*A, *pet*B, *pet*D) of tobacco. Introns may have regulatory functions in gene expression. When the introns are excised from the message, the coding sequences or exons are spliced together by ligases to give the functional mRNA transcript from which protein is made, but splicing in chloroplasts has not been demonstrated, perhaps because of the need for particular conditions or proteins.

Some 20 *rps*2 genes for 30S ribosomal proteins, three in IRs, making a total of 23 (four with introns), eight *rpl* genes for 50S ribosomal proteins and probably components of RNA polymerase (the *rpo*A, B and C genes for the α, β and β¹-subunits) plus initiation factor 1 (*inf*A) and DNA binding protein (*ssb*) have been identified in chloroplasts. As mentioned earlier, ribosomes of the chloroplast are sensitive to chloramphenicol and erythromycin and have clear homology with bacteria genes. An rRNA operon exists in all IRs examined in higher plants and cyanobacteria (the symbiotic ancestor of the chloroplast), suggesting its retention from the earliest symbiotic phase. Some 30 tRNA species occur as part of the protein synthesizing machinery, with seven repeated, a total

Table 10.1. Components of the chloroplast coded for chloroplast DNA genes and their size in amino acids coded for. Those genes which are polycistronically transcribed are shown as PC, those monocistronically transcribed as MC

Chloroplast	Gene	Gene product	Size	Transcription
ATP synthesizing	*atpA*	$CF_1\alpha$ subunit	507	
complex (CF_1–CF_0)	*aptB*	$CF_1\beta$ subunit	498	
	atpE	$CF_1\varepsilon$ subunit 1	133	PC
	*atpF**	CF_0 subunit III (=c)	184	
	atpH	CF_0 subunit IV (=a)	81	
	atpI	CF_0 subunit I	247	
Cytochrome	*petA*	cytochrome *f*	320	
b–f complex	*petB**	cytochrome b_6	215	PC
	*petD**	17 kDa polypeptide subunit 4 cytochrome *b6–f*	160	
Photosystem I	*psaA*	P700–Chl *a* protein	750	
	psaB	P700–Chl *a* protein	734	PC
	psaC	8 kDa 2 (4Fe–4S) protein	81	
	psaJ	4.9 kDa protein		
	psaI	4.0 kDa protein		
Photosystem II	*psbA*	D1 (32 kDa, Q_B) protein	353	MC
	psbB	47 kDa chl *a* protein (CP 47)	508	
	psbC	44 kDa chl *a* protein	473	
	psbD	34 kDa protein, D2	353	PC
	psbE	9 kDa cytochrome b_{559}	83	
	psbF	4 kDa cytochrome b_{559}	39	
Photosystem II	*psbG*	24 kDa polypeptide	284	
	psbH	10 kDa phosphoprotein	73	
	psbI	4.8 kDa reaction center	52	PC
	psbK	2.4 kDa polypeptide		
	psbL	5 kDa polypeptide	38	
Rubisco	*rbcL*	Large subunit	477	MC
NADH dehydrogenase	*ndhA*	ND1 ⎫		
	ndhB	ND2 ⎪		
	ndhC	ND3 ⎬ subunits of respiratory enzyme		
	ndhD	ND4 ⎪		
	ndhE	ND4L ⎪		
	ndhF	ND5 ⎭		
Ferredoxin	*frxA* ⎫			
	frxB ⎬ 4Fe–4S type ferredoxins			
	frxC ⎭			

Genes for gene expression and genetic apparatus

Ribosomal DNA	23S rDNA ⎫			
	16S rDNA ⎬ Components of the ribosomes			
	5S rDNA ⎪			
	4.5S rDNA ⎭			
Transfer RNA	30 different *trn* genes	Components of protein synthesis		
Ribosomal proteins	20 different *rpl* genes	30S + 50S ribosomal proteins		
RNA polymerase	*rpoA*	Subunit α		
	rpoB	Subunit β1		
	rpoC	Subunit β		
Initiation factor 1 proteins	*infA*	Components of protein synthesis		

* Denotes light activation.

of 37, sufficient to code for all tRNA functions. Ribosome binding sites on mRNA of some chloroplast genes have initiator sequences of the Shine–Dalgarno type.

Proteins encoded by cpDNA.
Some of the major proteins coded in the chloroplast genome are listed in *Table 10.1*. Aspects of their coding, synthesis and function are now considered.

*rbc*L **gene for Rubisco.** Rubisco is the major enzyme component of the chloroplast, forming 40–60% of the total soluble protein of leaves. The LSU is encoded by cpDNA in the *rbc*L gene, in a sequence of 1431 bp corresponding to 477 amino acids. There are no introns and there is considerable homology with the cyanobacterial gene (e.g. *Anacystis nidulans* with 472 amino acids). The LSU is highly conserved, with sequences from different species having more than 80% homology when the silent base changes in the cpDNA are allowed for. This is much greater than the homology (40% between organisms) for the nuclear-encoded SSU. In eukaryotes the single copy of *rbc*L is mono-cistronically transcribed in the chloroplast and is under post-transcriptional control, that is processing of RNA. The gene has a strong promoter region which is light inducible. There are very many more genes for the LSU than for the SSU and it is still unclear how the necessary stoichiometry of the LSU and SSU components is achieved. In C3 plants, expression of the Rubisco genes occurs only where chloroplasts are found, and the expression is much greater in leaves than stems. C4 plants have Rubisco genes in both mesophyll and bundle sheath cells but expression is restricted to the bundle sheath.

atp **genes for coupling factor.** The ATP synthase (CF_1–CF_0) complex has 6 of its nine subunits encoded by the B strand of cpDNA in all species, the rest by nDNA. The chloroplast genes are in two separate loci 40 kbp apart in the large single copy region, with those for β and ε (*atp*BE or β operon) in one locus and IV, III, I and α (*atp*IHFA or α operon) in the other. These are both conserved in clusters but the distance between them is variable, depending on species. The most abundant transcript of these genes is of *atp*H, as CF_0III (also called c, see Section 6.5.1) is the most abundant subunit in the complex. A polycistronic transcript of the operon is probably made and then cleaved into a mixture of mono-, di- and polycistronic products. The β operon is *ca* 700 bp upstream from *rbc*L and, as it is on the opposite strand, transcribed in the opposite direction. The stop codon of *atp*B (UAG) and the adenine base preceding it form the initiation codon of *atp*E (AUAG), so they form a single operon. The β operon promoter in spinach lies 454 bp upstream of the 5′ initiator codon for *atp*B and has two sequences, TTGACA and TGTATA, resembling other chloroplast promoters, although there are species differences. Knowledge of the regulation of transcription of these operons is poor; there are complex patterns of mono- and polycistronic transcripts. No evidence exists for close coordination of gene expression between the two operons in the chloroplast or with the nuclear genes for coupling factor subunits. There is considerable similarity in subunit structure and function and in organization of *atp* genes between the chloroplasts of higher plants, cyanobacteria and bacteria. A particular order of genes is probably not essential for function. Genes for α and β subunits are strongly conserved in different organisms, possibly because of their role in catalysis. Such similarity in the very fundamental aspects of coding for this essential enzyme supports the theory of the prokaryotic endosymbiont origin of chloroplasts.

psa **genes for photosystem I components.** The PSI complex has two subunit apoproteins (A1 and A2) of P700 coded in the *psa*A and *psa*B genes in the middle of the LSC region of cpDNA separated by 25 base pairs (in tobacco). They are without introns, have a single transcriptional start upstream of the *psa*A genes and are co-transcribed (polycistronic). This may aid the maintenance of the correct 1:1 stoichiometry in the complex. Regulation of expression occurs at translational and post-transcriptional levels with complex control by light and developmental stages. The proteins are *ca* 750 and 730 amino acids long (about 83 kDa). Gene *psa*C, the first photosynthetic component found in the SSC region of cpDNA, codes for a 9 kDa polypeptide of 81 amino acids, the apoprotein of Fe–S centers A+B of PSI. This *psa*C gene is co-transcribed with a gene (*ndh*D) for a subunit of NADH dehydrogenase. In addition, *psa*J and *psa*I code for 4.9 and 4.0 kDa proteins of the PSI complex in barley. Other components of PSI are encoded by the nuclear genome and are discussed later.

psb **genes for photosystem II.** Photosystem II has nine components coded in the cpDNA of tobacco but the 33, 24 and 18 kDa polypeptides of the water-splitting apparatus are encoded by nDNA. The chloroplast genes are located at six different sites within the large single-copy region. The D1 (also called the 32 kDa or Q_B) protein of the PSII reaction center is encoded by *psb*A and transcribed monocistronically as a 34 kDa precursor which is then processed to form the 32 kDa polypeptide; the coding sequence is 317 codons. The D1 protein binds herbicides, for example atrazine, and mutants resistant to the herbicide have single base alterations in the gene code related to specific changes in the amino acid sequence. The mRNA is stable in the dark (more so than the mRNA from *rbc*L) although the product is unstable and rapidly degraded in darkness. Indeed, the protein itself turns over very rapidly. Gene *psb*B codes for a 51 kDa polypeptide; it contains an intron and is upstream of genes for cytochrome b$_6$ (*pet*B) and petD. The four genes *psb*B, *psb*H, *pet*B and *pet*D form a single operon, called the *psb*B operon, transcription of which has been described; *pet*B and *pet*D contain introns which are removed from the primary transcript, the exons are spliced and the mRNA processed to give multiple smaller mRNAs. The *psb*C and *psb*D genes overlap by 53 bp; they code for the 44 kDa and D2 proteins of PSII. In addition, the *psb*E and *psb*F genes are coding sequences for the 9 and 4 kDa subunits of cytochrome b_{559} which are co-transcribed. Other PSII genes (*psb*G and I) have been identified as such by Western blotting but their function is not established.

pet **genes of the cytochrome b–f complex.** This complex has six polypeptide components involved in the electron transport pathway between PSI and PSII. The three genes for cytochrome *f*, cytochrome b_{563} and subunit IV (*pet*A, B, D) are contained in cpDNA. The other three, coded in the nDNA, include the Rieske Fe–S protein which has a vital role in the assembly of the complex. The precursor protein is made in the cytosol as a 26 kDa polypeptide which is transported into the chloroplast and then processed to the 19 kDa protein. *pet*A is 4 kbp downstream of *rbc*L on the A strand and separated from it by several open reading frames (ORFs). These are base sequences which could be transcribed, but with no identified protein product. *pet*A is co-transcribed with an associated ORF as a large polypeptide and the N-terminal 35 amino acids removed during processing. Cytochrome *f* has a hydrophobic C-terminal sequence (amino acids 250–271) which allows positioning in the thylakoid membranes. Genes *pet*B and *pet*D are close together and some 15 kbp from *pet*A, so it is likely that they are not polycistronic and therefore are under transcriptional control. The *pet*B and *pet*D genes are co-transcribed with *psb*B and *psb*H. Each gene has a single intron and very small initial exons (6 and 8 bp only) which are very highly conserved in all genomes. The sequences suggest that there are four and three hydrophobic, transmembrane segments in each protein, respectively.

Genes for NADH dehydrogenase.   Possible protein products of ORFs have been identified by comparing their base sequences with DNA sequences from other organisms (using a DNA library) for which products have been identified. The ORFs of cpDNA contain some six genes with strong homology to sequence components of NADH dehydrogenase from human mitochondria and *Chlamydomonas*. This enzyme functions in mitochondrial respiration, taking electrons from NADH and transferring them to ubiquinone. However, the enzyme has not been described in higher plant chloroplasts and may be either a relic of earlier respiratory activity there, or may function in chlororespiration. The genes are *ndh*A, D, E and F in the SSC region, *ndh*C in the center of the LSC region and *ndh*B in the inverted repeats. Single introns occur in *ndh*A and *ndh*B genes; the homology of these genes in several different higher plants and the liverwort genomes is strong.

Other plant chloroplast genomes.   Several other genomes have been studied in detail (Bogorad and Vasil, 1991a). The liverwort *Marchantia* has, in comparison with tobacco, a somewhat smaller cpDNA genome of 121 024 bp, but in general the cpDNA structure, the gene products, *et cetera*, are very highly conserved. As with tobacco, *Marchantia* cpDNA has two large IR sequences of 10 058 bp with a duplicated complement of genes; each contains genes for rRNAs. The IR sequences are separated by LSC and SSC regions of 81 095 and 19 813 bp, respectively. The cpDNA may code for up to 136 products, including all the rRNAs and tRNAs and proteins associated with the ribosomes of the chloroplast (four rRNA, 32 tRNAs, 19 ribosomal proteins and 25 RNA polymerase subunits). Twenty genes for polypeptides of photosynthetic membrane electron transport and other membrane components have been identified. Further gene sequences have been related, through sequence homology with other organisms, to 4Fe–4S-type ferredoxin from bacteria (two ORFs), and seven genes have strong homology with NADH dehydrogenase of human mitochondria (seven ORFs), as in tobacco. There is also a component of nitrogenase (one ORF); other genes resemble those genes coding for the permeases of *E. coli* membranes and an antenna protein of a cyanobacterium. Such similarity between these two very different organisms suggests a common origin of the cpDNA and considerable conservatism in the evolution of the genome.

There is rather more variation in cpDNA than the basic pattern of tobacco and *Marchantia* chloroplast DNA suggests. Legumes such as pea *(Pisum sativum)* and broad bean *(Vicia faba,* genome 123 kbp) have no large repeated sequences and are thus unique among land plants so far examined. *Vicia* cpDNA has only one set of rRNA genes, about 30 tRNA genes and encodes the LSU of Rubisco, the α, β, ε and subunit III(c) of ATP synthase as well as others described for the tobacco and liverwort genomes. In legumes, the gene *rp*l22 is nuclear but in other angiosperms it is chloroplastic. The reasons for this and the mechanism by which the transfer occurred (probably from cpDNA to nDNA) are obscure. Variation in the IR is the main cause of differences in size of the cpDNA in different species. Reasons for the particular structural features of cpDNA and the differences between species are largely unknown, although it is speculated that it may be related to processes of gene copying. Possibly, the IR was present in the cpDNA of the common ancestor of land plants (note *Marchantia)* and the major changes, such as the loss of the IR in legumes, occurred later.

10.4.3 Nuclear encoded chloroplast proteins

The presence of genes in the nuclear genome for many chloroplast components was shown by genetic analysis, for example, the distribution of defective pigmentation into the offspring of mutant plants. *Table 10.2* lists some of the proteins associated with the photosynthetic apparatus which are encoded by nDNA. Virtually all aspects of the chloroplast structure involve nuclear gene products (e.g. about half of the approximately 60 types of polypeptides in the light reaction components are nuclear encoded), raising interesting questions about the way proteins are made in one cell compartment and transferred to another and about co-ordination of chloroplast development. Indeed, the evolutionary history of this prokaryotic organelle in a eukaryotic environment is an intriguing subject, involving many unsolved questions about regulation and the mechanisms by which genes migrated from symbiont to host during evolution.

Protein synthesis. The protein synthesis system of plastids has components restricted to the chloroplast (rRNA and tRNA) and components which are nuclear (elongation and termination factors plus enzymes which modify RNAs) but some components appear in both compartments (ribosomal proteins and initiation factors). The locations of some genes for different components are shown in *Table 10.3*.

*rbc*S **genes.** The SSU of Rubisco is nuclear encoded by the *rbc*S multigene family in all eukaryotic chlorophytes (all green algae and land plants) so far examined. The number of genes differs with the organism, for example five in pea (*Pisum sativum*) and tomato (*Lycopersicon esculentum*), eight in *Petunia* and 13 in duckweed (*Lemna minor*), but there are only two *rbc*S genes at one locus in *Chlamydomonas*. The five *rbc*S genes in tomato are at three genetic loci on two chromosomes: *rbc*S-1 on chromosome 2 and *rbc*S-2 on chromosome 3, both genes encoding a single gene product, and *rbc*S-3 on chromosome 2, consisting of three genes arranged in tandem over a 10 kb region, *rbc*S-3A, -3B and -3C; they are monocistronically transcribed and differ in number and size of introns. Although these *rbc*S genes differ between and within species, the coding regions are highly conserved, for example 91–100% in the tomato genes when silent nucleotide substitutions are allowed for. The evolutionary pressure to conserve the functional protein was probably greater than the pressure to conserve the nucleotide sequences. Despite their similarity, there is substantial difference in the ability of these nuclear genes to produce products. In tomato, for example, the *rbc*S-3A and *rbc*S-1 genes are expressed in dark-grown leaves and change little with illumination. In contrast, *rbc*S-3B and -3C are hardly expressed in the dark but increase greatly in the light. In *Petunia* the eight genes differ 100-fold in expression.

cab **genes.** A chloroplast membrane component entirely encoded by nDNA is the light-harvesting complex of PSII (LHCII), composed of a major and three minor chlorophyll *a/b* binding proteins, namely CP29, LHCII (genes *Lhcb*1, *Lhcb*2, *Lhcb*3), CP26 and CP24 of

Table 10.2. Some chloroplast proteins which are nuclear encoded and regulated by light

Chloroplast component	Gene designated	Gene products	Light response	Photoreceptor
Rubisco	*rbcS* (multigene family)	Small subunit	+	R, B, UV
Rubisco activase	*Rca*		+	?
Electron transport	*petC*	Cytochrome *b/f* Rieske Fe–S protein		
	petE	Plastocyanin	+	R, B
	petF (Fed-1)	Ferredoxin NADP oxidoreductase	+	?
Photosystem II	*psbO*	33 kDa ⎤	+	R
	psbP	24 kDa ⎮ water-splitting	+	R
	psbQ	18 kDa ⎮ polypeptides	+	R
	psbR	10 kDa ⎦	+	R
Photosystem I	*psaD*	18 kDa ⎫	+	R, B
	psaF	17 kDa ⎬ polypeptides of		
	psaE	9.2 kDa ⎱ PSI complex		
	psaG	10.8 kDa ⎰		
	psaH	10.4 kDa ⎭		
Light-harvesting chlorophyll protein	*cab* (multigene family)	Polypeptides of the chlorophyll protein complex of PSII	+	R, B, UV
ATP synthase (coupling factor)	*atpC*	γ		
	atpD	δ		
	atpG	II (or b′)		
Glyceraldehyde	*GapA*		+	?
phosphate dehydrogenase	*GapB*		+	?
Ferredoxin 1	*Fed-1*		+	
Glutamine synthase 2	*GS2*		+	
Phytochrome	*Phy*		−	R
Protochlorophyllide reductase	*Pcr*		−	R
Nitrate reductase			+	(?R)
PEP carboxylase			+	?
Phospho-ribulose kinase			+	?
Flavonoid biosynthesis enzymes				
Chalcone synthase	*Chs*		+	R, B, UV
Phenylamoniumlyase	*Pal*		+	UV

R = red, B = blue, UV = ultraviolet.

29, *ca* 27, 29 and 21 kDa, with classical Mendelian inheritance. These four are encoded by *cab* genes which have eukaryotic 5′ and 3′ flanking regions of TATA and CAAT boxes and polyadenylation sites. They form a multigene family of 3–16 genes depending on the species. In *Petunia*, the 16 genes are grouped into five small families in which the genes are related but the differences between families are greater. A sequence of 31–37 amino acids forms a transit peptide, that is, a sequence of amino acids at the amino-terminal end of the polypeptide chain which enables the protein to be transferred into the chloroplast

Table 10.3. Compartments in which chloroplast protein translation components are coded

Translation system component	Genes in plastid	Nucleus
tRNA	+	−
rRNA	+	−
r-Proteins	+	++
Elongation factors	−	+
Termination factors	−	+
rRNA processing and modification enzymes	−	+
r-Protein modifying enzymes	−	+
Aminoacyl-tRNA synthases	−	+
mRNA maturation enzymes	−	+

+ = present; ++ = abundant; − = absent.

envelope and thylakoid membrane. The *cab* genes produce mRNA in the nucleus which is translocated to the cytosol where it associates with ribosomal components to form free polysomes on which the precursor protein is made. On release from the polysomes the polypeptides attach (in a nonenergy-dependent binding) to the outer chloroplast envelope membrane: some 3–5000 molecules per chloroplast, thus allowing rapid synthesis of the protein even if transport is slow. The protein binding site is probably specific, for example Rubisco and LHCP precursors have different binding characteristics, pH requirements and sensitivity to proteolysis.

Genes for other thylakoid proteins.   The apoprotein of the Rieske Fe–S centers, plastocyanin and ferredoxin-NADP oxidoreductase, are products of the *pet*C, *pet*E and *pet*F nuclear genes, respectively. The *pet*C product is made as a 26 kDa precursor, 7 kDa larger than the mature protein, on cytosolic ribosomes and then transferred into the chloroplast from the cytosol. From the deduced sequences of amino acids there are two highly conserved sequences, Cys–Thr–His–Leu–Gly–Cys and Cys–Pro–Cys–His, near the C-terminus of the protein which probably coordinate the iron–sulfur center and also sequences which code for very hydrophobic amino acids which may anchor the protein in the membrane.

psa **and** *psb* **genes.**   Photosystem I has five of its eight polypeptides encoded in the *psa*D, *psa*F, *psa*E, *psa*G and *psa*H genes. They are, respectively, subunits II, III and possibly IV and VI; II is the site for Fe–S centers, III is involved in the reduction of the PSI reaction center, and the others are of unknown function. All are synthesized as larger precursors which are modified ('processed') after entering the chloroplast. Photosystem II is characterized by having four of the polypeptides of the water-splitting complex, the 33, 24, 18 and 10 kDa polypeptides, coded in the nucleus by the *psb*0, *psb*P, *psb*Q and *psb*R genes. Each protein has two transit peptide sequences attached, one enabling transport across the chloroplast envelope and the other across the thylakoid enabling the water splitting complex to be assembled in the thylakoid lumen.

atp **genes.**   Coupling factor (CF_0–CF_1) has the γ, δ and II subunits nuclear encoded in the *atp*C, *atp*D and *atp*G genes, respectively. Suggestions have been made that transfer of these genes to the nucleus took place during evolution after the initial endosymbiosis. The precursor proteins are synthesized in the cytosol and transported into the chloroplast after which the transit peptide is removed. However, regulation of this process is poorly understood.

Nitrate and nitrite reductases. Nitrate assimilation in the chloroplast is intimately associated with utilization of the primary products of the light reactions (Chapter 7). Nitrite reductase (NiR) is in the chloroplast and nitrate reductase (NR) in the cytosol. Both NR and NiR are coded for by small gene families in the nucleus and are synthesized on cytosolic ribosomes; NR remains in the cytosol but NiR is transported into the chloroplast. Regulation of NR synthesis occurs at transcription, post-transcription and post-translation steps, and is complex with nitrate ions, sugars and metabolites involved in induction of transcription.

10.4.4 Chloroplast gene expression and protein synthesis

Some chloroplast genes are transcribed monocistronically (e.g. *rbc*L and *psb*A) whilst others are polycistronic (e.g. *atp*1-*atp*A and *atp*B-*atp*E), resembling prokaryotic genes. Upstream of the genes are sequences of DNA similar to promoters of prokaryotes. The initiation sites for transcription of tobacco *rbc*L, *psb*B and *atp*B-E genes have upstream sequences (probably promoters) very similar to the -10 and -35 promoters of bacteria. However, the presence of introns, particularly in tRNAs, is less characteristic of prokaryotes. The chloroplast has a complete system for transcription and translation of cpDNA genes. Half of the total leaf ribosomes are in the chloroplast and half of the total leaf protein is made in there. Thus, the chloroplast is the most active compartment for protein synthesis in mesophyll cells, much more so than the mitochondria. Chloroplast genes are transcribed by RNA polymerases, of which there are two classes in the chloroplast. One class is tightly bound to DNA, the other soluble; this could be important for gene regulation. Expression of chloroplast genes may be controlled at several steps in the process and regulation is complex; the references (e.g. Bogorad and Vasil, 1991a, b) should be consulted for this essentially molecular biological rather than photosynthetic problem.

Protein synthesis in chloroplasts is prokaryotic, for example ribosomes of chloroplasts are very similar to the prokaryotic 70S type in their subunit structure (50S and 30S) and sedimentation coefficients, initiation of chain elongation and the sequences and size of the rRNAs. Proteins of prokaryotic ribosomes have considerable homology with the products of chloroplast genes. The chloroplast DNA sequences have great homology with the sequences coding for ribosomal proteins in *E. coli*. These are composed of rRNA (with a G + C content of 55.8%) and about 60 proteins and are sensitive to prokaryotic antibiotics (e.g. chloramphenicol), but not to eukaryotic inhibitors (e.g. cycloheximide). Initiation of protein synthesis requires N-formylmethionine rather than methionine, and factors for initiation and elongation can be interchanged with those of bacteria and remain fully functional. However, there is not total similarity between chloroplast and bacterial ribosomes: chloroplasts are affected by both prokaryotic and eukaryotic ribosomal inhibitors. This is explained by the involvement of both nuclear and chloroplastic genomes in the synthesis of plastid ribosomes. Of the 60 higher plant ribosomal proteins, only about 20 are made in the chloroplast; it is assumed but not proven that the remainder are made in the cytosol. Chloroplast and nuclear genomes cooperate intimately in the synthesis of the machinery needed to form the chloroplasts components, and a chloroplast protein may be needed for synthesis of a nuclear encoded protein (see Roell and Gruissem, 1996; and Gray, 1996 for details).

10.4.5 Regulation of chloroplast protein synthesis

Protein synthesis by chloroplasts is driven by light and needs no additional energy. Low intensity light (15–30% of the energy needed for full photosynthesis) is effective and blue light is more effective than red light of equivalent energy. In darkness, protein synthesis will take place if ATP is supplied, but at only half the rate of that in saturating light. The rate can be considerably increased by supplying Mg^{2+} to overcome the depletion of Mg^{2+} resulting from supplying ATP alone. There is probably a requirement for electron transport but it is indirect; the synthesis of sufficient ATP will depend on the balance with NADPH reactions. Competition for ATP may occur between CO_2 assimilation and amino acid incorporation into proteins but factors which stimulate the turnover of NADPH and increase the rate of ATP production increase protein synthesis. Synthesis of chloroplast proteins takes place on ribosomes, including polysomes, bound to the thylakoid membranes by ionic charges and the nascent polypeptide chain. Binding is stimulated by light and the increase in stromal pH resulting from illumination. However, it seems that there is specificity in the function of the bound ribosomes, for example membrane-bound polypeptides such as the D1 and PSI reaction center proteins and subunit III(c) of coupling factor are made on thylakoid bound ribosomes, whereas Rubisco LSUs seem to be made on both the stromal ribosomes. Proteins from the ribosomes may be prematurely released by environmental stresses such as low oxygen, salts and darkness. Light affects the rate of gene transcription of protein accumulated for proteins involved in photosynthesis which are encoded in the nucleus, but has relatively little effect on the rate at which chloroplast mRNAs are transcribed or the amount accumulated. However, increasing light results in the rapid translation of many chloroplast mRNAs, so that expression of both nuclear and chloroplast proteins is increased and coordinated.

Chloroplasts originated as prokaryotic endosymbionts, as indicated by the 70S ribosomes and polycistronic messages without 5′ caps and polyadenylated tails. However, many of the original symbiont genes have transferred to the nuclear genome and their function has been replaced by new mechanisms in the chloroplasts. Proteins and RNA interact in regulation of translation in chloroplasts and respond to illumination. *Trans*-acting protein factors may regulate translation, for example if cytochrome *f* is not incorporated into the cytochrome b_6f complex, then it may slow translation of its own *pet*A mRNA. Mutations to nuclear genes affect the translation of specific chloroplast mRNAs, for example synthesis of the D2 reaction center protein of PSII from *psb*D mRNA which may be controlled by several nuclear protein factors. Also proteins which differentially bind at 5′ untranslated regions of the chloroplast mRNAs may regulate translation. They may be associated with chloroplast membranes and it is likely that they also respond to the energy status of the membranes and to NADPH and ATP by undergoing phosphorylation, for example.

Chloroplast genes may be activated by light (*Table 10.4*). Such 'photogenes' show large increases in transcripts and their protein products when illuminated; they include the D1 protein of PSII and *rbc*L for LSU of Rubisco. The latter is barely detectable in darkness and light greatly stimulates production of the mRNA. Expression of *rbc*L genes in the chloroplast is regulated primarily at the post-transcriptional level so that the stability of the chloroplast RNA and of the mechanism of gene expression during translation, together with promoter strength, may be important. Nuclear genes are also light activated, e.g. those for LHCII and the SSU of Rubisco. This important aspect of gene regulation in relation to the environment will be considered later. There is a strong regulation of gene expression depending on the tissue. Genes coding for photosynthetic components, such as the SSU of Rubisco, occur in tissues other than leaves of higher plants (e.g. stems and fruits, where they are expressed but to a greatly reduced extent, and in roots where they are negligible). Also, in C4 plants, genes may be present in the mesophyll and bundle sheath tissues but only expressed in one cell type; Rubisco, for example, only accumulates in the bundle sheath. Such tissue-specific regulation of expression is related to the control of DNA sequences which provides tissue specificity.

Very active photosynthetic cells are responsible for a large part of the plant's total protein metabolism, both *de novo* synthesis and rapid and substantial turnover. The system is very dynamic and environmental conditions leading to the production of active oxygen species and photoinhibition may increase degradation, repair and replacement of proteins. Much of the newly synthesized protein may be rapidly degraded by proteolysis if it is defective or if other components which are essential for the structures are not available. An example of protein which is rapidly broken down and resynthesized is the D1 protein of photosystem II. In photoinhibitory conditions (Section 5.7), particularly in bright light, the protein is rapidly turned over. In fact, the D1 protein is the most rapidly synthesized and degraded of all chloroplast proteins. Although a major product of protein synthesis, it never accumulates when chloroplast development is blocked as do some other components of the chloroplasts. The mechanism of proteolysis of D1 is not known.

Table 10.4. Light activated genes of the chloroplast

Gene	Product	Effect of light
psbD	Protein of PSII	+
psbK	Protein of PSII	Independent, constitutive
psbI	Protein of PSII	Independent, constitutive
rbcL	Rubisco large subunit	+
petA	Cytochrome *f* apoprotein	+
petB	Cytochrome b_6 apoprotein	Independent, constitutive
atpA	Coupling factor α	+
psaA	PSI apoprotein	+
psaB	PSI apoprotein	?
trnK	tRNA genes	Light independent, constitutive
trnG	tRNA genes	Light independent, constitutive
rps16	Ribosomal protein	Light independent, constitutive

10.5 Light receptors and gene activation

Regulation of gene expression by light occurs both in nuclear and chloroplast genes. Light regulates development through a system involving light perception, transduction of the energy into a signal and coupling of the signal to gene expression and regulation of products. There are four main photoreceptors in plants:

(i) phytochrome, which is regulated by red/far-red light;

(ii) the blue/UV-A absorbing photoreceptor (often called 'cryptochrome') which is also involved in phototropism and stomatal opening;

(iii) a UV-B receptor, important in inducing plant protective responses to damaging UV-B radiation; one of its more important features is stimulation of genes responsible for chalcone synthase production, a key point in the synthesis of flavonoids which are major protective pigments against UV-B radiation;

(iv) protochlorophyllide reductase.

Plastid genes dependent on light regulation are *psb*A for the D1 (32 kDa Q_B) protein of PSII, *rbc*L (LSU of Rubisco) and *pet*A, the cytochrome *f* apoprotein. Nuclear genes which are increased include the *cab* genes encoding the light-harvesting *a/b*-binding protein of the LHCII complex, *rbc*S for the Rubisco SSU (in most species examined but not barley) and the ferredoxin gene *Fed*-1. However, gene expression for protochlorophyllide reductase (NADPH:protochlorophyllide oxidoreductase; Section 8.12) and phytochrome is decreased. Species differ greatly in response to light. For example, *cab*1 gene expression is very sensitive, the mRNA being greatly increased by extremely dim light, which gives only 1% conversion of inactive phytochrome (P_r) to the active form (P_{fr}) and is saturated by light flux at which only 3% of the P_{fr} is produced. Interestingly the *phy*A and *Pcr* genes are also very sensitive to very low fluxes but their transcripts are decreased. The very low light response is shown by barley (*Hordeum vulgaris*) and *Arabidopsis* but not by oat (*Avena sativa*); there are differences in expression of the individual genes within a gene family and *cab*1 in wheat shows both low and normal light responses and in pea some genes show no low light response. There is, at present, no simple explanation of these complex differences. They are probably related to the ecological behavior of plants, as the phytochrome system may confer advantages in regulation of development. The sensitivity of plants to light flux and to the changed spectrum may be regarded (teleologically) as an advantage when seedlings grow towards the soil surface or in deep shade. Dim light enriched towards the far-red end of the spectrum predominates and signals that the plant must extend into bright light in order to photosynthesize fully. It is an advantage to prevent development of a competent photosynthesis apparatus (because of the drain on resources in the seed or other storage organ) until light is available and then a very rapid response is needed to achieve full competence.

10.5.1 Phytochrome

Phytochrome is of the greatest physiological significance for plants, determining development and thus ability to respond to environment; it is rapidly induced by low-intensity light and is sensitive to spectral quality, principally the red/far-red ratio. It is a chromoprotein with a 120–127 kDa apoprotein and a linear tetrapyrrole chromophore. It exists in dark-grown tissues in a dominant form (P_r) which is physiologically inactive but can absorb red light at 665 nm absorption maximum and is then converted to P_{fr}, the physiologically active form. Only very brief (1 min) exposure to red light of low energy is necessary (*Figure 10.2*). This is sufficient to trigger response of light-sensitive genes. Equilibrium is achieved between P_r and P_{fr} depending on light quality; it is constant under steady-state illumination but changes in the spectrum, for example due to growth of vegetation around an organ or plant, result in a growth response. When P_{fr} absorbs far-red light (735 nm) it reverts back to the inactive form. Thus, the effect of red light can be reversed by immediate application of far-red light; equilibrium between the two forms is altered quickly by the red/far-red ratio, permitting regulation of photomorphogenesis and chloroplast structure, leading to such differences as those between sun and shade leaves (Section 12.4.2). This rapid, low-energy red/far-red reversibility provides a means of differentiating phytochrome effects from other light effects. There are, however, more complications for different forms of P_r and P_{fr} existing in tissues grown in darkness or in light; they are distinguished by molecular characteristics and have different effects on gene expression.

Phytochrome itself is nuclear encoded by the *phy* genes: *phy*A codes for the I type phytochrome and *phy*B and *phy*C coding sequences are like the type II phytochromes

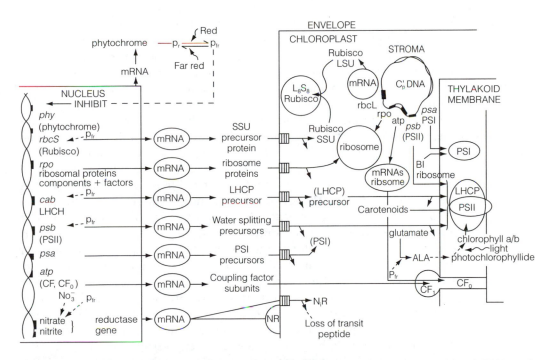

Figure 10.2. Scheme of the interactions between nuclear and chloroplast genomes (nDNA and cpDNA, respectively) in terms of their gene products and their incorporation into the chloroplast stromal and thylakoid compartments. The role of phytochrome in regulation of gene expression is indicated. In addition, the interaction of nitrate ions in the regulation of gene expression for nitrate and nitrite reductases is shown. Details of the processes are discussed in the text

(based on sequence homology). Phytochrome provides an interesting example of the complex regulatory processes which occur in plants. The *phy* genes are autoregulated, that is small exposure to light and formation of P_{fr} decreases expression of the gene and formation of phytochrome; in rice and oats the response is large but the effect in *Cucurbita is* much smaller. Thus, a regulatory cycle with very subtle balancing between the amount of receptor, the response of selected genes and light quality is possible, allowing rapid and large developmental and morphogenetic responses to sudden illumination. Phytochrome operates at the level of gene expression and the effects on the development of the chloroplast are many and well described (some will be considered later). However, details of the mechanisms by which light quality is transduced into regulation of gene activity are not well understood. Direct interaction with DNA is unlikely; P_{fr} has not been found in the nucleus. Amplification of the signal is most likely, for the rapid and far-reaching effects would need more than the few P_{fr} molecules produced in very low light.

Light-regulated gene promoters in plants are known, and their structures and the ways in which the signal from the primary light receptors is transduced are being clarified. Flowering plants have light-inducible genes for proteins involved in, for example, the light-induced synthesis of PSII components, Rubisco SSU *rbc*S and flavanoids such as anthocyanins, which have very strong homology with those in ferns and mosses which are not responsive to light signals. From the three light receptors, signal tranduction results in a control sequence being initiated which ultimately results in stimulation (or sometimes repression) of gene transcription by a complex of *cis*-acting elements and protein transcription factors in the promoter. There is no single control system, promoter structure or group of regulatory proteins regulating function in all organisms, so that the process is very complex, perhaps of multiple origins and selected under different ecological conditions. How the systems evolved is not yet clear. The 'triggering' reactions caused by phytochrome are strongly exhibited by tissues grown in the dark but induction depends on the gene. The transcription rates of *cab* and *rbc*S genes and that for GS2 (a plastidic glutamine synthetase) increase rapidly over 24 h following illumination whereas products of *Fed*-1 (a gene for ferredoxin) increase rapidly in the first 2 h and then remain constant. However, sensitivity to far-red light is retained for longer by *Fed*-1 than by *cab*, *rbc*S and GS2. There is much evidence of differences in the ability of tissues to respond to the light, e.g. production of anthocyanins by P_{fr} occurs only in subepidermal cells of mustard seedlings. Competence of the system is possibly determined by production of a 'plastid factor', which is needed for expression of nuclear genes but is stopped by chloramphenicol, although it seems unlikely to be a protein itself. There is much evidence for the interaction of the phytochrome and blue light receptors in regulation of gene expression in tissues grown in the light and transferred to darkness for a period. For example, abundance of *rbc*S and *cab* mRNAs was unaffected by the phytochrome equilibrium in pea tissues, despite the marked response of dark-grown material, but blue light did induce gene expression, although it was reduced by applying far-red light. Perhaps such complex regulatory responses influenced by light quality allow full expression of tissue characteristics and thus enable the plant to become very efficient in the normal environment. The interpretation is that blue light receptors control the red light effects, i.e. there is synergism between red and blue.

10.5.2 Light and the organization of thylakoid membranes

Phytochrome has major effects on the expression of a number of genes coding for components of chloroplasts; this influence extends to the development and organization of thylakoids and chloroplasts. However, there is as yet no fully accepted time course or understanding of the mechanisms; a scheme is given in *Figure 10.2*. Constitutively expressed genes allow the tissue to develop to the point where further changes depend on light. There are regulatory steps in the synthesis of chlorophyll *a* and *b* and in the formation of carotenoids (the xanthophylls violaxanthin, neoxanthin and lutein; Chapter 3). Very importantly, expression of the *psb* genes of the nucleus and chloroplast, and especially the *cab* nuclear genes, are strongly regulated. Without the latter, no PSII complex would be made. With light and P_{fr} formation, *cab* and other nuclear gene products are made in the cytosol and transported into the chloroplast stroma. There further processing by proteases may take place. If energy is available from ATP (or from the thylakoid proton energy gradient), the polypeptides are inserted into the thylakoid membranes. Assembly into the light-harvesting complex requires the correct carotenoid and, of course, chlorophyll *a* and *b*. Without chlorophyll, no complexes are made and the *cab* gene products are rapidly broken down. Concomitantly, P_{fr} stimulates synthesis of the water splitting and reaction center polypeptides; these are assembled in the membrane and form the active complex. However, this simple statement hides a considerable degree of uncertainty about the undoubtedly complex mechanisms. Similar considerations apply to the other multicomponent complexes in the membrane. Thus, photosystem I is assembled from polypeptides, which are both nuclear and chloroplast coded. In darkness, chloroplasts contain small amounts of the products of the *psa*A and *psa*B chloroplast genes (the A1 and A2 protein of the core of PSI). On illumination, this hardly changes, but polypeptide II, a product of the nuclear gene, is rapidly synthesized. This is followed by component III and thereafter the other proteins. Polypeptide II may act as a core for assembly of the other units. The products of the *psa*A and *psa*B genes are synthesized on mRNA which may be at low abundance in the dark in some plants (e.g. maize, *Zea mays*) and subsequently increases in the light. In other cases, there may be a large amount of mRNA but products do not increase markedly in the light. Transcripts of these developmentally very important components may decrease as the leaves age but gene transcription decreases even faster (see Jenkins, 1991). Synthesis of the coupling factor complex involves the transcription of nuclear and chloroplastic genes, transport of the former into the stroma and assembly into the transmembrane segment (CF_0) and the stroma unit (CF_1). Evidence for some light-regulated control of ATP synthase exists, but there is rather poor understanding of how the different groups of genes are coordinated in their expression and how the final structure and stoichiometry of the complex is achieved *in vivo*.

Another type of control mechanism, which is only partly analyzed, involves a signal from the chloroplast to the nucleus, a 'chloroplast factor', which 'tells' the nucleus that the chloroplast is 'ready' to receive the protein. In the case of nitrate reductase, this factor, active phytochrome (P_{fr}) and nitrate are all necessary for NR gene expression. Without light, energy for nitrate reduction is not available in the chloroplast, so this mechanism avoids draining the plant's material resources in the construction of the system until light and nitrate ions are available; when both are, it is essential to have the light-transducing and nitrate-reducing components fully functional. Many of the photosynthetic system components, including those encoded by *cab*, NR and to a smaller extent *rbc*S, *rbc*L, *psb*A, *psb*O and *psa*D genes, exhibit diurnal rhythms in amount and activity probably related to the combined effects of changes in light intensity and spectrum and the inherent patterns which are under complex control in different types of plants at many different levels of activity. It is still unclear how the balance between the chloroplast and nuclear genomes is regulated. There are many questions: is there one- or two-way exchange of regulatory factors made by chloroplasts and nucleus, are common signals responsible for regulating both compartments? How is the content of Rubisco LSU and SSU genes regulated? The genes are in very different doses yet how is the stoichiometry of the products fixed? There is evidence that decreased production of LSU may lead to accumulation of the SSU, triggering breakdown of the SSU and thus regulating the amount.

10.5.3   Regulation of nuclear gene expression by light and sugars

Nuclear gene expression in plants is regulated by a multistep series of controls in which *cis*-acting DNA sequences, control elements, and *trans*-acting gene products (probably proteins which bind to specific DNA sequences including the promoter regions) determine the initiation of transcription and are affected by light. Also, expression is inhibited by accumulation of sugars in higher plants. It has been found that nontranscribed regulatory sequences confer the photoregulatory features; the transcriptional regulatory sequences are in the promoter and enhancer sequences 5' to the transcription initiation site. There are binding sites for nuclear factors which are essential for light responses. As an example of the mechanism by which light controls gene expression, the *rbc*S gene from pea was fully expressed in petunia, with the 970 bp 5' flanking sequence determining the response to light; a -35 to -2 promoter was essential for transcription with sequences further upstream required for maximum gene expression in light. This effect of enhancer elements is seen in *rbc*S gene expression, where two regions controlling expression (between base pairs -373 and -204) affect photoregulation and tissue specificity. In *Petunia,* an enhancer 3' to the start of coding also affected gene expression. It has been shown that small blocks of sequences in these regions determine the great differences in degree of expression in the petunia genes. It seems that there are positive, negative and reiterated DNA sequences in the enhancer regions so that a very subtle control of gene transcription is possible, and indeed likely, given the many environmental and plant factors which affect gene expression.

Regulation of transcription seems the most common method of controlling mRNA for nuclear encoded proteins, but for chloroplast genes translational processing is most common. For example, *rbc*S mRNA decreased in seedlings of soybean, but transcription decreased even faster. However, the opposite effect was seen in mature soybean plants. Translational and post-translational controls are very important in synthesis and regulation of chloroplast-encoded components, allowing more rapid and flexible response to environment. Light probably affects translation, protein folding and transport and protein degradation among other factors. Translational control applies to LSU together with the SSU of Rubisco in pea; their synthesis stops when light-grown plants are placed in darkness, probably because the ATP supply is interrupted. The RNAs for both the LSU and SSU remain bound to the polysomes in the chloroplast stroma and cytosol, respectively, but are inactive (although if cultured *in vitro* they are functional). In dark-grown plants, no polysome RNA complexes are made, suggesting that initiation is prevented. Another example of light regulation of translation is given by the mRNA (transcribed from *psa* genes) for LHCP polypeptides. Transcription is blocked in darkness at the level of chain elongation and synthesis of chlorophyll *a* removes the block.

Regulation of gene expression by sugars has been analyzed in considerable depth, as it is a mechanism by which the composition and activity of the photosynthetic carbon assimilation may be regulated (in part) with the growth of the plant and thus with the environment (Jang and Sheen, 1997; Smeekens and Rook, 1997). In particular, light 'sets the scene' for development but other conditions, especially the supply of nutrients, will determine how the photosynthetic system develops. The internal state of the plant is reflected in the accumulation of carbohydrates. Active photosynthesis, with rapid growth, results in less sucrose and starch accumulating than if growth is slowed (e.g. by low temperatures or poor nutrition). Sugars repress gene expression and are thus a key element in regulating development of the photosynthetic system. The structure of the system depends on synthesis of proteins, and the main regulation occurs in the accumulation of mRNA transcripts, with protein accumulation then following. Small concentrations of metabolizable sugars at the site of expression stimulate (or 'derepress') transcription, but large concentrations inhibit. Repression of nuclear and plastid genes for a wide range of proteins involved in many aspects of photosynthesis has been shown by feeding isolated leaves or leaf discs sucrose or glucose, for example these sugars at 175 mM inhibit *rbc*S expression and accumulation of the Rubisco small subunit. Inhibition of production of the triosephosphate translocator is also inhibited. Artificially increasing carbohydrates in leaves (e.g. by preventing translocation through cold-girdling or by increasing photosynthesis with elevated CO_2) also has the same effect. The mechanism is conserved across a wide range of eukaryotes. Glucose is the main regulator of gene expression and involves hexokinase as the glucose sensor, although several other sugars may act in this way and have a role under adverse conditions, although there is no direct link between gross accumulation of carbohydrates and gene repression; possibly there are other factors which interact and control regulation.

10.6 Protein transport into chloroplasts

The mechanisms by which proteins synthesized in the cytosol or chloroplast stroma are incorporated into the stroma and thylakoid membranes are understood in some detail (see Robinson *et al.*, 2000). Those encoded in the plastid and destined for the stroma require no further transport, but if they are to be incorporated into the thylakoid membrane or lumen, then they are targeted directly into the particular compartment during or after synthesis. Nuclear encoded proteins are synthesized in the cytosol and imported into the chloroplast by a single import system in the envelope, but those for the thylakoid are modified before entry into the membranes, probably by four different mechanisms. Proteins synthesized in the cytosol and destined for the chloroplast stroma and thylakoid membranes (intrinsic proteins) or the lumen must be transported across the chloroplast envelope (two membranes) into the stroma, then those destined for the thylakoid membrane must enter the lipid bilayer (perhaps requiring specific mechanisms) whilst those for the lumen must pass across the thylakoid (three membranes to cross). There is a complex system of protein trafficking to transport the proteins coded by the nuclear genome and synthesized in the cytosol, and the proteins encoded in the chloroplast genome and synthesized in the chloroplast. Once in the final compartment, the proteins must combine into the complexes, as most of the membrane components are multimeric and made from proteins encoded in nucleus and chloroplast. The basic mechanism is probably derived from the prokaryotic (cyanobacterial) endosymbiont cell, much modified by later evolution as genes were transferred from symbiont to the host nucleus.

10.6.1 Protein transport from cytosol to stroma

Movement of proteins from the cytosol to stroma requires transport across the outer (cytosolic) and inner (stromal) chloroplast envelope membranes; a summary of the structures and mechanism is given in *Figure 10.3*. There is a single 'translocon' system for nearly all chloroplast proteins synthesized in the cytosol. The outer membrane has three integral proteins, Toc159 (or Toc36 plus Toc86), Toc75 and Toc34, forming a complex, and the inner membrane three integral proteins, Tic110, Tic22 (also called Toc25) and Tic20 (or 21), plus

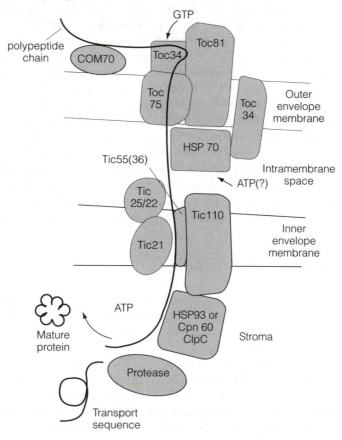

Figure 10.3. The translocation system of the chloroplast envelope for transport of proteins from the site of synthesis in the cytosol to the stroma, details are given in the text

possibly Tic55, also forming a complex. Detailed analysis of the proteins and the gene sequences and their homologies in bacteria and cyanobacteria (principally *Synechocystis*) suggests that the translocon is of both cyanobacterial and eukaryotic origin with extensive modification of the system during chloroplast evolution in the algae and higher plants, as the components have only low homology to bacterial or other transport systems. The system functions in the following way: the protein with transit sequences for the envelope binds to the outer membrane, probably to a molecular chaperone (heat-shock-protein-type protein family, com70) and then to Toc159 and, in an ATP requiring step, enters Toc75. The Toc and Tic components are in close arrangement and the protein moves through Toc75, perhaps via another chaperone (hsp70) in an energy-requiring step, to Tic20 and Tic22 (perhaps between them), into the stroma. As the protein emerges it is processed by stromal chaperones (the hsp60 homolog cpn60 and the chloroplast ClpC homolog), as indicated in *Figure 10.4*, associated with the stromal side of the Tic complex, specifically to Tic110. The chaperones bind to the unfolded protein and, with the energy of ATP, the structure of the mature protein is established. Finally, a stromal peptidase (protease) removes the transit sequence and the mature protein is incorporated into the stroma or then proceeds to the next stage for transport into the membrane or into the lumen.

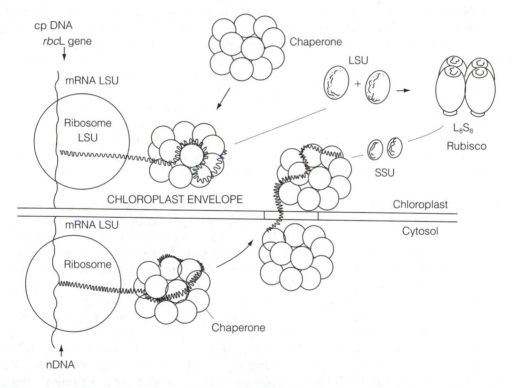

Figure 10.4. Highly schematic illustration of the role of a molecular chaperone, a protein of many subunits, in permitting the correct folding of the primary amino acid chain after synthesis in the ribosome. The situation for Rubisco large subunit (LSU) and small subunit (SSU) illustrates the role the molecular chaperones may have in transporting proteins between cellular compartments, the correct folding of the subunits and their assembly into the mature L8S8 Rubisco protein

10.6.2 Transport into the thylakoid

To enter the thylakoid lumen, proteins made in the cytosol (e.g. the products of psbQ, psbP and psbO, the 16, 23 and 33 kDa polypeptides of the water-splitting complex) cross the envelope as described. In the stroma, the envelope transit peptide is removed by peptidase. Then the thylakoid targeting domain reacts with the thylakoid transport mechanisms. There are two mechanisms for lumen proteins, allowing passage into the lumen where the transit sequence is removed and the mature protein structure formed. Two groups of lumen-proteins exist, one (e.g. 33 kDa water-splitting and plastocyanin) requires ATP and a carrier mechanism, called Sec-A, which is a stromal prokaryotic component which reacts with a translocase in the membrane. The other group (e.g. the 16 and 23 kDa water-splitting proteins) depends on the ΔpH across the membrane, and probably a distinct translocase system, but not SecA. This was shown by experiments on isolated chloroplasts and thylakoids, with and without ATP and ΔpH, and by competition studies where proteins within a group were found to compete with each other for transport but not with proteins from the other group. The structure of the transit sequences provides the targeting signals for the different mechanisms despite strong similarity between the two groups. The structure is prokaryotic with three distinct domains, an amino-N-terminal (N-) enriched in basic and hydroxylated amino acids and few acidic residues for entry into the stroma, a hydrophobic core (H-) enriched in serine, threonine, lysine and arginine, and a more hydrophobic carboxy-terminal (C-) domain. This may form a β-strand and has a role in the processing in the stroma. The composition and structure regulates transport across the thylakoid and entry into the lumen. The signal is given by the presence of arginine in the region between the N- and H-domains, together with the charge in the H and, particularly, the basic amino acid residues (lysine) in the C-domain. Probably this dual transport for lumen proteins is a consequence of symbiosis (SecA is present in cyanobacteria), but the others have no counterpart so may have evolved later, to accommodate proteins which, because of co-factors of secondary or tertiary structure, could not be transported by the Sec system.

Intrinsic, membrane-spanning proteins or parts of complexes with multiple membrane-spanning, hydrophilic, amino acid sequences, are transported by a third system. Light-harvesting chlorophyll a/b protein (LHCP), with three membrane-spanning sequences, is synthesized with only the envelope transit sequence, so the signals required for correct insertion into the membrane are derived from the mature protein. Transport into the membrane depends on a stromal protein (54CP) in a signal recognition particle, or SRP (perhaps part of a ribosomal complex) and GTP, not ATP. The complex recognizes and binds to a sequence of hydrophobic amino acids specific to the protein, and so not recognized by the ATP and Sec system proteins, and allows the protein to enter across the water–membrane interface and insert into the membrane lipid. A fourth system is used by proteins with a single membrane span, for example CF_0 subunit II and subunits W and X of PSII, with N-terminus in the lumen. Despite similarities in the transit peptide with Sec-transported proteins, the proteins integrate without Sec, SRP and GTP or ATP. Probably the structure of the protein allows insertion into the hydrophobic membrane, followed by reorganization of the secondary/tertiary form and removal of the signal segment by specific peptidases. The complexity suggests that during evolution the existing mechanisms of the cyanobacterial symbionts and the host were used and new systems devised as chloroplast genes were transferred to the nucleus and so protein import had to increase, at the same time as maintaining all the conditions required for efficient assembly of the photosynthetic system.

The mechanisms at the molecular level by which proteins recognize specific sequences and by which the large molecular components (often of a hydrophilic nature) pass the hydrophobic lipid membrane are now being analyzed. Transit peptidases have been partially purified and described but their regulation remains obscure. Once in the stroma the protein is assembled, in the case of Rubisco LSU and SSU, into the holoenzyme. Proteins destined for thylakoid complexes are incorporated into the membrane; details of the mechanism are unclear, particularly the signaling mechanisms. These ensure that the different components of complexes, often produced by very different synthetic pathways (e.g. chl *a* and *b* and xanthophylls), are available in the correct stoichiometry. Many interesting features of the translocation of proteins and processing remain to be answered and also many intriguing questions about the evolution of such a complex system. Was the transit peptide 'tacked onto' a pre-existing nuclear encoded enzyme, enabling it to enter the symbiont before the symbiont's own gene for the enzyme was eliminated in the course of evolution? Were the characteristics of the cpDNA product (e.g. inability of the polypeptide to unfold and refold if translocated) the reason why some proteins remained encoded in cpDNA and others moved to the nucleus? Perhaps the regulatory processes in the cytosol would have been disturbed by particular sequences of proteins which thus remained in the cpDNA because of the selection pressure, even if there was a positive advantage in terms of genome organization and expression for cpDNA genes to migrate into the nucleus. Better understanding of transport mechanisms into chloroplasts may provide a way of effectively directing foreign and modified proteins into the organelle to achieve a particular photosynthetic response. Attempts to incorporate foreign Rubisco, e.g. from *Anacystis*, into higher plant chloroplasts or chimeric types of proteins chloroplasts have largely been unsuccessful because the translocating and organizational aspects of system organization have not been capable of forming a correctly folded protein, resulting in precipitation in the cell.

10.6.3   Protein folding and molecular chaperones

Proteins, once made on the mRNA in the form of the primary amino acid chain, must undergo folding to achieve the correct secondary and tertiary structures which are essential for function. Also, if part of a complex, they must be combined with other components, such as prosthetic groups and metal ions. Multiunit proteins, such as Rubisco, are then assembled in the chloroplast stroma to form the active holoenzyme (*Figure 10.4*). Folding was assumed to be 'spontaneous', based on electrostatic charge and size and shape of the primary chain. However, with increasing use of molecular engineering to produce foreign proteins in cells, it became clear that often newly formed proteins would not fold and assemble properly even if the primary structure was correct. An example was given in the previous section of the problem of expressing foreign Rubisco in higher plant chloroplasts. A similar effect was observed when higher plant Rubisco was expressed in *Escherichia coli*. Only inactive, denatured protein was made which precipitated out in the bacteria. The reason for this is the absence of the machinery required to fold and assemble the enzyme correctly. In chloroplasts it was observed that folding of newly synthesized LSUs of Rubisco required brief, noncovalent binding with a large (720 kDa), soluble, oligomeric protein composed of 12 subunits of two types, α and β, of molecular mass 61 and 60 kDa, respectively. This Rubisco LSU binding protein has an α_6 β_6 structure and is nuclear encoded. It is made as a higher molecular mass precursors in the cytosol before transport into the chloroplast. It is 50% homologous to a protein of *E. coli* called cpn60 or GroEL protein, which is required for folding one of the head proteins of phage virus. The concept is that the unfolded polypeptides of the large and small subunits of Rubisco are 'chaperoned' by the binding protein (*Figure 10.3*) giving the folded form. The LSU polypeptide, in the presence of the chaperone, forms a dimer; dimers may associate, giving the L_8 structure of Rubisco to which the folded small subunits bind, thus forming the active L_8S_8 holoenzyme. There are many types of proteins involved in the folding of proteins in different cells. They may recognize the structure of the partly folded protein and bind to it in such a way that some deleterious but spontaneous arrangements are prevented whilst others can proceed correctly. This prevents synthesis of inactive proteins and their precipitation in cells. Chaperonins are possibly also involved in the transport of proteins from cytosol to chloroplast; one or more chaperonins may bind to the newly synthesized protein chains during or immediately after transcription and transport it to the membrane, and once in the stroma it is picked up by another chaperonin and the transit peptide removed. If the correct chaperonins are not available (as in a foreign cell) then active enzyme cannot be made or transported. Chaperonins are important in protein folding and DNA replication and appear to protect cell development from environmental stresses, for example, heat stress proteins (hsp) of different molecular mass (e.g. hsp70 and hsp60) with such function appear in temperature- and water-stressed plants.

10.7 Chloroplast development

In flowering plants, formation of the components of chloroplasts and their integration into a functional structure is a complex, very well regulated process which depends upon environmental signals, particularly light. This 'photomorphogenesis' (control of development by light) requires low-energy radiation and depends on light quality (Section 10.5.1). Regulation is thus through information, that is using light as a signal, rather than for provision of energy-rich assimilates from photosynthesis. In darkness, the photosynthetic system does not develop as in plants grown in light, instead it undergoes skotomorphogenesis (*skoto* is Greek for darkness). Light is needed to start development of chloroplasts and associated metabolic systems, for example chlorophyll formation. Chloroplasts do not arise *de novo* but from proplastids which in developing cells are of a generalized nature, *ca* 1 μm diameter with a double-bounding membrane and perforated internal membranes. Proplastids may develop into plastids of several types, including chloroplasts, leucoplasts and chromoplasts and this largely depends on whether the tissue is exposed to light or darkness. Chloroplasts develop through stages in which the early thylakoids become increasingly perforated and differentiated into the stacked and unstacked regions of the mature chloroplast. During this phase of growth, the plastids enlarge to 5–10 μm diameter. The sequence of events is shown most clearly in monocotyledonous plants where the basal leaf meristem gives rise to daughter cells which grow and mature as they pass into the upper mature part of the leaf blade. Thus, there is a developmental sequence related to age (the above sequence may take only 6 h at 25°C) and to distinct environmental stimuli such as light. Both cells and chloroplasts divide: chloroplast division has been long observed by microscopy in the form of a dumb-bell shaped chloroplast. The relative rates of the two processes determine the number of chloroplasts per cell.

In young meristematic cells, development under the influence of light involves formation of gene products and their assembly into structural and biochemical components of the cell. During normal development in light, the very young meristematic cells contain Rubisco and chlorophyll in small amounts and as development proceeds these increase 6- and 80-fold, respectively, by maturity. Both Rubisco LSU and SSU increase in parallel in a tightly coordinated fashion, but the mechanism is obscure. There is a large complement of 70S and 80S ribosomes per plastid in young cells; about 60% of the mature state. The α- and β-subunits of coupling factor appear together very early in development before the thylakoids are fully developed or stacked. However, LHC is only detected during the phase of rapid granal assembly, leading to mature chloroplasts. It is interesting that the mRNA for LHC peaks earlier than that for Rubisco. During the later formation of the thylakoids, there is massive accumulation of the lipids of the membranes with the appearance of Δ^3–transhexadecenoic acid in phosphatidylglycerol. In darkness, the proplastids form etioplasts which lack chlorophyll and other pigments but possess a prolamellar body, a circular lattice of interconnecting membrane tubules derived from the inner plastid envelope with a few single membranes extending from it. The prolamellar body is largely composed of the protochlorophyllide oxidoreductase enzyme protein with protochlorophyllide and specific lipids of thylakoids (e.g. MGDG and DGDG). The prolamellar body may be an artifact of interrupted development when light is excluded or a normal feature, if only very transiently, in normal environmental conditions. It is as if development is blocked at this stage in darkness. When the tissue is illuminated, the prolamellar bodies develop into the thylakoid system with both stacked and unstacked membranes. Their formation is accompanied by chlorophyll synthesis, increase in proteins and activation of pre-existing ones and the incorporation of these into the system. Fully developed etioplasts have coupling factor components, Rubisco and other PCR cycle enzymes, but lack the functional light-harvesting complexes of the light-grown tissue; chlorophyll synthesis is required. When etioplasts are illuminated chlorophyll synthesis occurs and a functional chloroplast may result from the breakdown and re-assembly of the prolamellar body, with formation of photosystems, development of the energy-transducing system and lastly synthesis of enzymes of the PCR cycle.

A most important process in development of thylakoids is the formation of chlorophyll Section 8.12). Protochlorophyll(ide) captures light and is converted to chlorophyll(ide), the holochrome protein acting as a protochlorophyllide reductase (NADPH:protochlorophyllide oxidoreductase). Protochlorophyllide reductase is loosely bound to the membranes of etioplasts and acts by forming a photoactive ternary complex with protochlorophyllide and NADPH, which on illumination transfers H to the porphyrin, giving the chlorophyllide. There is a characteristic shift in absorption from 630 to 638–650 nm resulting from binding of the protochlorophyllide to the complex and this is again shifted to 678 nm by light (resulting in one reduction step) and to 684 nm as the second light-independent reduction takes place, before chlorophyllide is released (absorption at

672 nm); these events take place with characteristic time scales. Protochlorophillide reductase and its mRNA are abundant in the prolamellar body of dark-grown chloroplasts and are actively synthesized in darkness. However, within 30 min of illumination the amount of protein and activity of the enzyme decreases to only 10% of that in darkness, with loss of translatable mRNA. Probably regulation operates via phytochrome, which controls the amount of mRNA, although there is the possibility that substrate availability is an important factor regulating enzyme activity. The absorption spectrum of chlorophyll in the developing membranes *in vivo* changes from 684 to 672 nm with the protochlorophyllide to chlorophyllide conversion, reflecting different binding of the pigments to proteins or other forms of aggregation or disaggregation, for example the pigment absorbing at 672 nm may be unbound chlorophyllide. The spectral changes occurring after 1 min of bright illumination allow conversion of protochlorophyllide (absorption at 650 nm) to chlorophyllide (at 683 nm) followed by a further transition to 673 nm. This is called the Shibata shift and may correspond to the time course of phytol addition to the substrate. Normal development of primary thylakoids (which are perforated double membranes) can only proceed to maturity if the conversion of protochlorophyll(ide) to chlorophyll *a* takes place, otherwise development is halted and the plastids degenerate.

10.8 Turnover and replacement of proteins in the chloroplast

Synthesis of proteins is not restricted to developing chloroplasts but is important in the mature system. The most obvious example is the D1 protein of PSII, which is the most rapidly turned-over protein in the chloroplast. Damage is related to regulation of the energy balance of reaction centers (Anderson and Aro, 1997; Section 5.7). If the quinone-binding site on D1 cannot accept an electron then the probability of photodegradation increases. D1 is damaged between the fourth and fifth membrane spans where a particular amino acid sequence, found in other rapidly turned-over proteins, occurs. The damaged polypeptide is removed and replaced in a reaction requiring ATP and involving the protein ubiquitin which 'labels' the damaged center so that it may be identified and removed. Removal may not be by ordinary proteolysis but breakdown products have been observed. Details of the process are not clear, for example, how the PSII complex remains stable without the D1 protein and for how long and where replacement occurs. The protein is made on ribosomes in the stroma and may enter into the reaction center directly, the damaged PSII migrating to the stromal thylakoids from the granal lamellae or first entering the stromal membranes and then migrating in the membrane to the damaged complex. The latter seems more likely as CP43 and CP47 only occur in the granal membranes; D1 may be linked to palmitoyl groups which allow diffusion and correct insertion of the protein into the complex. Some 2 h are required for incorporation into PSII. Environmental factors, such as bright light, which causes the damage, and cold which decreases the rate of protein synthesis, increases photoinhibition.

References and further reading

Andersson, B. and Aro, E.-M. (1997) Proteolytic activities and proteases of plant chloroplasts. *Physiologia Plant.* **100**: 780–793.

Andersson, B., Ponticos, M., Barber, J., Koiviemi, A., Hagman, Å., Salter, A.H., Dan-Hui, Y. and Lindahl, M. (1994) Light-induced proteolysis of photosystem II reaction centre and light-harvesting complex II proteins in isolated preparations. In *Photoinhibition of Photosynthesis* (eds N.R. Baker and J.R. Bowyer). BIOS Scientific Publishers, Oxford.

Argüello-Astorga, G. and Herrera-Estrella, L. (1998) Evolution of light-regulated plant promoters. *Annu. Rev. Plant Physiol. Mol. Biol.* **49**: 522–555.

Baker, N.R. and Bowyer, J.R. (eds) (1994) *Photoinhibition of Photosynthesis.* BIOS Scientific Publishers, Oxford.

Bartley, G.E., Coomber, S.A., Bartholomew, D.A. and Scolnic, P.A. (1991) Genes and enzymes for carotenoid biosynthesis. In *Cell Culture and Somatic Cell Genetics of Plants,* Vol. 7B, *The Photosynthetic Apparatus: Molecular Biology and Operation* (eds L. Bogorad and I.K. Vasil). Academic Press, San Diego, CA, pp. 331–345.

Bogorad, L. and Vasil, I.K. (1991a) *Cell Culture and Somatic Cell Genetics of Plants,* Vol. 7A, *The Molecular Biology of Plastids.* Academic Press, San Diego, CA.

Bogorad, L. and Vasil, I.K. (1991b) *Cell Culture and Somatic Cell Genetics of Plants,* Vol. 7B, *The Photosynthetic Apparatus: Molecular Biology and Operation.* Academic Press, San Diego, CA.

Durnford, D.G and Falkowski, P.G. (1997) Chloroplast redox regulation of nuclear gene transcription during photoacclimation. *Photosynth. Res.* **53**: 229–241.

Feifelder, D. (1987) *Molecular Biology* (2nd Edn). Jones and Bartlet, Boston, MA.

Gatenby, A.A., Viitanen, P.V., Speth, V. and Grimm, R. (1994) Identification, cellular localization, and participation of chaperonins in protein folding. In *Molecular Processes of Photosynthesis,* Vol. 10, *Advances in Molecular and Cell Biology* (ed. J. Barber). JAI Press, Greenwich, pp. 355–388.

Gillham, N.W. (1994) *Organelle Genes and Genomes.* Oxford University Press, Oxford.

Gray, J.C. (1996) Regulation of expression of nuclear genes encoding polypeptides required for the light reactions of photosynthesis. In *Oxygenic Photosynthesis: the Light Reactions* (eds D.R. Ort and C.F. Yocum). Kluwer Academic, Dordrecht, pp. 621–641.

Gruissem, W. and Tonkyn, J.C. (1993) Control mechanisms of plastid gene expression. *Crit. Rev. Plant Sci.* **12**: 19–55.

Hartman, F.C. and Harpel, M.R. (1994) Structure, function, regulation and assembly of D-ribulose-1, 5-bisphosphate carboxylase/oxygenase. *Annu. Rev. Biochem.* **63**: 197.

Heins, L., Cullinson, I. and Soll, J. (1998) The protein translocation apparatus of chloroplast envelopes. *Trends Plant Sci.* **3**: 56–61.

Huang, A.H.C. and Taiz, I. (eds) (1991) *Molecular Approaches to Compartmentation and Metabolic Regulation.* American Society of Plant Physiology, Rockville.

Jang, J.-C. and Sheen, J. (1997) Sugar sensing in higher plants. *Trends Plant Sci.* **2**: 208–214.

Jenkins, G.I. (1991) Photoregulation of plant gene expression. In *Developmental Regulation of Plant Gene Expression, Plant Biotechnology,* Vol. 2 (ed. D. Grierson). Blackie, Glasgow, pp. 1–41.

Kapoor, S. and Sugiura, M. (1998) Expression and regulation of plastid genes. In *Photosynthesis: a Comprehensive Treatise* (ed. A.S. Raghavendra). Cambridge University Press, Cambridge, pp. 72–86.

Keegstra, K. and Cline, K. (1999) Dissecting the protein import and routing systems of chloroplasts. *Plant Cell* **11**: 557–570.

Kessler, F. and Blobel, G. (1996) Interaction of the protein import and folding machineries of the chloroplast. *Proc. Natl Acad. Sci. USA* **93**: 7684–7689.

Kloppstech, K. (1997) Light regulation of chloroplast genes. *Physiol. Plant.* **100**: 739–747.

Koch, K.E. (1996) Carbohydrate-medulated gene expression in plants. *Annu. Rev. Plant Physiol. Mol. Biol.* **47**: 509–540.

Kuhlemeier, C. (1992) Transcriptional and post-transcriptional regulation of gene expression in plants. *Plant Mol. Biol.* **19**: 1–14.

Leon, P., Arroyo, A. and Mackenzie, S. (1998) Nuclear control of plastid and mitochondrial development in higher plants. *Annu. Rev. Plant Physiol. Mol. Biol.* **49**: 453–480.

Maliga, P. (1994) Isolation and characterization of mutants in plant cell culture. *Annu. Rev. Plant Physiol.* **35**: 519–542.

Martin, W. and Herrmann, R.G. (1998) Update on gene transfer from organelles to the nucleus. Gene transfer from organelles to the nucleus: how much, what happens, and why? *Plant Physiol.* **118**: 9–17.

Pratt, L.H., Senger, H. and Galland, P. (1990) Phytochrome and other photoreceptors. In *Methods in Plant Biochemistry,* Vol. 4, *Lipids, Membranes and Aspects of Photobiology* (eds J.L. Harwood and J.R. Bowyer). Academic Press, London, pp. 185–230.

Reumann, S. and Keegstra, K. (1999) The endosymbiotic origin of the protein import machinery of chloroplastic envelope membranes. *Trends Plant Sci.* **4**: 302–307.

Robinson, C. and Mant, A. (1997) Targeting of proteins into and across the thylakoid membrane. *Trends Plant Sci.* **2**: 431–437.

Robinson, C., Woolhead, C. and Edwards, W. (2000) Transport of proteins into and across the thylakoid membrane. *J. Exp. Bot.* **51** (GMP Special Issue): 369–374.

Rochaix, J.-D. (1997) Chloroplast reverse genetics: new insights into the function of plastid genes. *Trends Plant Sci.* **2**: 419–425.

Roell, M.K. and Gruissem, W. (1996) Chloroplast gene expression: Regulation at multiple levels. In *Oxygenic Photosynthesis: the Light Reactions* (eds D.R. Ort and C.F. Yocum). Kluwer Academic, Dordrecht, pp 565–587.

Roy, H. and Gilson, M. (1997) Rubisco and the chaperonins. In *Handbook of photosynthesis* (ed. M. Pessarakli). Marcel Dekker, New York, pp. 295–304.

Roy, H. and Nierzwicki-Bauer, S.A. (1991) Rubisco: genes, structure, assembly, and evolution. In *Cell Culture and Somatic Cell Genetics of Plants,* Vol. 7B, *The Photosynthetic Apparatus: Molecular Biology and Operation* (eds L. Bogorad and I.K. Vasil). Academic Press, San Diego, CA, pp. 347–364.

Schnell, D.J. (1998). Protein targeting to the thylakoid membrane. *Annu. Rev. Plant Physiol. Mol. Biol.* **49**: 97–126.

Shinozaki, K., Hayashida, N. and Sugiura, M. (1989) *Nicotiana* chloroplast genes for components of the photosynthetic apparatus. In *Molecular Biology of Photosynthesis* (eds Govindjee *et al.*). Kluwer Academic, Dordrecht, pp. 1–25.

Sigler, P.B., Xu, Z., Rye, H.S., Burston, S.G. Fenton, W.A. and Horwich, A.L. (1998) Structure and function in GroEL-mediated protein function. *Annu. Rev. Biochem.* **67**: 581–608.

Smeekens., S. and Rook, F. (1997) Sugar sensing and sugar-mediated signal transduction in plants. *Plant Physiol.* **115**: 7–13.

Staub, J.M. and Deng, X.-W. (1996) Light signal transduction in plants. *Photochem. Photobiol.* **64**: 897–905.

Stern, D.B., Higg, D.C. and Yang, J. (1997) Transcription and translation in chloroplasts. *Trends Plant Sci.* **2**: 308–315.

Sugita, M. and Sugiura, M. (1996) Regulation of gene expression in chloroplasts of higher plants. *Plant Mol. Biol.* **32**: 315–326.

Sugiura, M. (1991) Structure and replication of chloroplast DNA. In *Frontiers in Molecular Biology: Molecular Biology of Photosynthesis* (eds B. Anderson, A.H. Salter and J. Barber). IRL Press, Oxford, pp. 58–74.

Thompson, W.F. and White, M.J. (1991) Physiological and molecular studies of light regulated nuclear genes in higher plants. *Annu. Rev. Plant Physiol. Plant Mol. Biol.* **33**: 432–466.

Vermaas, W.F.J. and Ikeuchi, M. (1991) Photosystem II. In *Cell Culture and Somatic Cell Genetics of Plants*, Vol. 7B, *The Photosynthetic Apparatus: Molecular Biology and Operation* (eds L. Bogorad and I.K. Vasil). Academic Press, San Diego, CA, pp. 26–112.

Weihe, A. and Börner, T. (1999). Transcription and the architecture of promoters in chloroplasts. *Trends Plant Sci.* **4**: 169–170.

Carbon dioxide supply for photosynthesis

11.1 Introduction

Carbon dioxide, the major substrate for photosynthesis, is supplied from the medium in which an organism lives: water for aquatic bacteria, algae and some higher plants and the atmosphere for terrestrial plants. Only transport of carbon dioxide from the atmosphere to higher plant leaves is considered here, although the basic principles are applicable to other situations. To understand the factors controlling the fluxes of CO_2 the relationships between mass and volume of gases, the effects of temperature and pressure and the basis of the expression of gas composition are briefly considered, as it is often essential to interconvert volume, mass and pressure of a gas in determining rates of photosynthesis using gas exchange techniques.

11.2 Gas laws

For an ideal gas, and employing SI units, the relation between the number of moles of gas (n), volume (V, m^3) and pressure (P) in pascals (1 Pa = 1 N m^{-2}) at a temperature (T, in kelvin) is given by the ideal gas equation:

$$PV = nRT \tag{11.1}$$

The molar gas constant (R) is 8.31 m^3 Pa mol^{-1} K^{-1}. Standard atmospheric pressure is 101 325 Pa = 1.013 bar = 760 mmHg. A mole of gas at 273K (0°C) and 101 325 Pa occupies 0.0224 m^3 (22.414 dm^3 or liters in the frequently employed but non-SI unit) and contains Avogadro's number of particles (6.023 × 10^{23}).

Concentrations of a component gas in a mixture of gases may be expressed in several different ways: number of moles per cubic meter (molar concentration) or mass per unit volume (mass concentration) may be employed. Also the ratio of the number of moles of the component gas as a proportion of a total number of moles, the mole fraction, is used. Mole fraction does not change when P or T are altered, although the molar and mass concentrations do. Another basis of expression frequently used in photosynthetic studies and gas analysis is the volume of the component gas per volume of total gas. From equation (11.1) it is clear that, for a given gas mixture, the volume/volume and mole fractions, as well as the pressure of the component compared to the total pressure of the mixture, are identical and will change with, for example, T in the same way. Many units of volume concentration are used and may be confusing, e.g. $m^3\ m^{-3}$, $1\ 1^{-1}$ or $\mu l\ l^{-1}$; the latter is equivalent to 1 volume per million volumes and this is the frequently used (volume) parts per million, abbreviated to vpm, vppm or ppm. Volume units are also expressed as volume %, e.g. 1 vpm = 0.0001 volume %. However, SI units of volume concentration are $m^3\ m^{-3}$ or preferred prefixes. Air contains about 360 $cm^3\ CO_2\ m^{-3}$, equivalent to 360 vpm or 0.036% CO_2 and 0.21 $m^3\ O_2\ m^{-3}$ or 210 000 vpm or 21 (volume)% O_2. Volumes of gas may be converted to mass and mole from equation (11.1); air at 0°C and 101 325 Pa pressure contains 36 Pa CO_2 or 706 mg $CO_2\ m^{-3}$ or 16 mmol $CO_2\ m^{-3}$ and 300 g or 9.4 mol $O_2\ m^{-3}$.

The contribution of a gas to a mixture of gases may be described by the partial pressure. Dalton's law of partial pressures states that in a mixture of gases (which do not interact) the partial pressure of each gas is the pressure which it exerts if occupying the volume alone and the total pressure is the sum of the partial pressures. For air (ignoring rare gases)

$$P_{total} = P_{O_2} + P_{N_2} + P_{CO_2} + P_{H_2O} \tag{11.2}$$

Partial pressure of CO_2 in air at standard T and P is currently 36 Pa (or in non-SI units 360 μbar) and that of O_2 is 21.3 kPa (or 210 mbar). Partial pressure changes with T and P [equation (11.1)] and therefore must be corrected for atmospheric pressure, altitude and temperature when making comparisons. As mentioned the mole fraction is equal to the partial pressure divided by the total pressure. *Table 11.1* gives some conversion factors for units expressing the composition of CO_2 in air under standard conditions.

It is important to appreciate the great difference in concentration between CO_2 and O_2 in the atmosphere; O_2 is 600 times more concentrated than CO_2, which favors even inefficient oxygenation reactions compared with carboxylation reactions, for example, RuBP oxygenase compared with carboxylase. Plants must accumulate CO_2 from very dilute concentration and, at the same time, function at large O_2 concentration with large gradients of water vapor pressure between leaf and atmosphere. A leaf at 23°C with the

Table 11.1. Conversion factors for CO_2 and O_2 concentrations in air at 20°C and 101 315 Pa

---

1 μl $CO_2\ l^{-1}$ = 41.6 × 10^{-6} ml $CO_2\ m^{-3}$ = 1.83 x 10^{-3} g $CO_2\ m^{-3}$
1 mg $CO_2\ m^{-3}$ = 0.0227 mmol m^{-3} = 0.554 μl $CO_2\ l^{-1}$
1 ml $CO_2\ m^{-3}$ = 44 g m^{-3} = 24.4 $cm^{-3}\ l^{-1}$
21% O_2 = 210 $cm^3\ O_2\ l^{-1}$ = 0.21 $m^3\ O_2\ m^{-3}$
1 mol $O_2\ m^{-3}$ = 32 g m^{-3} = 24.4 $cm^{-3}\ l^{-1}$

Pressure
1 μl l^{-1} = 1.01 μbar = 0.101 Pa
340 μl $CO_2\ l^{-1}$ = 340 μbar = 34.4 Pa
21% O_2 = 210 mbar = 21.3 kPa

---

internal air saturated with water vapor (100% relative humidity) has a vapor pressure of about 2.8 kPa compared with 1.5 kPa of air at 25°C and 50% relative humidity; this corresponds to a gradient of water content of about 12 g m^{-3}. Thus, leaves must absorb CO_2 whilst limiting loss of water vapor. Stomata function as 'control valves' which regulate these conflicting interests (Section 11.5).

11.3 Solubility of CO_2 and O_2 in water

Photosynthesis consumes forms of CO_2 dissolved in water, as enzymatic reactions proceed only in the aqueous state. Ribulose bisphosphate carboxylase uses dissolved CO_2, and PEP carboxylase consumes bicarbonate ions. Solubility of CO_2 and O_2 depends on solvent temperature, partial pressure of the gas, and the chemical nature of gas and solvent. For dilute solutions, Henry's law states that at constant temperature the volume or mass of gas dissolved in a given volume of liquid is directly proportional to the pressure. *Table 11.2* gives the solubility of CO_2 and O_2 in water. For air, 6.51 cm^3 of O_2 dissolve in 1000 cm^3 or 0.291 mol m^{-3} (2911 µM). For CO_2 at a partial pressure of 34 Pa and with solubility of 0.888 m^3 CO_2 m^{-3} water, 0.29 cm^3 dissolve in 1000 cm^3 or 12 mmol CO_2 m^{-3} (12 µM) (see *Table 11.3*). Gases are less soluble, generally, at warm temperatures than cold (*Table 11.2*). The solubility of gases is inversely proportional to the concentration of solutes. In complex biological fluids, proteins and salts decrease the solubility of all gases similarly.

Table 11.2. Solubility of CO_2 and O_2 in water at 101 325 Pa pressure of the gas as a function of temperature

T(°C)	mol CO_2 m^{-3}	m^3 CO_2 m^{-3}	mol O_2 m^{-3}	m^3 O_2 m^{-3}
0	76.4	1.71	2.17	0.049
10	51.4	1.19	1.68	0.038
15	43.1	1.02	1.50	0.034
20	36.5	0.87	1.36	0.031
25	31.1	0.76	1.23	0.028
30	26.7	0.67	1.12	0.026
40	21.2	0.53	1.01	0.023

Table 11.3. Solubility of CO_2 and O_2 in water when in equilibrium with air containing 32 Pa (320 µbar) of CO_2 and 21 kPa (21%) O_2 at atmospheric pressure as a function of temperature

T(°C)	µM CO_2	cm^3 CO_2 (m^3 water)$^{-1}$	µM O_2	m^3 O_2 (m^3 water)$^{-1}$	Molar ratio, O_2/CO_2
0	23	515	458	0.0102	19.9
10	16	371	356	0.0079	22.2
20	12	290	291	0.0066	24.3
25	10	244	263	0.0059	26.3
30	9	223	245	0.0055	27.2
40	7	179	216	0.0048	30.9

After Sesták *et al.* (1971).

11.3.1 Carbon dioxide

In pure water over 99% of CO_2 is in solution, but some is hydrated to carbonic acid, H_2CO_3, which dissociates to give the bicarbonate ion, HCO_3^-, and H^+, decreasing pH:

$$CO_2 \leftrightarrows CO_2 + H_2O \leftrightarrows H_2CO_3 \leftrightarrows H^+ + HCO_3 \qquad (11.3)$$
$$\text{(gas)} \quad \text{(dissolved)} \quad \text{(solution)} \quad \text{(solution)}$$

Formation of carbonate ions, CO_3^{2-}, from $HCO_3^- \leftrightarrows CO_3^{2-} + H^+$ is not considered further. If H_2CO_3 or HCO_3^- are removed by chemical reactions then the apparent solubility of CO_2 changes. As H_2CO_3 in solution is only 1/400 of the other components, it may be neglected. The pH of solutions of CO_2 in water depends on the molar concentration of CO_2, the gas phase CO_2 partial pressure, the dissociation constant (pK) and bicarbonate ion concentration. This is expressed in the Henderson–Hasselbach equation:

$$pH = pK + \log [HCO_3^-] - \log [CO_2] \qquad (11.4)$$

The pH determines the balance between CO_2 dissolved and the bicarbonate ion concentration; H^+ ions drive the reaction in favor of CO_2 (by the law of mass action) so that in acid solutions there is little bicarbonate (acids are used to remove bicarbonate from solution and produce CO_2). Changes in CO_2 and bicarbonate concentration are important to photosynthetic organisms, algae, for example, grow in acid or alkaline waters differing greatly in temperature. With increasing acidity and temperature the concentrations of bicarbonate ions and dissolved CO_2 decrease, and with them the supply of substrate for photosynthesis. Also, within photosynthesizing tissues the cellular fluids, such as the chloroplast stroma, have a variable pH which, together with temperature, influences the CO_2 concentration. The solubilities of CO_2 and O_2 and their ratio in equilibrium are shown in *Table 11.3*. The solubility of O_2 is not affected by pH so the CO_2/O_2 ratio rises greatly with increased alkalinity. As temperature increases, the solubility of O_2 decreases less than that of CO_2 so the ratio of O_2/CO_2 increases substantially.

11.3.2 CO_2 and bicarbonate equilibria and photosynthesis

The chloroplast stroma is alkaline during illumination and bicarbonate ions predominate, but Rubisco uses CO_2 as substrate, not HCO_3^-. This is shown by supplying the enzyme with $^{14}CO_2$ or $H^{14}CO_3$ at high pH and low temperature (10°C), where HCO_3^- formation from CO_2 is very slow. With $^{14}CO_2$ the reaction can proceed and ^{14}C is incorporated into 3PGA but $H^{14}CO_3^-$ is not used. By adding the enzyme carbonic anhydrase, which increases the equilibration rate greatly, $H^{14}CO_3^-$ is used as well, as it is converted to $^{14}CO_2$. During photosynthesis, in order to avoid starving Rubisco of substrate CO_2, the rate of supply of CO_2 must match the rate of reaction. The rate of conversion of bicarbonate to CO_2 in alkaline conditions is slow, so that both the CO_2 concentration in solution and the rate of supply to Rubisco could limit assimilation at high pH. However, the rate of conversion is increased 100-fold in tissues by carbonic anhydrase; as CO_2 is depleted it is rapidly produced from bicarbonate. Conversely, if CO_2 is available in solution but HCO_3^- is needed (by PEP carboxylase) then the rate of HCO_3^- formation is also increased by carbonic anhydrase.

Carbonic anhydrase. Carbonic anhydrase, a protein of 180 kDa molecular mass and containing zinc, is found in three distinct forms in photosynthetic tissues, often in very large amounts, particularly in the chloroplast and at cell membranes (see Coleman, 2000). It also occurs in nonphotosynthetic tissues and in animals. The enzyme, which has an extremely large turnover number (10^6 s^{-1}), facilitates the diffusion of CO_2 by speeding up the formation of HCO_3^- in the cytosol and maintaining a large gradient of CO_2 concentration. Also, carbonic anhydrase allows HCO_3^- to act as a buffer (albeit small) to provide CO_2 if the supply temporarily fails. In the chloroplast, carbonic anhydrase prevents depletion of CO_2 and serves to maintain a high partial pressure of CO_2 at the active site of Rubisco, although it is below that of the intercellular spaces due to other limitations (Section 11.4.4), and minimizes the effects of low atmospheric CO_2. Plants or algae grown in high concentrations of CO_2 contain less carbonic anhydrase than those grown in low concentrations. Also, chloroplasts of C3 plants contain more carbonic anhydrase than C4 plants, which have a CO_2 concentrating system, although the enzyme is active in the C4 mesophyll. Algae, particularly in alkaline natural waters, have carbonic anhydrase at the cell surface which increases the rate of supply of CO_2 to the cells.

11.3.3 Carbon dioxide concentrating mechanisms

Metabolic systems which increase the effective CO_2 concentration for Rubisco are familiar from discussion of the C4 and CAM mechanisms (Chapter 9). The advantages are apparent even though CO_2 diffuses to Rubisco active sites in leaves mainly in air and only over short distances in the liquid phase of the cell. In aquatic algae and higher plants, CO_2 and bicarbonate ions must diffuse through a water film around the surfaces of cells in tissues. As diffusion of CO_2 in air is 10^4 times greater than in water, the problem of CO_2 supply in aquatic environments is much greater than in the aerial environment. In acid water, the total carbon content is small and the bicarbonate–CO_2 equilibrium lies far towards CO_2.

Aquatic organisms such as cyanobacteria, green algae and aquatic higher plants (e.g. *Elodea*), are limited by the carbon supply for photosynthesis. In some organisms active transport systems which increase the CO_2 concentration in cells have developed; their effect is similar to that of C4 photosynthesis. Cyanobacteria often grow in environments where the CO_2 concentration is 5–20 times smaller than the K_m (CO_2) of Rubisco. The concentrating mechanism is an active metabolic pump which transports bicarbonate ions and leads to internal concentrations 40–1000-fold greater than in the medium. The pump in the cyanobacterium *Anacystis* is a 42 kDa polypeptide on the cytoplasmic membranes, using energy from photosynthesis and ATP. An exchange of bicarbonate and CO_2 with H$^+$ and Na$^+$ probably occurs. Carbonic anhydrase may be combined with the pump to maintain the required rate of supply of substrate for photosynthesis under adverse conditions of pH, *et cetera*. Rubisco within the cell is enclosed in a protein shield (in which carbonic anhydrase may also be involved) with the small subunits of Rubisco possibly forming part of the shield. This carboxysome (*Figure 11.1*) increases the CO_2 within the shield; bicarbonate is pumped into the shield and converted to CO_2 (carbonic anhydrase speeding up the processes), thus increasing the CO_2 in the carboxysome so that Rubisco can function efficiently. Such mechanisms are very important adaptations to extreme environments and may be induced by the availability of CO_2.

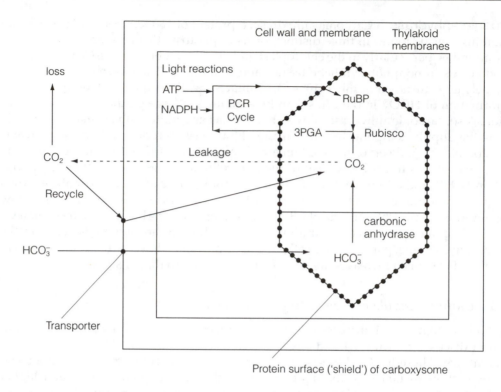

Protein surface ('shield') of carboxysome

Figure 11.1. Carboxysome structure and function in photosynthesis in cyanobacteria. The carboxysome functions as a CO_2 concentrating mechanism allowing Rubisco to function efficiently in an environment at alkaline pH where the main source of carbon is bicarbonate (HCO_3^-) ions. Carbonic anhydrase and Rubisco are restricted to the carboxysome. Carbonic anhydrase ensures rapid conversion of HCO_3^- to CO_2. The protein surface to the carboxysome decreases leakage of CO_2

11.3.4 Carbon dioxide supply to photosynthetic cells in leaves

Carbon dioxide in the chloroplast stroma is removed by the RuBP carboxylase reaction and a gradient of CO_2 develops across the chloroplast envelope, cytosol, cell membranes and walls to the intercellular spaces and, via the stomata, to the ambient air (*Figure 11.2*). The gradient is the driving force for CO_2 diffusion, but the rate at which this occurs to the reaction site depends on the conductances (reciprocal of resistance) to CO_2 diffusion in the gas and liquid phases in the leaf and atmosphere and on external CO_2 concentration. In the atmosphere, CO_2 diffuses toward and O_2 away from the leaf during illumination and the fluxes are reversed in darkness; water vapor diffuses away from the wet internal surfaces in the leaf to the dry atmosphere under most conditions. In the turbulent air, gas concentrations are effectively uniform, due to rapid mixing by mass flow. In the intercellular gas spaces and also in the liquid spaces in the cell wall and cytosol, the movement of gases is by diffusion, not by mass mixing. In the 'boundary layer' surrounding the cell or leaf there is incomplete mixing by mass flow and gases move partially by diffusion. Diffusion is the

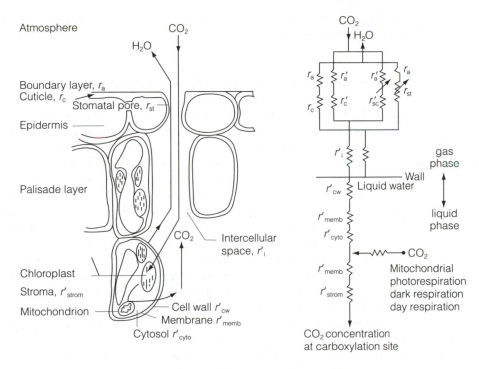

Figure 11.2. Pathway for water vapor loss from, and carbon dioxide entry into a leaf in the light, shown as an electrical analog with resistances denoting sites of restricted diffusion, i.e. small conductances (conductance is the reciprocal of resistance) related to leaf structure. r', resistance to CO_2; r, resistance to H_2O

movement of molecules of a substance from higher to lower concentration and diffusion coefficients depend on molecular species, the medium, and temperature and pressure. It is very rapid over short distances and faster in gases than liquids, for example diffusion of CO_2 in air is 0.16 cm^2 s^{-1} and in water 0.16×10^{-4} cm^2 s^{-1} at 15°C. Graham's law states that the diffusion coefficient for a gas in air is approximately equal to the reciprocal of the square root of the molecular mass. Thus, H_2O (mass 18) and CO_2 (mass 44) have, in air, diffusion coefficients of 0.257 and 0.16 cm^2 s^{-1}, respectively, so that the ratio of DH_2O/DCO_2 is approximately 1.6. Thus, it is possible to convert conductances for water vapor to those for CO_2 by this factor. This applies to diffusion in air. In the boundary layer over a leaf with the transition from diffusion to turbulence, simple diffusion does not apply and the rate of diffusion of water vapor is approximately 1.37 times that of CO_2. When the stomata are closed, water vapor and CO_2 diffuse through the plant surfaces, where the cuticle is the main barrier; in this complex medium the diffusion of the two gases is very different with CO_2 more restricted, so DH_2O/DCO_2 may increase greatly. Diffusion coefficients increase for all gases by approximately 1.3 between 0 and 50°C so the ratios are almost constant over the physiological temperature range. The magnitude of the diffusion coefficients and their ratio are important in calculating fluxes and for determining the diffusion of one molecular species from measurements of another. Diffusion of gases and ions in the liquid phases of cell walls and cytosol is particularly important as it slows the flux of materials and may limit the rate of photosynthesis.

11.4 Measurement of leaf water vapor and CO_2 exchange

Fluxes of CO_2, O_2 and H_2O between a leaf and surrounding atmosphere may be measured in several ways both separately and, more usefully, in combination to provide understanding of their interaction and dependence on the plant characteristics and environmental conditions. Physiological measurements are generally made by enclosing a leaf of known projected surface area (L, m²) (*Figure 11.3*) in a transparent chamber through which air (or another gas) flows (mol s⁻²). The air is stirred vigorously to maximize boundary layer conductance and keep it constant. Water vapor pressure is measured by a suitable sensor, for example dew point hygrometer, capacitative resistance sensors or infra-red absorption in the air entering and leaving the chamber and the water vapor flux (mol m⁻² s⁻¹) is calculated from the differences in water vapor pressure, flow rate and leaf area. The mean water vapor in the chamber air is calculated from the inlet and outlet humidity and vapor pressure in the intercellular spaces is determined by measuring the leaf temperature and assuming that air is saturated at that temperature. The boundary layer is estimated by using wet replica leaves or from the leaf energy balance. This basic approach is also used in porometers which rapidly measure water vapor flux from the leaf surface, leaf temperature and humidity of the air to compute conductance. Carbon dioxide exchange may be measured at the same time as water vapor using infrared analyzers to determine the concentration of CO_2 in the air entering and leaving the leaf chamber; from this, the flow rate and leaf area, the rate of CO_2 exchange (photosynthesis in the light, respiration in the dark under normal atmospheric CO_2 and O_2 pressures) is determined. Infrared gas analysis depends on the absorption of infrared radiation by CO_2 over the range 2.5–25 µm. The main peaks of CO_2 absorption are at 2.7, 4.2 (principal band) and 17 µm. There are several designs of infrared analyzer, one type has two tubes, with windows at each end made of fused silica or calcium fluoride which are transparent to infrared radiation. Radiation from a source is passed through the end windows along the tube to a detector; CO_2 molecules in the tubes absorb radiation and so radiation reaching the detector is inversely proportional to the number of CO_2 molecules in the light path. Differences in radiation absorption caused by differences in CO_2 concentration in the tubes are detected by the instrument and may be calibrated to provide a measure of CO_2 concentration of air entering and leaving a leaf chamber. Water vapor also absorbs infrared radiation (in three major bands at 2.5–2.8 µm, which overlaps with one CO_2 band, 5–8 and 18–28 µm). As water vapor is at much greater concentration than CO_2 in the normal atmosphere, it must be removed or special optical filters used to block the water absorption bands. Infrared analyzers are sensitive to differences in partial pressure of 0.01 Pa CO_2 over a range of 0–100 Pa, enabling CO_2 exchange to be routinely measured.

Oxygen pressure may be measured by paramagnetic detectors or zirconium chemical cells, but these methods are generally too insensitive to detect the small fluxes in O_2 from leaves against the very large background of O_2. In small volume, closed-chamber polarographic oxygen electrodes are sufficiently sensitive to determine the exchange but have not been generally applied to open gas exchange systems.

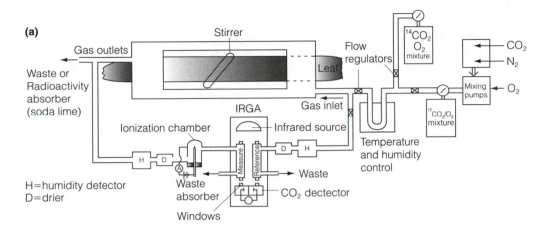

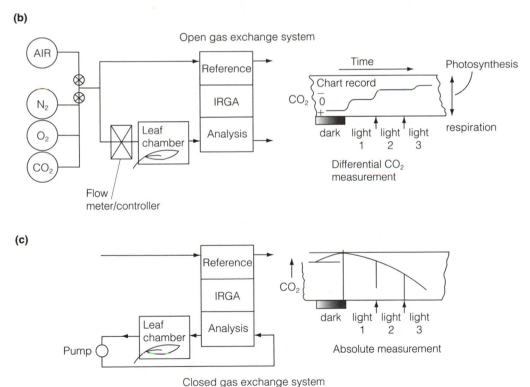

Figure 11.3. (a) Leaf chamber and 'open' gas exchange system for measurement of photosynthesis under steady-state CO_2 concentration with regulated O_2 and humidity. Addition of $^{14}CO_2$, premixed in a pressurized cylinder, and ionization detector enables $^{14}CO_2$ uptake to be measured at the same time as $^{12}CO_2$ and the leaf may also be sampled for analysis of radioactive assimilation products. (b) Open and (c) closed gas exchange systems

Techniques of gas exchange. The type of system shown diagrammatically in *Figure 11.4b* is an open system, as the gas flowing into the chamber, is of constant composition and no recirculation of the gas occurs. The rate of CO_2 assimilation and the composition of the atmosphere are constant, providing that light and other features of the plant's environment are not altered. The difference between the CO_2 in the gas entering and leaving the chamber is used to calculate the exchange rate. By altering the CO_2 in the gas stream entering the chamber, CO_2 response curves can be constructed and also, by

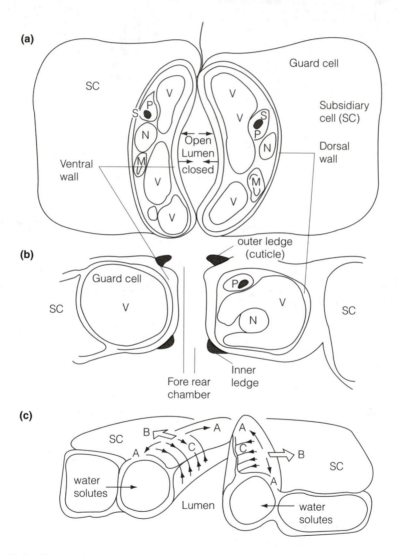

Figure 11.4. Semi-schematic view of a dicotyledonous plant stomatal complex: (a) in plan view indicating cell contents; (b) in section. V, vacuole; N, nucleus; P, plastid with S, starch; M, mitochondrion. SC, subsidiary cell. (c) Forces in the stomatal apparatus generated when water and solutes enter the guard cell. Microtubules in the ventral walls, orientated as indicated (C), coupled with the thickenings prevent expansion into the lumen but allow it to expand into the subsidiary cells (SC) because the walls are thin and not braced longitudinally. Expansion is possible in direction A

changing irradiance, light responses can be determined. Closed gas exchange systems, in contrast (*Figure 11.4c*), recirculate the air over the leaf so the composition of the atmosphere changes with time and the leaf experiences different humidity and CO_2 and its rate of photosynthesis decreases with CO_2 depletion from which the photosynthetic rate is calculated. This type of system may also be used to determine, for example, the CO_2 compensation point by allowing the leaf to come to an equilibrium with respect to CO_2 in the atmosphere; it is difficult to seal closed systems against the inflow of CO_2 from outside and allowances must be made for this. Yet another approach for measuring photosynthesis is to use a closed system but to inject gas containing CO_2 into the chamber to maintain a gas phase of required, constant composition; the rate of injection of CO_2 provides a measure of photosynthesis. This method rests on the use of sensitive, accurate flow monitors and controllers which respond to the infrared analyzer. Maintenance of constant humidity in closed gas exchange systems is more difficult than in open systems and is achieved by circulating part of the chamber air through a desiccant.

Additional methods used in photosynthesis studies. Radiotracer studies.   Measurement of photorespiratory CO_2 efflux may be made by allowing a leaf in an open gas exchange system to reach steady-state net photosynthesis with $^{12}CO_2$ and then switch to a gas of identical composition except that it contains $^{14}CO_2$ of known specific activity (^{14}C/total C ratio). The depletion of $^{14}CO_2$ from the gas in the first 15–60 s after passage over the leaf (measured with a flow-through ionization chamber or scintillation detector of small volume) allows the gross flux of CO_2 into the leaf to be calculated from the ^{14}C depletion and the specific activity. During this short period ^{14}C enters the PCR cycle and the glycolate pathway but very little ^{14}C is evolved in photorespiration (Chapter 8), hence allowing gross photosynthesis to be determined.

Integrating sphere.   The Ulbrich or integrating sphere is a method for determination of the number of photons required for the fixation of a given amount of CO_2, which requires accurate estimation of the absorption of photons by the leaf, independent of reflection or transmission of radiation (Section 13.5), to give the true quantum yield. This device has a highly reflective inner surface (coated with Eastman white reflecting paint, for example) which allows the absorbed light to be measured without the effect of reflectance or transmittance. Light enters the chamber from light guides and is reflected to give diffuse radiation. A chamber enabling the CO_2 exchange of the leaf to be measured as described earlier is placed in the integrating sphere and photosynthesis measured over a range of irradiances within the sphere. The irradiance at the sphere surface is determined by quantum sensors and the difference between the irradiance measured with and without the leaf in the chamber is used to determine the net photon absorption. This is then related to CO_2 exchange close to the light compensation point. From this, the true quantum yield is obtained. Details of techniques may be obtained from Šesták *et al.* (1971), Pearcy *et al.* (1989) and Hashimoto *et al.* (1990).

11.4.1   Pathway of water vapor flux and stomatal conductance

Evaporation from the water-saturated cell surfaces determines the vapor pressure in the intercellular spaces, which is assumed to be the saturated vapor pressure of pure water at the bulk leaf temperature, as mesophytic leaves are essentially isothermal throughout.

Even under extreme wilting, the vapor pressure may be saturated. For a leaf in a gas exchange chamber (*Figure 11.4*), with air of controlled, unsaturated humidity entering and passing over the surface (as in a steady-state open gas analysis system), the rate of transpiration (E, mol H_2O m^{-2} s^{-1}) for a leaf of plan area L (m^2) is given by the difference in the water vapor content of air entering and leaving the chamber times the flow rate. With a molar flow into the chamber of v_e (mol s^{-1}) and a mole fraction of water m_e (mol H_2O mol^{-1}) and v_o and m_o the flow and mole fraction of the air leaving the chamber:

$$E = v_o m_o - v_e m_e / L \tag{11.5}$$

A correction is required for the addition of water vapor to the air. The corresponding correction for exchange of CO_2 by a photosynthesizing leaf is negligible.

Conductance of the pathway for water vapor for evaporating water surfaces to the bulk atmosphere is complex, as *Figure 11.3* shows. Assuming that it is made up of the stomatal conductance (g_s), cuticular conductance (g_c) and a boundary layer conductance (g_a) the total conductance of the leaf is $g_1 = g_s + g_c + g_a$. The g_1 may be calculated as the constant of proportionality between the vapor concentration gradient from the inside of the leaf to the bulk air and the transpiration rate. This is a form of Fick's law where the flux is proportional to the gradient and inversely proportional to the resistance (the reciprocal of conductance) of the path. Thus:

$$g_1 = E(1 - m)/(m_i - m_a) \tag{11.6}$$

where $m = (m_i + m_a)/2$ and m_i and m_a are the mole fractions of water vapor in air inside the leaf and in the ambient air, respectively. The units of conductance are mol m^{-2} s^{-1}. To measure the boundary layer conductance a replica leaf surface (e.g. wet paper) with no stomata or cuticle is substituted. However, it is difficult to determine the true conductances on a complex structure like the leaf where stomata are distributed often non-uniformly on both sides; if the boundary layer conductance is large then such errors are relatively small.

Boundary layer.   Boundary layer thickness and therefore conductance at the leaf surface varies with wind speed, surface dimensions and characteristics such as hairiness. At wind speeds of 0.5–10 m s^{-1}, g_a is 0.2–4 mol m^{-2} s^{-1}. The effects of g_a on transpiration and CO_2 assimilation depend on its magnitude relative to other conductances in the system. In general, g_a is large compared to g_s (2 cf. 0.4 mol m^{-2} s^{-1} with open stomata) and so has little control over water or CO_2 flux. The boundary layer during measurement is minimized and kept constant by vigorous stirring.

Cuticular and stomatal conductance.   The waxy cuticle is a very effective barrier to water loss with small conductance (g_c, 0.005 mol m^{-2} s^{-1} or smaller) but it also prevents the entry of CO_2. Desiccation is a major limitation to growth of terrestrial plants, which have not developed a material allowing diffusion of CO_2 but no water vapour. Stomatal pores in the cuticle offer a high conductance pathway (g_s) for CO_2 flux into the leaf, but allow H_2O to escape so plants lose water during CO_2 assimilation. Plants appear to be in an evolutionary impasse, with large surface area required for light and CO_2 capture but losing water as a consequence. Cuticular and stomatal pathways operate in parallel and both are in series with the boundary layer (*Figure 11.3*). Stomata occur on leaf surfaces in variable numbers (0–3 × 10^8 m^{-2}) on both surfaces (amphistomatous) or only on upper (ad-) or lower (abaxial) surfaces. The ratio of total stomatal pore area to leaf surface is about 1% so that water vapor diffusion out of, and CO_2 diffusion into, the leaf occur only through a very small part of the leaf area; the maximum g_s of mesophytic C3 crop plants is 1.2–0.4 mol m^{-2} s^{-1} for H_2O vapor, but many trees and C4 plants generally have smaller maximum stomatal conductances than C3 plants. The stomatal pores are bounded by guard cells which regulate the width and area of the aperture via changes in turgor pressure, controlling CO_2 entry and, possibly of more importance for land plants, water vapor loss. Conductance of closed stomata is of the order of 0.01 mol m^{-2} s^{-1} and probably approaches cuticular conductance. As the maximum g_s is much greater than g_c the efflux of water and influx of CO_2 occurs via the pores. Conditions which encourage stomatal opening – adequate water supply, bright light and high humidity – stimulate rapid photosynthesis; some aspects of stomatal physiology are discussed in Section 11.5.

11.4.2 Carbon dioxide exchange and conductances

The pathway for CO_2 entry and exit from the inside of the leaf is shown in *Figure 11.3*. In darkness, respiration in the mitochondria produces CO_2 which diffuses from the cells into the intercellular spaces and thence into the atmosphere. In the light, CO_2 is removed from the chloroplast stroma by the Rubisco carboxylation reactions and a gradient of CO_2 develops across the chloroplast envelope, cytosol, plasmalemma and wall to the intercellular spaces. Atmospheric CO_2 diffuses into the leaf through the stomata. Respiratory CO_2 contributes to the carbon supply to the chloroplast, or may (depending on the relative resistances of the transport pathways) diffuse into the intercellular spaces. Determination of the fluxes and concentrations of CO_2 is important for understanding the regulation of photosynthetic rate.

The net CO_2 assimilation rate (A) of leaf area L (m^2) is determined in an open gas exchange system (*Figure 11.4*) from the difference in CO_2 content of the air entering and leaving the chamber and the flow rates:

$$A = v_o C_o - v_e C_e / L \tag{11.7}$$

where v_e and v_o are the flow of air (mol s^{-1}) and C_e and C_o are the mol CO_2/mol air entering and leaving the chamber, respectively. Then A has units of mol CO_2 m^{-2} s^{-1}. The same considerations about the differences in flow into and out of the chamber apply as in discussion of water vapor exchange. The conductance of the leaf surface and the rate of photosynthesis determine the concentration of CO_2 in the intercellular spaces C_i. It is important to know this value because the CO_2 concentration at the cell surface is the effective supply of CO_2 which the assimilating cells experience within the leaf (see Chapter 4). Thus, it is possible to relate metabolic processes to CO_2 availability more directly than if related to atmospheric CO_2 (C_a). Calculation of C_i is from:

$$C_i = C_a - A/(g_s' + g_c' + g_a') \tag{11.8}$$

where g_s', g_c' and g_a' are the stomatal, cuticular and boundary layer conductances to CO_2. These conductances to CO_2 are calculated from the corresponding conductances to water vapor from the ratio of the diffusion coefficients by the factor 1.6. However, as discussed earlier, $g_a' = g_a'/1.37$. The CO_2 concentration of the intercellular, as opposed to substomatal, cavities depends on the geometry of the mesophyll air spaces and the conductance of the pathway; this is generally large (4–0.8 mol m^{-2} s^{-1}) and probably constant except in severely wilted leaves. Leaves with large photosynthetic rate have relatively large internal spaces, allowing rapid diffusion of gases and a large cell surface to leaf surface ratio (20–40 in many mesophytes) which minimizes the gradients and provides large surface area for CO_2 absorption.

11.4.3 CO_2 partial pressure at the Rubisco active site

Extending the Fick's law concept to the flux of CO_2 from the cell surface to Rubisco, we may consider the movement as across a series of 'resistances' in the liquid phases of the cell wall, cell membrane, cytosol, chloroplast membranes and stroma to the enzyme active site. These, in combination, may be expressed as a mesophyll conductance, which is difficult to estimate because the CO_2 concentration at the enzyme site (C_c) is uncertain. It was once assumed that the CO_2 concentration in the chloroplast was the same as the compensation concentration but this cannot be so for A is then zero. Another view was that the conductances of the liquid phase (which are essentially determined by metabolic processes) were large so that the chloroplast CO_2 was very close to the intercellular CO_2, although clearly some gradient must exist otherwise CO_2 would not diffuse to Rubisco. Also, the rate of CO_2 assimilation is not a linear function of C_i or of the activity of Rubisco, but reaches a plateau at higher rate of A, which may be explained by a 'draw-down' of CO_2 at the active site at high A, even if C_i is large. These studies were on plants grown with different nitrogen supply and the mesophyll conductance was calculated as about 0.45–0.5×10^{-5} mol m^{-2} s^{-1} Pa^{-1}, which would result in C_c about 6 Pa below, that is for a leaf with A of 30 μmol m^{-2} s^{-1} at C_a of 36 Pa, and g_s of 0.4 mol m^{-2} s^{-1}, C_i would be 25 Pa and C_c 19 Pa. Confirmation of the conductances and C_c was provided by analysis of carbon isotope ratios ($^{13}C/^{12}C$) and fluxes, based on the differences in discrimination due to diffusion in the aqueous phase of the cell and in the Rubisco CO_2 fixation process. The rate of assimilation was changed by altering the photon flux. The conductance between the active site and substomatal cavities was estimated as 0.02–0.06 mol m^{-2} s^{-1} Pa^{-1} over the range of about 0–40 μmol m^{-2} s^{-1} A. Thus, the conductances from the intercellular spaces to the Rubisco sites is variable. However, this means that C_c remains rather constant as A changes, with C_c/C_i about 0.7 or C_c/C_a of 0.5. The mesophyll conductance increases with the nitrogen content of leaves and with the area of the chloroplast appressed to cell walls in contact with the intercellular spaces, although the dependence of conductance on anatomy is probably rather variable also reflecting complex biochemical processes that are involved and may change without affecting the structure. Details are given in Evans and von Caemmerer (1996) and von Caemmerer (2000).

The CO_2 partial pressure in the chloroplast may also be assessed by measuring CO_2 exchange rates at different CO_2 partial pressures (Laisk and Oja, 1998). When the CO_2 is measured as a function of C_a (on the linear part of the response curve), extrapolation back to zero CO_2 gives the rate of respiration from the leaf. Then C_i can be calculated from equation (11.8) (see *Figure 11.2* and Section 12.2.5), allowing the effect of stomata to be removed and giving the respiration from the cells. By correcting for the solubility of CO_2 in the wet cell walls, the concentration of CO_2 at the wall is obtained. From the changing slope of the lines, the effects of different resistances of parts of the pathway may be calculated.

The current evidence is that the CO_2 partial pressure in the chloroplast and at the Rubisco active sites, related by $A = g_1(C_i - C_c)$, is some 30% below that of the intercellular space under normal atmospheric conditions, that is C_c/C_i of 0.7 or C_c/C_a of 0.5 (von Caemmerer and Evans, 1991). The relation between A and C_i is discussed at length in Chapter 12. Here we may mention that the carboxylation efficiency (μmol CO_2 assimilation m^{-2} s^{-1} Pa^{-1}) may be calculated from the slope of the relation between A and C_i when A is zero. If the liquid phase conductances to CO_2 in the cell were infinite then this would be a measure of metabolic efficiency, for example enzyme reactions and diffusion of metabolites in the cell. Because C_i is not the chloroplast CO_2 concentration, then the relation includes the supply of CO_2. Carboxylation efficiency is related to metabolism, with C3 plants having smaller efficiency than C4 plants, and it depends on the amount of Rubisco in the tissue, for example nitrogen deficiency decreases both Rubisco amount and activity and also the efficiency in C3 and C4 plants.

11.5 Stomatal structure and function

Stomata play a particularly important role in the life of terrestrial plants. These microscopic pores (C_a 15–30 μm long in *Vicia faba*) occur in the outer epidermal surfaces of leaves and CO_2 diffuses into the tissue and H_2O vapor diffuses out through them. Stomata are chemically and osmotically driven, turgor-operated valves which respond rapidly to environmental conditions, particularly light, CO_2, water vapor and temperature. They also respond to conditions within the plant, particularly the water status and metabolic events such as the supply of abscisic acid (related to water and other stresses). Specialized guard cells (*Figure 11.5*) which contain chloroplasts, mitochondria and large continuous vacuoles, bound the pore and are capable of movement, increasing the pore area when conditions are conducive to CO_2 fixation, for example in light, and when the water supply is adequate. They thus regulate CO_2 and H_2O fluxes and by decreasing the latter, protect the plant from desiccation and are very important in plant survival strategies. Stomata were probably a prerequisite for plant invasion of dry land as they are found in fossils of *Rhynia*, a Psilophyte of the middle Devonian period 365 million years ago, present day examples of which have similar stomatal structure. The number of stomata per unit area of leaf surface (stomatal density), and as a proportion of the number of epidermal cells (stomatal index) is not a fixed value, but changes with conditions. Of particular interest is the effect of CO_2 concentration; as this has increased over the last 200 years, so both stomatal density and index have decreased. The effect has been shown experimentally. Also, stomatal numbers in fossil leaves from the Quaternary period are correlated with CO_2 fluctuations during the glacial–interglacial cycles, which are known from ice cores. Leaves from the Tertiary and Paleozoic also show relations to CO_2 partial pressure based on palynological (pollen) evidence which can be used to derive temperature and CO_2 data. In the early

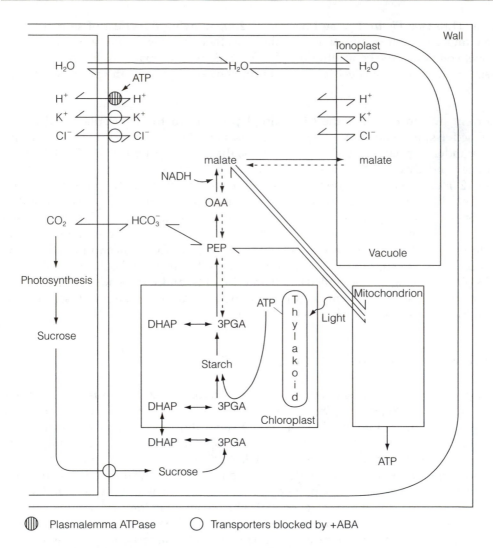

Figure 11.5. Simplified scheme of ion fluxes and generation of organic metabolites in a stomatal guard cell which leads to accumulation of osmotica in the vacuole, increased turgor and pore opening

Devonian, atmospheric CO_2 was 10–12 times greater than present, similar to that at the end of the Carboniferous before rising again in the Permian (*Figure 1.3*) (Berner, 1993; Beerling and Woodward, 1996).

The form and structures of a stomatal complex are illustrated in *Figure 11.4a*. The kidney-shaped guard cells of bean are typical of many dicotyledonous plants, whereas dumb-bell shapes are characteristic of grasses and other monocotyledons. Guard and subsidiary cell structure determines how movement of the guard cells and pore aperture are regulated. Guard cells are firmly anchored at each end and linked to the subsidiary cells by rather thin walls (*Figure 11.4b*). In contrast the outer and inner ventral walls are thick with cuticular ridges; when the pore is closed these provide an effective seal. The cell walls are constructed of a cellulose matrix with microfibrils running around the cell diameter of the outer and ventral walls (*Figure 11.4c*). An analogy is that of radial steel-belted motor car tyres. When the guard cells absorb water via the thin walls connecting them to the subsidiary cells, which is the pathway for rapid exchange of solutes and water, they swell and the forces generated result in the dorsal wall expanding into the subsidiary wall. However, because the ventral wall is thickened it does not expand longitudinally (pore and guard cell length remain almost constant between opening and closing) although the outer and inner more dorsal regions can expand between the tubules. Thus, the guard cell bows and the pore opens, the aperture increasing from zero to some 10–15 µm. During opening, guard cell volume may increase by 40% from the volume of 4 pl per guard cell when closed. The first phase of opening (Spannungsphase) involves inflation of the guard cells and a large increase in turgor. Osmotic potentials increase by 4–6 MPa with opening (compared to a range of 1–2 MPa in epidermal and mesophyll cells generally) and water potentials decrease similarly, thus 'pulling' water from the epidermis and increasing turgor greatly. Turgor generates the forces which distort the guard cell walls and open the pore.

What is responsible for the changes in turgor? Experimental determination of the amounts of ions and solutes in guard cells has provided the basis of the model in *Figure 11.5*. Clearly, metabolism regulates the processes, which involve energy and ion and organic solute fluxes resulting in marked changes in concentration between the closed and open states. These determine the passive water movement into the guard cell. During opening in bean, the K^+ ion concentration rises some 6–10-fold (by 300 mM, i.e. K^+ increases from 0.3 pmol per guard cell when closed to 2.5 pmol when open). This occurs within 0.5–2 h, so large K^+ fluxes – of the order of 30 pmol cm^{-2} s^{-1} – occur. For every 1 µm increase in aperture, the K^+ increases by 25 mM. Chloride ions also increase six-fold. Stomatal closure is faster than opening (15 min) and the reverse fluxes are correspondingly rapid. During opening the membrane potentials in guard cells become more negative and there is a flux of protons (H^+) from the guard cells so that their pH rises from 5.1 to 5.6.

Associated with these changes are those involving organic compounds. Guard cells contain starch, amounts of which change with stomatal activity; closed stomata contain abundant starch and open stomata contain much less. From this early observation arose the theory of starch hydrolysis as providing the osmoticum required for turgor generation and, therefore, opening. It is now known that malate also increases greatly, 10-fold, with opening, from 50 to 500 mM. Changes in malate are related to the observed large amounts and activities of PEP carboxylase. However, there is little Rubisco or other PCR cycle enzymes and reduced amounts of chlorophyll, suggesting that guard cell chloroplasts are not very active in CO_2 fixation. In the light they probably do not synthesize starch, but the products go to make malate. In darkness guard cells contain relatively little osmoticum and have low turgor, the walls are not stretched and the pore is closed. The relatively large content of starch in the plastids probably results from synthesis using malate which is abundant in the light and is converted to starch in darkness.

On illumination of stomata (when water is adequate) starch is remobilized using ATP from the thylakoid reactions or from mitochondria (blue light may stimulate their metabolism) to form PEP via glucose and fructose-6-phosphates and 3PGA and DHAP. PEP reacts with HCO_3^- (catalyzed by PEP carboxylase), giving OAA, which is reduced to malate by malate dehydrogenase, using NADH. Malate enters the vacuole and is osmotically active. At the same time K^+ and Cl^- are pumped into the cytosol via a plasmalemma ATPase and then into the vacuole via the tonoplast. Protons are exported from the cell. These processes generate additional osmotic potential and the turgor required for opening during the 'motorische phase'.

Regulation of stomatal operation is complex; clearly it must be flexible to be effective at maintaining plant homeostasis in the face of multiple environmental stimuli. Light plays a major role. Blue light, at low intensity, stimulates the fluxes of ions and increases metabolism; it probably operates via flavins and may set the system so that it can respond to other signals including additional response to white light. Carbon dioxide also influences stomatal behavior: large partial pressures decrease stomatal aperture whereas low pressures increase it, but the CO_2 effect is probably indirect. CO_2 is clearly involved in the regulation of HCO_3^- concentration and thereby the PEP carboxylase reaction and generation of malate. CO_2 probably operates indirectly on the mechanism, so that internal CO_2 of the leaf is maintained approximately, but not absolutely, constant under many conditions. Optimization of net photosynthesis and water loss is shown by stomata, but quantitative aspects of the regulatory mechanisms are still to be analyzed. Both the light and CO_2 responses are overridden by the water balance of the tissue. Water stress, with decreased leaf turgor, prevents opening, probably because the necessary guard cell turgor cannot be generated. There is also rapid hormonal regulation of stomatal movement. Abscisic acid and farnesol, both sesquiterpenoids, are present in normal tissues and may regulate stomatal aperture, but in water-stressed plants their synthesis is greatly increased either via the C15 pathway from mevalonic acid or by the C40 route via carotenoid breakdown. If $+ABA$, increases above a threshold, it blocks the K^+ transporter on the plasmalemma and prevents opening, thus acting as a long-term 'off switch': this is important for plant survival when water loss exceeds supply.

Nutrition also affects stomata, for example inadequate phosphate decreases stomatal opening and reduces sensitivity to other environmental stimuli, but the mechanisms are not understood. Atmospheric humidity also regulates stomatal aperture; dry air frequently causes pore closure, possibly because loss of water from the subsidiary cells upsets the turgor balance of the guard cells. There are many intriguing questions still remaining about the mechanisms of stomatal regulation and control by the environment, including what determines the behavior of CAM plant stomata which open in darkness and close in the light.

11.5.1 Stomatal control of water and CO_2 exchange

Regulation of leaf gas exchange involves a balance between assimilation and water loss. For a given ambient CO_2, humidity and energy balance, light determines the assimilation rate and C_i decreases until a steady state is achieved between stomatal conductance, water loss and CO_2 uptake. If the leaf loses water and turgor decreases then stomatal conductance decreases and with it C_i and also A if it is not on the saturation part of the response curve. Evidence from measurement of chlorophyll a fluorescence of water-stressed leaves suggests that with mild stress but very small stomatal conductance, C_i may decrease to the compensation point. However, C_i may not drop so low, because of the effects of water deficits on R_{day} or on the capacity for CO_2 assimilation. If C_i is saturating for A then stomatal closure may occur with reduction in water loss but no effect on A. Under conditions favorable for many plants, for example abundant water, light and moist air, stomata may open so that C_i rises and A increases. There is a complex, dynamic control system which maintains CO_2 fixation whilst preventing excessive water loss because of the feedback from turgor and ABA. These factors are probably genetically determined by evolutionary selection of the mechanisms. For example, plants of very droughted environments may have inherently small assimilation rates (limited metabolic capacity) and operate with small stomatal conductance, whereas plants of well-watered, nutrient-rich environments may have large capacity and productivity but lose water and therefore be susceptible to drought conditions. Stomatal closure for prolonged periods is unusual in mesophytes normally growing in good environments, possibly because it increases photorespiration, induces metabolic imbalance in supply of assimilates related to demand and may cause damage. Thus, stomata may regulate, indirectly, energy balances of cells, providing protection for the biochemical mechanisms as well as regulating water and carbon fluxes.

11.6 Stable isotopes in photosynthetic studies

11.6.1 Carbon isotopes and photosynthetic discrimination

Atmospheric CO_2 is composed of the stable isotopes of carbon: ^{12}C is 98.9% and ^{13}C 1.1%. The thermodynamic and kinetic properties of compounds containing the different isotopes differ due to mass, and they participate to different extents in physical and chemical reactions, that is, there is discrimination. Thus, the proportion of isotopes in products of the reactions changes. The $^{13}C/^{12}C$ ratio gives important information about reactions and can be used as a tracer in carbon metabolism. The ratio is measured by mass spectrometry of CO_2 (organic matter is first oxidized) and compared with a standard of known ratio, usually a belemnite (fossil shell) earth from the Peedee formation in South Carolina, USA and called the PDB standard. The ratio of unknown to standard isotope distribution is $\delta^{13}C$:

$$\delta^{13}C(\%_o) = \frac{(^{13}C/^{12}C) \text{ unknown substance} - (^{13}C/^{12}C) \text{ standard}}{(^{13}C/^{12}C) \text{ standard}} \times 1000 \qquad (11.9)$$

Negative $\delta^{13}C$ shows ^{13}C enrichment; atmospheric CO_2 is $-7\%_o$). Plant material $\delta^{13}C$ ranges from -10 to $-40\%_o$; C3 plants have -22 to $-34\%_o$ (mean $-27\%_o$); C4 plants -10 to $-18\%_o$ (mean $-13\%_o$) and CAM plants are in the range of C3 and C4 plants. These differences arise as Rubisco discriminates more strongly ($-28\%_o$) against $^{13}CO_2$ than PEP carboxylase ($-9\%_o$) due to the chemical reaction; Rubisco uses CO_2 and PEPc the HCO_3^- ion. In C4 plants, CO_2 is concentrated in the bundle sheath and $^{13}CO_2$ is fixed more effectively than from the atmosphere. CAM plants use both PEPc and RuBP carboxylase to assimilate CO_2, hence their wide range of $\delta^{13}C$ values. Measurement of $\delta^{13}C$ provides confirmation of the metabolic origin of organic carbon; it suggests that very old samples of organic carbon were derived from photosynthesis some 3.5–4 billion years ago, thus setting this process very early in earth's biological history. $\delta^{13}C$ discrimination also shows that ferns and bryophytes from the Carboniferous period had C3 photosynthesis. From isotopic analysis it is possible to distinguish between sucrose from sugar cane (C4) and sugar beet (C3) as well as between artificial and natural vanilla (produced by an orchid). Isotope discrimination is affected by the physiological state of plants, for example, small stomatal conductance increases the utilization of CO_2 from the intercellular spaces and decreases discrimination against ^{13}C. Small stomatal conductance also reduces water loss so that discrimination is related to water use efficiency. By determining $\delta^{13}C$ composition (which is a relatively rapid method requiring little tissue preparation) for total plant dry matter accumulated over long periods, it is possible to select plants with high water use efficiency.

11.6.2 Oxygen isotope exchange

The metabolism of O_2 has been studied by use of isotopes, ^{18}O and ^{16}O and mass spectrometry, to assess the exchange of O_2 between the different pools of metabolites involved in photorespiration, CO_2 assimilation *et cetera* within leaves, as well as of O_2 exchange between the atmosphere and the leaf. There is one source of O_2 production in plants – from water splitting in photosynthesis, where approximately 1 mol O_2 is released per mol of CO_2 assimilated. So by enclosing a leaf or cells in a chamber with an atmospheric O_2 isotopic composition (enriched in $^{18}O_2$) different from that of water in the tissue (enriched in ^{16}O), the change in isotope ratios (increase in $^{16}O/^{18}O$) can be used to determine O_2 release from the chloroplast. There is no isotopic discrimination in the water-splitting reaction. However, there are several processes consuming O_2 at the same time as it is released: (1) principally photorespiration (in oxygenation of RuBP, and in the respiration of NADPH produced in the glycolate oxidase reaction in the peroxisomes); (2) in mitochondrial respiration in the light, either acting as terminal acceptor linked to glycolysis and the TCA cycle or only to consumption of reductant, for example transferred by the malate shuttle from chloroplasts; and (3) Mehler peroxidase reaction.

Rates of O_2 consumption of the different pathways are difficult to distinguish under normal conditions. The photorespiratory use can be estimated by comparing normal and low O_2; alternatively Rubisco characteristics may be used to calculate the flux of O_2 (Chapter 12). From the rates of Rubisco carboxylation and oxygenation, V_c and V_o, and the known transfer of four electrons per O_2 evolved, and with CO_2 compensation point without mitochondrial respiration, Γ_*, and CO_2 pressure, C, the total flux of electrons for CO_2 assimilation and photorespiration is

$$J_A = 4V_c + 4V_o = (4 + 8\Gamma_*/C)\,V_o$$

and the O_2 evolution rate is therefore $J_A/4 = (1 + 2\Gamma_*/C)\,V_o$. As the stoichiometry between electron transport, proton transport and ATP synthesis is not established, neither is the exact rate of O_2 production due to the activities of the processes. Assuming

$3H^+$/electron transported, $4H^+$/1ATP, then the O_2 evolution rate is $(1 + 2.33\Gamma_*/C) V_o$. For $2H^+$/electron and $3H^+$/ATP, then the O_2 evolution rate is $(1.125 + 2.625\Gamma_*/C)$, and for $3H^+$/electron and $3H^+$/ATP then the O_2 evolution rate is $(0.75 + 1.75\Gamma_*/C)$. Thus there is uncertainty about the rate of O_2 evolution. There is also electron transport associated with nitrate reduction, a process which varies with leaf age and differs between species.

Oxygen uptake by oxygenation of RuBP uses $\frac{1}{2}O_2$/RuBP plus $\frac{1}{2}O_2$ for the glycolate oxidase reaction, or $1.5V_o=3\Gamma_*V_c/C$. From isotopic exchange the carboxylation and oxygenation rates of RuBP by Rubisco have been estimated and also the value of C_c *in vivo*, with assumptions about the rates of dark respiration and the Mehler reaction. These rates correlate very well with estimates from fluorescence (see Section 5.6). The consumption of O_2 by mitochondrial respiration in the light is not established and neither is that by the Mehler–peroxidase reaction. The latter is thought to be negligible from some studies, more significant in others.

11.7 Global carbon, CO_2 and O_2

It is perhaps useful here to outline the global CO_2 and O_2 environment in which plants function. This is also very important from the view-point of alterations to CO_2 in the atmosphere in the last 150 years and the consequences for global temperature, and thus many other processes.

11.7.1 Carbon

Carbon is held in organic compounds in living and dead biomass of plants and other organisms, particularly forests (boreal, temperate and tropical), in peat and in soil humus of the terrestrial biosphere and also in the oceans and deposited on ocean floors (*Figure 11.6*). There are, however, much larger quantities of carbon of plant origin in coal, oil, oil shales and natural gas. Yet, even these quantities of carbon are insignificant compared to the huge sedimentary deposits of carbonates in the ocean sediments and in rocks. Also, the oceans hold some 40 000 Gt (equivalent to 10^{15} g) of carbon as CO_2 and bicarbonate ions, which is the major store of readily remobilized carbon and an enormous buffer for the world's atmospheric CO_2. Oceans take up 2 ± 0.8 Gt carbon per year, based on modeling studies calibrated with tracers, such as the man-made chlorofluorocarbons and radiocarbon (^{14}C) from atomic weapons. The atmosphere acts as a pool linking the carbon fluxes between the solid forms of carbon. There are very large fluxes between the oceans, continents and atmosphere; changes in their magnitude, even if quite small, affect the partial pressure of CO_2 in the atmosphere. One of the major exchange processes for carbon is in photosynthesis, with a flux of 100 Gt per year of which 40 Gt is to the oceans and 60 Gt per year to the terrestrial vegetation. Standing terrestrial biomass is of the order of 620 Gt carbon. However, these large fluxes fixing carbon are almost offset by the respiration of the biomass and soil. Consequently, the terrestrial carbon sink may be small (0.3 ± 1 Gt carbon per year). The rate of exchange between the pools is very different; from vegetation to atmosphere the time scale is tens to hundreds of years, from soil to air probably an order of magnitude slower, between shallow seas and deep water of the order hundreds of years but from these to ocean sediments requires thousands of years.

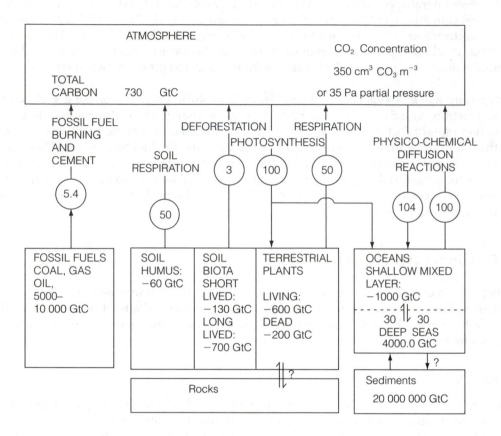

Figure 11.6. Simplified scheme of the main pools and fluxes, ⑤, of carbon between the atmosphere, biosphere and geosphere. The size of each is given in gigatonnes of carbon (GtC; 1 GtC = 10^{15} g C). The fluxes are also in GtC. The CO_2 concentration in the atmosphere is shown as volumes per volume (equivalent to volume parts per million volumes (vpm)) or as partial pressure (Pa). Data are derived from the literature cited

11.7.2 CO₂

Currently the atmosphere holds 760 Gt of carbon as gaseous CO_2 and this has increased by some 20–30 Gt since the mid-nineteenth century. Between 1980 and 1999 the increase was 3.6 ± 0.2 Gt carbon per year. In terms of partial pressure of CO_2, this represents an increase from 27 Pa (270 µl l^{-1} or vpm) in 1850 to about 33 Pa in 1970; in mid 1998 it was 36 Pa. Measurements made since 1957 at Mauna Loa in Hawaii, an elevated site unaffected by vegetation and surrounded by oceans, and more recently at different sites around the world, show that CO_2 is increasing rapidly (*Figure 11.7*) and has accelerated from about 0.7 Pa per decade early this century to about 1.5 Pa per decade between 1973 and 1988; rates of increase in the 1990s have been about 1.7 Pa per decade. The Mauna

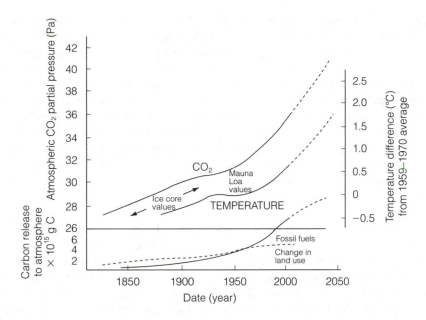

Figure 11.7. The trends in atmospheric CO_2 partial pressure and global temperature since 1850 and projected to 2050. These changes (which are based on experimental evidence and climate modeling) are subject to uncertainty and debate. Correlation of increased atmospheric CO_2 with use of fossil fuels is firmly established; the link to climate change is discussed in the text and the literature cited

Loa data show a seasonal cycle of atmospheric CO_2 (amplitude about 0.8 Pa) which is related to the large CO_2 production and very low rates of photosynthetic uptake in the Northern hemisphere (which has a larger terrestrial land area than the Southern) during the winter and the converse in summer. The current increase in atmospheric CO_2 is largely, if not exclusively, the result of increased human activity including burning fossil fuels and removal of forests; some 11 000 km^2 per year of tropical forest is currently being destroyed. This is equivalent to 0.4–2.5 (average 1.6 ± 1) Gt carbon per year released by burning and land use changes. The carbon in terrestrial ecosystems is about one-third in standing biomass (forests) and two-thirds in soils. Forests sequester (store) carbon when young and growing but once they mature (100–200 years) the uptake and loss of C is in equilibrium. Soils hold carbon that may be much slower to turn over. The tropical regions were strong sources of CO_2 in the 1980s but may have been nearly neutral in the 1990s. Temperate carbon sinks in forests were probably positive in the 1990s due to forest regrowth in Europe and North America (mainly due to land-use changes but with some effect of increasing CO_2). However, over the last century, industrialization based on exploitation of fossil carbon reserves, coupled with the rapid growth of human population and demand for agricultural land, rapidly increased emission of CO_2. Simultaneously, accumulation of carbon in standing vegetation decreased, particularly in the tropics. Deforestation also allows soil organic matter to be oxidized, as does drainage of peat lands and marshes and their use in agriculture. All these processes contribute to increased atmospheric CO_2. The rate of consumption of fossil fuels plus the production of cement has accelerated in recent decades (*Figure 11.7*) and currently they release 5.5 ± 0.5 Gt carbon per year. There is little indication of the rate decreasing in the near future, nor will it as the developing economies and populations grow further. Indeed, increased consumption is likely and therefore further rapid increase in atmospheric CO_2. However, there is carbon sequestration in temperate forests of some 0.5 ± 0.5 Gt carbon per year and uptake by other terrestrial processes may remove 1.4 ± 1.5 Gt carbon per year (note the great uncertainty in these values). The role of photosynthesis in removing CO_2 from the atmosphere is clear, and it is likely that the increased CO_2 will stimulate photosynthesis, thus partially absorbing the gas and increasing organic carbon. It is unclear if photosynthetic organisms will have the capacity to keep increasing their accumulation, particularly if water and nutrient shortages restrict the ability of plants to grow effectively.

Increased absorption by the atmosphere, predominantly by CO_2, of long-wave radiation emitted by the surface (terrestrial and oceans) of the globe as a consequence of the 'greenhouse effect', will increase temperature. This will stimulate respiration of plants and microorganisms according to a Q_{10} (the ratio of the rates of a process caused by a temperature rise of 10°C) of about 2, thus offsetting the gains from enhanced photosynthesis. Attempts to calculate the components of the total carbon balance are fraught with difficulty; indeed a large fraction of the CO_2 evolved cannot be accounted for. Possibly, it is being sequestered by the temperate and boreal forests, in roots, soil humus, *et cetera*. It is unlikely to be entering the oceans where plant growth is severely inhibited by deficiencies of nutrients.

11.7.3 O_2

Oxygen in the atmosphere is another component of the plant's environment that has large effects on metabolism. The pool of atmospheric oxygen is very large compared to that of carbon dioxide. However, it is possible to estimate the change in it by measuring the ratio of O_2/N_2, as the N_2 pool is so large it can be assumed to be virtually constant. Oxygen has low solubility in water, so changes in O_2 reflect the balance between oxidation of carbon by burning and respiration, and the release in photosynthesis. The O_2 concentration is decreasing slightly faster than the CO_2 is increasing. From these analyses, the uptake of CO_2 by the oceans was 1.9 ± 0.8 Gt carbon per year between 1989 and 1994. There is a terrestrial net carbon gain of 1.8 ± 1.1 Gt carbon per year, which is in agreement with that previously estimated.

11.7.4 Effects of CO_2 on world climate

Carbon dioxide absorbs infrared radiation; short-wave radiation warms the earth and plants, which re-emit long-wave infrared radiation. This is absorbed by CO_2 and the atmosphere heats up, producing the 'greenhouse' effect mentioned earlier, so-called because it is analogous to heating in greenhouses (glass houses) caused by poor transmission of infrared compared to short-wave radiation by glass. Other gases also absorb infrared radiation effectively, for example methane is 20 times more efficient than CO_2 at absorbing infrared. Currently methane production, for example in anaerobic paddy fields, urban refuse disposal and by ruminants, is increasing rapidly. Increased atmospheric

CO_2 and methane are expected to increase the atmospheric temperature. The extent of global warming is hotly debated, but much evidence suggests that surface temperatures are increasing. Over the last century, temperatures have risen by approximately 0.7°C and some of the hottest years of this century have occurred in the 1980s and particularly the 1990s. These observations may also be linked with changes to world climate with, for example, increased droughts in many areas, such as the Sahel. Factors controlling climate are complex and some feedback processes may reduce potential climate change. For example, a warmer atmosphere increases evaporation, resulting in more clouds which have a very large albedo, thus reflecting radiation back into space and so decreasing the input of energy. The role of vegetation in these complex processes cannot be modeled accurately with current methods, although it is clearly important to place vegetation and photosynthetic processes in their correct, central role between the geosphere and atmosphere. Mathematical modeling of carbon fluxes in the earth from earliest times (Berner 1993, suggest that 100 million years ago the atmosphere contained 20–30 times more CO_2 than today. However, by 60 million years ago there was probably only twice the present concentration. Atmospheric CO_2 has fluctuated considerably, for example, there was an increase of CO_2 some 40 million years ago associated with its release by volcanic activity caused by sea-floor spreading and continental drift. In the much more recent past, the last ice age for example, the CO_2 content of the atmosphere was probably very small. Thus, plants have faced substantial, if slow, changes in CO_2 and climate during their evolutionary history (Beerling and Woodward, 1996). Changes in climate may affect natural vegetation and agriculture and may seriously disrupt the stability of ecosystems; it is therefore important to assess the changes.

References and further reading

Badger, M.R. and Spalding, M.H. (2000) CO_2 acquisition, concentration and fixation in cyanobacteria and algae. In *Photosynthesis: Physiology and Metabolism* (eds R.C. Leegood, T.D. Sharkey and S. von Caemmerer). Kluwer Academic, Dordrecht, pp. 369–397.

Beerling, D.J. and Woodward, F.I. (1996) Palaeo-ecophysiological perspectives on plant responses to global change. *Trends Ecol. Evol.* 20–23.

Berner, R.A. (1993) Paleozoic atmospheric CO_2: importance of solar radiation and plant evolution. *Science* **261**: 68–70.

Bolin, B. (ed.) (1981) *Carbon Cycle Modelling.* Scientific Committee on Problems of the Environment (Scope 16), John Wiley Chichester.

Bro, E., Meyer, S. and Genty, B. (1995) Heterogeneity of leaf CO_2 assimilation during photosynthetic induction. In *Photosynthesis: from Light to Biosphere,* Vol. V. (ed. P. Mathis). Kluwer Academic, Dordrecht, pp. 607–610.

Coleman, J.R. (2000) Carbonic anhydrase and its role in photosynthesis. In *Photosynthesis: Physiology and Metabolism* (eds R.C. Leegood, T.D. Sharkey and S. von Caemmerer). Kluwer Academic, Dordrecht, pp. 353–367.

Coombs, J., Hall, D.O., Long, S.P. and Scurlock, J.M.O. (eds) (1985) *Techniques in Bioproductivity and Photosynthesis* (2nd edn). Pergamon Press, Oxford.

Ehleringer, J.R., Hall, A.E. and Farquhar, G.D. (eds) (1993) *Stable Isotopes and Carbonwater Relations.* Academic Press, San Diego, CA.

Evans, J.R. and von Caemmerer, S. (1996) Carbon dioxide diffusion inside leaves. *Plant Physiol.* **110**: 339–346.

Farquhar, G.D., Ehleringer, J.R. and Hubrick, K.T. (1989) Carbon isotope discrimination and photosynthesis. *Annu. Rev. Plant Physiol. Plant Mol. Biol.* **40**: 503–537.

Field, C.B., Ball, T. and Berry, J.A. (1989) Photosynthesis: principles and field techniques. In *Plant Physiological Ecology, Field Methods and Instrumentation* (eds R.W. Pearcy, J. Ehleringer, H.A. Mooney and P.W. Rundel). Chapman and Hall, London, pp. 209–253.

Griffiths H. (1998) *Stable Isotopes: The Integration of Biological, Ecological and Geological Processes.* BIOS Scientific, Oxford.

Hashimoto, Y., Nonami, H., Kramer, P. J. and Strain, B.R. (1990) *Measurement Techniques in Plant Science.* Academic Press, San Diego, CA.

Keeling R.F., Piper S.C. and Heimann M. (1996) Global and hemispheric CO_2 sinks deduced from changes in atmospheric O_2 concentration. *Nature* **381**: 218–221.

Laisk, A. and Oja, V. (1998) *Dynamics of Leaf Photosynthesis: Rapid-response Measurements and their Interpretations.* Techniques in Plant Sciences. CSIRO Publishing, Collingwood.

Luo, Y. and Mooney, H.A. (eds) (1999) *Carbon Dioxide and Environmental Stress.* Academic Press, San Diego, CA.

Mannion, A.M. (1991) *Global Environmental Change.* Longman Scientific and Technical, Harlow.

Mansfield, T.A., Hetherington, A.M. and Atkinson, C.J. (1990) Some current aspects of stomatal physiology. *Annu. Rev. Plant Physiol. Plant Mol. Biol.* **41**: 55–75.

McElwain, J.C. and Chaloner, W.G. (1995) Stomatal density and index of fossil plants track atmospheric carbon dioxide in the Palaeozoic. *Ann. Bot.* **76**: 389–395.

Outlaw, W.J. Jr and Harris, MJ. (1991) Water stress, stomata, and abscisic acid. In *Impact of Global Climate Changes on Photosynthesis and Plant Productivity* (ed. Y.P. Abrol). Oxford and IBH Publishing, New Delhi, pp. 447–461.

Parkhurst, D. (1994) Diffusion of CO_2 and other gases inside leaves. *New Phytol.* **126**: 449–479.

Pearcy, R.W., Ehleringer, J., Mooney, H.A. and Rundel, P.W. (eds) (1989) *Plant Physiological Ecology, Field Methods and Instrumentation*. Chapman and Hall, London.

Reddy, K.R. and Hodges, H.F. (eds) (2000) *Climate Change and Global Crop Productivity*. CABI Publishing, Wallingford.

Robinson, J.M. (1994) Speculations on carbon dioxide starvation, late Tertiary evolution of stomatal regulation and floristic modernisation. *Plant Cell Environ.* **17**: 345–354.

Schlesinger, W.H. (1991) *Biogeochemistry. An Analysis of Global Change*. Academic Press, San Diego, CA.

Šesták Z., Čatský J. and Jarvis, P.G. (eds) (1971) *Plant Photosynthetic Production, Manual of Methods*. Junk, The Hague.

Slater, R.J. (ed.) (1990) *Radioisotopes in Biology. A Practical Approach*. IRL Press, Oxford.

Syvertsen, J.P., Lloyd, J., McConchie, C., Kriedemann, P.E. and Farquhar, G.D. (1995) On the relationship between leaf anatomy and CO_2 diffusion through the mesophyll of hypostomatous leaves. *Plant Cell Environ.* **18**: 149–157.

von Caemmerer, S. and Evans, J.R. (1991) Determination of the average partial pressure of CO_2 in chloroplasts from leaves of several C3 plants. *Aust. J. Plant Physiol.* **18:** 287–305.

von Caemmerer, S. (2000) *Biochemical models of leaf photosynthesis*. Techniques in Plant Sciences, CSIRO Publishing, Collingwood.

Willis, K.J. (1996) Plant evolution and the ancient greenhouse effect. *TREE* **11**: 277–278.

Woodward, F.I. (1987) Stomatal numbers are sensitive to increases in CO_2 from pre-industrial levels. *Nature* **327**: 617–618.

Zeiger, E., Farquhar, G.D. and Cowan, I.R. (eds) (1987) *Stomatal Function*. Stanford University Press, Stanford, CA.

Photosynthesis by leaves

Chapter

12

12.1 Introduction

Carbon dioxide assimilation of leaves is of the greatest importance for production of the materials necessary for growth and reproduction of higher plants. It underpins almost all terrestrial biology and ecology, including agriculture, which serves human requirements. Therefore measurement of photosynthesis by leaves in relation to environmental conditions is very important. A central problem in ecology and agriculture is to estimate the potential photosynthesis which is possible with the available light energy and temperature in particular environments, and to determine what photosynthetic production will actually be when other factors, such as water and nutrient supply, are taken into consideration. Also, it is important to know the maximum photosynthesis and production which could be achieved by particular plants, given the best combination of environmental conditions. The discrepancy between the potential and actual photosynthesis may be more readily decreased if the mechanisms and factors determining photosynthetic productivity are understood. Currently there is much effort given to understanding how photosynthetic rates are related to composition of the leaf tissue, and to assessing the effects of genetically modifying the activities of enzymes *et cetera*. These approaches provide understanding of the mechanisms by which the photosynthetic system interacts with environment and help to explain the effects produced by modifying environment, for example, by altering light and CO_2, and by changing nutrition. Carbon dioxide assimilation depends on the development of an effective integrated system of individual biochemical and biophysical processes, which were discussed at length in earlier chapters, and is a complex function of the photosynthetic system's interaction with environmental conditions. This chapter considers the nature of the responses of net photosynthesis (A) of leaves of C3 and C4 plants to the principal environmental factors – light, CO_2, temperature, nutrition and water – and relates them to the metabolic and physiological structures and functions of the photosynthetic system. Despite the complexity and great number of species and environmental combinations which have been described, certain features of the photosynthetic system are common to all plants, particularly within the C3 type, and are key to understanding their function and efficiency in given environments. The approach draws on understanding of the basic mechanisms described in earlier chapters and interprets responses of leaf photosynthesis using Rubisco kinetics and RuBP supply, based on the model analysis of Farquhar and von Caemmerer (1982) as summarized by von Caemmerer (2000).

12.2 Response of C3 photosynthesis to environment

In the next sections, the effects of photon flux and CO_2 on photosynthesis are considered in some detail. The effect of other environmental factors (temperature, nutrition and water) on CO_2 assimilation in leaves is considered later. The processes determining short-term CO_2 assimilation in leaves include: (1) electron transport leading to synthesis of ATP and NADPH; (2) synthesis of RuBP; (3) rate of diffusion of CO_2 to the active sites of Rubisco; (4) CO_2 assimilation by Rubisco; and (5) consumption and transport of assimilates out of the chloroplast, cell and leaf. In addition, the respiratory losses from leaves are an important part of the carbon balance. These complex problems of regulation in a multistep process are considered in greater detail later in this chapter. Before describing and analyzing the photosynthetic responses, the use of models of photosynthesis is discussed to provide a basis for later interpretation.

12.2.1 *Metabolic models of photosynthesis*

Early studies of the relation between CO_2 assimilation and characteristics of the plant attempted to correlate A with leaf composition and structure, for example pigment or protein content, cell surface area and stomatal frequency (see Heath, 1969). However, this approach could not explain or simulate the process and led to the concept that the rate of photosynthesis by the intact leaf is the result of integrated cellular biophysics (e.g. diffusion of cell metabolites or substrates), biochemistry (e.g. enzyme activities) and physiology (e.g. control of CO_2 supply by stomatal conductance). It is now accepted that A depends on the properties of the combined energy capturing and transducing systems (light reactions, electron transport, etc.) in the thylakoids, acting in series with the processes taking place in the chloroplast stroma (e.g. enzymology of the PCR cycle) together with those consuming assimilates (e.g. respiration and growth), as well as on the factors regulating CO_2 supply to the chloroplast. These processes are considered to be 'balanced' to achieve optimization of the rates of each step so that they are not greatly in excess or deficient. This 'systems approach' is developing rapidly, based on the quantitative analysis of the different components and formulation of mathematical simulation models of photosynthetic systems. Use of genetically modified plants with altered amounts and activities of enzymes is also rapidly and substantially enlarging understanding of interactions in the photosynthetic system. This approach may be regarded as 'mechanistic modeling'.

A related method of considering the system is that of metabolic flux control analysis, which examines the effect of changing enzyme activity on the output, in this case CO_2 assimilation under different conditions. Application of metabolic flux control analysis has shown how subtle regulation is, with particular components of greater or lesser importance, depending on the environment (Section 7.4.3). Although flux control analysis suggests that control of A is not due to a single limiting factor, it is also clear that the major features of photosynthesis are largely determined by electron transport leading to regeneration of RuBP, and Rubisco characteristics. Thus, flux control analysis is less mechanistic than the dynamic models based on characteristics of specific components.

Some 'models' are mathematical relationships, but with little mechanistic basis, describing the rate of photosynthesis in relation to one or more environmental factors, for example curves of A as a function of radiation and of CO_2. These are valuable for calculation of photosynthetic productivity of natural vegetation and in crop production for practical purposes, for example, but treat the mechanism as a 'black box'. Statistical correlations are used similarly, for example regressions of A on N content of leaves (see Thornley and Johnson, 1990, reference in Chapter 13), but are not considered further here. Mechanistic models should provide, in the long term, better understanding and predictive capacity than generalized 'black-box' or statistical approaches.

12.2.2   Carbon balance of leaves

Net CO_2 exchange (A) of leaves is the difference between the uptake of CO_2 in gross photosynthesis (A_g) and the production of CO_2 by photorespiration (R_{pr}) from the glycolate pathway and respiration by other processes, predominantly the tricarboxylic acid (TCA) cycle. Carbon dioxide produced by the TCA cycle is often called 'dark' respiration (R_{dark}) as it is the form of respiration occurring in darkened leaves and other organs. However, TCA cycle respiration proceeds in the light, although the rates are not easily measured and the contribution is rather ill-defined. It is considered to be the same mechanism but not necessarily to operate at the same rate or with the same stoichiometries. To distinguish this respiration from other sources of CO_2 it is called 'day' respiration (R_{day}). To further complicate the terminology, 'dark respiration' of nonphotosynthetic plant parts occurs in the light and dark (see Section 8.9, Chapter 8 for discussion and literature). The net uptake of CO_2 (A) of leaves in the light is:

$$A = A_g - (R_{pr} + R_{day}) \tag{12.1}$$

Over a day–night cycle the net assimilate production by a leaf in the light would be decreased by R_{dark} of the leaf, so total net CO_2 assimilation over the diurnal cycle is:

$$A = Ag - (R_{pr} + R_{day}) - R_{dark} \tag{12.2}$$

Changes in any one of these components alter the net carbon assimilation. Other losses of carbon from leaves (e.g. isoprene emission) are small compared with the net carbon assimilation and are neglected. Integrated over a sufficient period, such as the growing season of a crop, even small differences in A can have significant effects on productivity.

12.2.3   Rubisco models in leaf assimilation

Models based on Rubisco characteristics provide a mechanistic analysis of carbon assimilation using well-established parameters. The basic features of Rubisco reactions (and the symbols) have been discussed in Chapter 7. They are repeated here because they are so important in understanding photosynthesis of leaves (*Figure 12.1*).

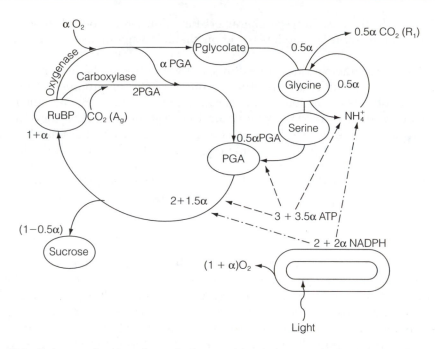

Figure 12.1. Scheme of carbon fluxes in the combined photosynthetic carbon reduction and photorespiratory cycles in a C3 leaf. A_g is the gross CO_2 assimilation, R_{pr} the CO_2 release in photorespiration, and α the ratio of oxygenase/carboxylase activity. The rate of carbon export from the combined cycles, shown as a flux to sucrose, is $1 - 0.5\alpha$. Compare this figure with the reactions of the glycolate pathway in *Figure 8.5*

The rate of carboxylation, V_c, at saturating RuBP (W_c) in the presence of O_2 is:

$$V_c = W_c = V_{c_{max}} C/C + K_c (1+O/K_o) \tag{12.3}$$

which simplifies at small partial pressure of CO_2 to:

$$W_c = V_{c_{max}}/K_c (1 + O/K_o)C \tag{12.4}$$

The rate of oxygenation, V_o, at saturating RuBP (W_o) in the presence of CO_2:

$$V_o = W_o = V_{o_{max}} O/O + K_o (1 + C/K_c) \tag{12.5}$$

K_c and K_o are the Michaelis–Menten constants for the Rubisco carboxylase and oxygenase respectively. The ratio $V_c/V_o = (V_{c_{max}} K_o/K_c V_{o_{max}})C/O$ describes the ratio of carboxylation to oxygenation and is an important characteristic of carbon fluxes in carbon metabolism by the combined PCR and photorespiratory cycles. The term $(V_{c_{max}} K_o/K_c V_{o_{max}})$ is the relative specificity factor (alternatively called the CO_2/O_2 specificity), $S_{c/o}$, of Rubisco, and is the rate of the two reactions at equal CO_2 and O_2 partial pressures, C and O, at the Rubisco active site in the chloroplast (Sections 7.2 and 11.4.4). The specificity factor is larger in higher plants (77–90 mol/mol in C3 plants and 70–76 in C4 plants) than in green algae (*ca* 60), cyanobacteria (48) and photosynthetic bacteria (10–20). However, the high $S_{c/o}$ is related to a slower rate of catalysis (k_{cat}), which is 3.5 mol CO_2 (mol enzyme sites)$^{-1}$ s^{-1}. Carboxylation is 2000–3000 times greater than oxygenation at equal partial pressures ($S_{c/o}$ 2000–2500 bar/bar in C3, 1500–2000 bar/bar in C4 plants) but *in vivo*, with small CO_2 and large O_2 partial pressures, actual ratio of carboxylation to oxygenation is much smaller, around 4:1.

The relationship $V_c/V_o = (1/S_{c/o})\,O/C = \alpha$. If R_{day} is zero, then A is also zero when α is 2, which occurs when the chloroplast has a CO_2 partial pressure of Γ^*. The Γ^* (Section 12.2.6) is the CO_2 partial pressure in the absence of 'day' respiration (R_{day}), and $\Gamma^* = 0.5\,O/S_{c/o}$ and $\alpha = 2\Gamma^*/C$. Therefore, $A = (1 - \Gamma^*/C)V_c - R_{day}$ and by substituting in the equation for V_c the RuBP saturated (also called the Rubisco limited) rate of assimilation is:

$$A = (C - \Gamma^*)V_{c_{max}}/C + K_c\,(1 + O/K_o) - R_{day} \tag{12.6a}$$

Alternatively,

$$A = V_c - 0.5V_o - R_{day} = V_c\,(1 - 0.5\alpha) - R_{day} \tag{12.6b}$$

Analysis of the combined carboxylation/oxygenation cycles shows that the rate of RuBP synthesis is $(1 + \alpha)V_c$, and the rate of 3PGA synthesis $(2 + 1.5\alpha)V_c$. From knowledge of the reactions the rate of ATP consumption in the combined cycles is $(3 + 3.5\alpha)V_c$ and rate of NADPH use is $(2 + 2\alpha)V_c$. It is not yet certain what the stoichiometry of ATP and NADPH synthesis is in relation to electron transport. Either 3 or (more probably) 4H$^+$ are required per ATP synthesized (see Chapter 7). The number of protons moved by electron transport depends on whether the whole chain (2H$^+$/e$^-$), whole chain plus Q-cycle (3H$^+$/e$^-$), cyclic electron transport through PSI (1H$^+$/e$^-$) or cyclic electron transport through PSI plus Q-cycle (2H$^+$/e$^-$) is used. This then affects the ATP supply: with 3H$^+$/ATP the ATP production of the different modes of electron transport mentioned is 1.33, 2, 1.33 and 2.66, respectively, or with 4H$^+$/ATP, 1, 1.5, 1 and 2, respectively. This uncertainty is most important for understanding the system and for simulating its behavior. It may be that the processes of electron transport and ATP and NADPH synthesis are not rigidly linked, but operate flexibly according to demand and conditions.

For whole chain electron transport, with 2H$^+$ transported per e$^-$ and assuming 3H$^+$/ATP, then the total e$^-$ flow, J, required for CO_2 assimilation and photorespiration is

$$J = (4.5 + 5.25\alpha)V_c \tag{12.7}$$

Too little ATP is produced for the reactions and alternative sources are required, for example from e$^-$ transport to O_2 in the Mehler reaction. With whole chain electron transport and the Q-cycle but 4H$^+$/ATP, $J = (4 + 4.66\alpha)V_c$ and generates more ATP. Also, cyclic e-flow through PSI results in increased electron flow. The differences in stoichiometry are important, as the ATP supply is crucial to the system function. A conservative estimate of whole chain electron transport required is $J = (4 + 4\alpha)V_c$.

The RuBP (electron transport) limited rate of A, A_j, may be derived from the flux of electrons needed to support carboxylation but is uncertain because of the stoichiometries of electron and proton transport and ATP synthesis:

$$A_j = (C - \Gamma^*) \, J/4C - 8\Gamma^* - R_{day} \tag{12.8}$$

Rates calculated from this equation generally do not saturate with CO_2 at partial pressures where measurements approach saturation. If transgenic plants with much smaller activities of Rubisco are used, then the model predicts their photosynthesis well, indicating that other processes in leaves limit the rate of CO_2 assimilation at high CO_2, probably RuBP synthesis.

A further consideration is the rate of export of assimilation products, A_p, from the chloroplasts. From *Figure 12.1*, and taking into account the rate of exchange of inorganic phosphate with the triosephosphate from the chloroplast, T_p:

$$A_p = (C - \Gamma^*)(3T_p)/(C - (1 + 3\alpha/2) \, \Gamma^*) - R_{day} \tag{12.9}$$

The rates of CO_2 assimilation will be limited by whichever is the smaller of A_j and A_p. Derivation of the equations is given by von Caemmerer (2000) and references therein. The kinetic constants of Rubisco are summarized in *Table 12.1*. From these relationships, combined with the factors determining the CO_2 supply to the chloroplast (Chapter 11) the carbon metabolism of C3 leaves may be explained with considerable precision.

12.2.4 Photosynthesis and irradiance

Chlorophyll in leaves absorbs light very effectively (Chapter 13) even at very large photon fluxes. However, the efficiency with which electron transport is driven depends on the capacity of the rest of the photosynthetic system to use the electrons. At small photon fluxes efficiency is large, but not at large photon fluxes. The rate of J saturates progressively as the photon flux increases, because of the limitations imposed by the structure, composition and turnover of the electron transport components, particularly the rate of electron flow through plastoquinone; J is principally a function of incident radiation, I_o, and of e^- and H^+ transport through the plastoquinone pool. Eventually, at very large photon flux J is saturated, and the rate is referred to as J_{max}. The relation between irradiance, rate of electron transport, J, and CO_2 assimilation varies with plant species and growth conditions. Genetically determinant (i.e. obligate) shade plants or shade-grown plants, with less PQ and reduced electron transport capacity compared to those grown in bright light, have smaller J_{max} than genetically determinant sun-plants or those grown in the bright light. Thus photosynthesis of low-light plants may be limited by electron transport in bright light. As the rate of NADPH synthesis is a fundamental determinant of A, and the reduction of $NADP^+$ by two electrons depends on the concentration of the electron carriers (e.g. the cytochrome b/f complex), acceptors and $NADP^+$-ferredoxin reductase activity, so genetic limitation of these components will limit A. With adequate acceptor and enzyme the rate depends on J, but as their concentrations decrease so the rate is determined by the step with the smallest capacity. Measurements suggest that electron transport and enzyme activity are unlikely to limit

Table 12.1. Some Rubisco kinetic values and chloroplast characteristics required for calculation of CO_2 assimilation in C3 and C4 leaves (derived from literature cited; see von Caemmerer, 2000)

Parameter	Units	C3	C4
K_c	μM	13	26
	Pa	38	65
K_o	μM	414	362
	Pa	33	45
$V_{c_{max}}$	μmol m^{-2} s^{-1}	80	60 (depends on species, nutrition etc.)
$V_{o_{max}}$	μmol m^{-2} s^{-1}	20	—
$V_{o_{max}}/V_{c_{max}}$		4	—
$V_{P_{max}}$	μmol m^{-2} s^{-1}	—	120
$S_{C/O}$	mol mol^{-1}	77–90	80
k_{cat}	mol CO_2 (mol sites)$^{-2}$ s^{-1}	3	3.8
R_{day}	μmol m^{-2} s^{-1}	0.01–0.02 $V_{c_{max}}$	0.01 $V_{c_{max}}$
J'_{max}	μmol m^{-2} s^{-1}	2 $V_{c_{max}}$ (approx 160)	2 $V_{c_{max}}$ (approx 400)
g_i	μmol m^{-2} s^{-1}Pa^{-1}	30	—

NADPH production and it is, therefore, mainly a question of the rate of recycling NADP$^+$.

ATP production is more complex, because three modes of electron flow coupled to photophosphorylation provide flexibility. In linear electron flow, J and ATP synthesis are related, as 3–4H$^+$ are consumed per ATP and 2H$^+$ are transported into the thylakoid per electron. In cyclic photophosphorylation one electron gives 1H$^+$ and synthesis of one ATP is related to one-third of cyclic electron flow, as 3e$^-$ gives 3H$^+$ and one ATP. ATP synthesis depends on the concentrations of ADP and CF1 enzyme active sites; as with the rate of NADPH synthesis, the rate of ATP production depends on the minimum value of J, ADP and CF1. There is probably adequate enzyme capacity, except for plants grown in low light exposed to bright light. Also, it seems likely that the capacities for ATP and NADPH synthesis change in parallel. A large concentration of ATP + ADP is needed to maintain large rates of RuBP synthesis by ribulose-5-phosphate kinase, so inadequate ATP synthesis severely impairs A. Thus, A depends on RuBP regeneration, which in turn depends on the efficiency of the PCR cycle, the amount of 3-PGA available and the supply of NADPH and ATP which are a function of light.

The relation between incident photon flux and electron transport is important when linking available energy to photosynthesis. Photon flux absorbed by PSII, I_{II}, is calculated by the empirical relation:

$$\theta J^2 - J(I_{II} + J_{max}) + I_{II}J_{max} = 0 \qquad (12.10)$$

where I_{II} is the photon flux captured by PSII and J_{max} is the maximal electron transport rate; θ describes the curvature of the relationship of light absorbed to incident I_o (θ is typically 0.6–0.8). I_{II} is related to I_o by allowing for the absorptance (about 0.85) of the leaf, and for the wavelengths of light which are not used in photosynthesis ($f = 0.15$) and assuming 50% of the energy is absorbed by each photosystem, so $I_{II} = I_o \times$ absorbtance $(1 - f)/2$. The solution of the equations for J is:

$$J = \frac{I_{II} + J_{max} - \sqrt{(I_{II} + J_{max})^2 - 4\theta I_{II} J_{max}}}{2\theta} \tag{12.11}$$

12.2.5 Light responses

The photosynthetic rate per unit of leaf area (A) of C3 plants is related to incident photosynthetically active radiation (PAR), I_o, as shown by the light response curve (*Figure 12.2a*). Of course, there are many different response curves, depending on species, growth conditions and the environment during their measurement. However, there is basic similarity in curves, even when C3 and C4 plants are compared. In darkened leaves, CO_2 is released by respiration (R_{dark}), then follows a linear phase of increasing A as I_o rises from zero, gradually changing through a transition phase of progressively decreasing A with rising I_o to a final phase of no increase in A (the plateau) despite increasing photon flux. This type of light-response curve is often used in modeling photosynthesis of vegetation and is then described by a mathematical function to give the best representation of the response, without incorporating mechanistic understanding. The rectangular hyperbola (see Eqn 13.4). is often used.

Other developments of this type incorporate leaf characteristics determining photon absorption, stomatal conductance *et cetera*, providing a more mechanistic assessment of the response.

In darkness, A is zero and R_{dark} is released from the leaf. The rate is not constant; after darkening a C3 leaf, which had been photosynthesizing in air, there is rapid CO_2 evolution lasting for a few minutes, which is from photorespiration, R_{pr} (shown by its O_2 sensitivity). This is called the 'post-illumination burst'. The rate of respiration then drops within a minute or two to a smaller value as R_{pr} ceases. R_{day} may be low in the light if conditions during photorespiration slow it. Then R_{day} may predominate and determine CO_2 release. However, during the transition from light to dark the R_{day} changes to R_{dark}, with metabolism becoming dominated by mitochondrial processes, and R_{dark} rises again (for a period of many minutes) before dropping slowly (over many hours) during the period of darkness. The rate after an hour or more is often regarded as the true R_{dark}. Thus, the rate of R_{dark} may not be the same as R_{day}, although parts of the same metabolic system, including the mitochondria, are involved. Probably, there are progressive changes between dark and light and with the rate of A, size of metabolite (e.g. carbohydrate) pools *et cetera*. The rate of R_{dark} increases with warmer temperatures (approximately doubling for every 10°C rise in temperature – see later discussion of Q_{10}) and is probably greater when leaves contain large amounts of carbohydrate. The different mechanisms of respiration and the use of electrons from other parts of metabolism, for example by the 'alternative pathway' in mitochondria which 'burns off' carbohydrates without coupling the process to ATP synthesis, help to regulate cell metabolism. Then

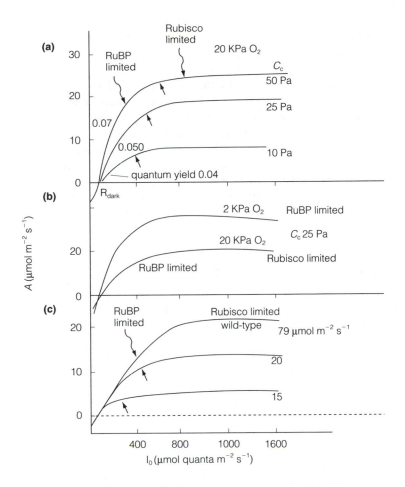

Figure 12.2. Photosynthetic rate (A) of C3 leaves in response to photon flux, I_o. (a) Generalized A/I_o response curves as a function of chloroplast CO_2 partial pressure (C_c) at atmospheric O_2 (21 kPa). The quantum yields are indicated on the curves. A_{max} denotes the light and CO_2 saturated rate of photosynthesis. The arrows show the changeover from RuBP to Rubisco limitation. (b) Effects of O_2 partial pressure on A in relation to I. In 2 kPa O_2 A is always RuBP limited, but at 20 kPa O_2 it is RuBP limited at low light and Rubisco-limited at high light. (c) A of tobacco with wild-type activity or transgenic with decreased activity (shown on the curves). Arrows indicate the progressive increase in the I_o at which transition from RuBP to Rubisco limitation occurs. (Based on information in von Caemmerer (2000))

mitochondria use reductant and perhaps reduce the energy load on the photosynthetic system (Section 8.9). In darkness, with no photosynthetic electron transport and without the PCR cycle operating, the energy balance of cells is much less complex than in the light. Different parts of the mitochondrial respiratory processes may be involved in light and dark, so that reductant and ATP production may be linked to CO_2 production by glycolysis and TCA cycle respiration very flexibly, allowing fine regulation of metabolism and conditions in the cell. There is relatively poor understanding of the way that CO_2 and energy fluxes in respiration are regulated (see Chapter 8) and the mechanisms and consequences of different forms and rates of respiration for effective photosynthesis under a range of conditions require further analysis.

On illumination, photosynthesis starts (induction is considered in Section 7.5) and the rate of CO_2 emission from the leaf falls, because CO_2 is assimilated and possibly because R_{dark} decreases, although in the normal atmospheric CO_2 and O_2 concentrations R_{pr} rises. At a particular photon flux, called the light compensation point, there is no net CO_2 exchange, that is CO_2 evolution and assimilation are in balance. The gross A_g equals the respiratory losses ($R_{dark} + R_{day} + R_{pr}$) so A is zero. Light compensation occurs at about 50 µmol m^{-2} s^{-1} of incident PAR in 36 Pa CO_2 and 21 kPa O_2 (photorespiratory conditions) and decreases with nonphotorespiratory conditions of saturating CO_2. Although called the compensation 'point', it varies with many factors, and should not be regarded as a fixed value. Shade plants, for example, generally have smaller compensation points than sun plants (20 compared to 80 µmol quanta m^{-2} s^{-1} because of their relatively greater ability to capture photons at low I_o). The exact light compensation point depends on A_g, R_{pr}, R_{dark} and R_{day}, which are affected by different conditions, R_{pr} by O_2 and CO_2 partial pressures and temperature, and R_{dark} and R_{day} by temperature. Other environmental factors, such as nutrition and water stress which affect one or other of these components, also influence the light compensation point.

With further rise in photon flux, the rate of A increases linearly, up to about 300–400 µmol quanta m^{-2} s^{-1} (*Figure 12.2a*) as electron transport is proportional to the photons captured and there is a fixed stoichiometry of electron transport to ATP and NADPH synthesis and consumption of these for CO_2 assimilation. At low photon flux, in non-limiting CO_2, the slope of the curve at $A = 0$ is CO_2/I with units of µmol CO_2 m^{-2} s^{-1}/µmol quanta m^{-2} s^{-1} or mol CO_2 quantum^{-1}, and is called the quantum yield. It is a measure of the photochemical efficiency, ϕCO_2, and shows the ability of the photosynthetic system to utilize photons for CO_2 fixation at the greatest efficiency. The 'true quantum yield' is calculated for photons absorbed by the antenna pigments. If light incident upon the leaf is used as a measure of photon flux then only an 'apparent quantum yield' is obtained; this is not a measure of the true efficiency of the system because light reflected and transmitted by the tissue is included in the energy supposedly utilized for CO_2 assimilation. The methods of measuring photosynthetic rates and quantum yield are described in Chapter 11. In C3 leaves in low O_2 or high CO_2 partial pressure (chloroplast CO_2 or C_c of 40 Pa), which effectively stop photorespiration, ϕCO_2 is about 0.07–0.08 or about one CO_2 per 12–14 photons absorbed, and in air ϕCO_2 is about 0.04–0.05, that is one mol CO_2 is fixed per 20–25 photons absorbed. True quantum yield for different species is very constant at 0.08 for a very wide range of plants under standard conditions. Quantum yield depends on the condition of measurement. For example, for gas exchange with incident light from above a leaf, the quantum yield may be slightly larger than with light from below, probably because the optical conditions affect the micro-light environment which cannot be easily determined by methods used to estimate the total number of photons absorbed. Leaves absorb most of the incident light and the response is a weighted average of the responses of different tissues. Hence, the leaf's response to conditions is very different from that of an individual chloroplast or of suspensions of chloroplasts in biochemical studies *in vitro*. Within leaves, there is great heterogeneity of cell types and distributions. Also, light distribution differs greatly in leaves, for example palisade mesophyll cells and chloroplasts receive more light when illuminated from above than spongy mesophyll chloroplasts next to the lower epidermis. The CO_2 environment is also nonuniform because of the different locations and rates of uptake and production. Thin leaves are rather isothermal, but not thick leaves (e.g of succulent plants). Such differences may be of considerable long-term ecological importance.

Light compensation point. Fluxes of carbon and partial pressures of CO_2 around the light compensation point, considered above in qualitative terms, are described by Rubisco kinetics. The rate of CO_2 assimilation is zero at the light compensation point; A at low irradiance is given by:

$$A = (C - \Gamma^*)J/(4C + 8\,\Gamma^*) - R_{day} \tag{12.12}$$

where C is the CO_2 partial pressure, and Γ^* the CO_2 compensation point without day respiration, R_{day}. So the rate of A per unit of quanta absorbed (i.e. the quantum yield) giving the initial slope of the light response curve (dA/dI_o) as

$$dA/dI_o = (C - \Gamma^*)/(8C + 16\Gamma^*)(1 - f)\ \text{absorbance} \tag{12.13}$$

However the efficiency predicted from the model is large, 0.09, compared with the experimental values (which are often measured at small but not zero photon flux), as the model does not include the other processes using energy in metabolism. Analysis based on Rubisco kinetics shows the initial slope to be limited by the capacity of RuBP regeneration from electron transport-driven ATP and NADPH synthesis, so that turnover of Rubisco active sites is limited. That Rubisco has the capacity (and is active), is shown by studies where the RuBP pool is increased transiently by removing CO_2 for a few seconds, then supplying CO_2 again. The rate of A greatly exceeds the steady-state values for a very short time because the large RuBP pool provides substrate before being depleted. Quantum yield is decreased by low CO_2 (*Figure 12. 3a and b*) because of the oxygenase reaction.

Maximum rate of assimilation, A_{max}. As I_o increases further, the rate of increase in A is no longer linear, as the mechanism determining A becomes saturated and slows. The photon flux at which the slope departs from linearity is very species- and environment-dependent. Eventually, a plateau, A_{max}, is reached, that is the rate becomes constant, at large photon flux. At O_2 of 20 kPa and C_c of about 50 Pa, the A_{max} is around 20–40 µmol $m^{-2}\ s^{-1}$, and is also very species dependent.

Processes determining A_{max} in leaves as the photon flux increases are potentially synthesis of ATP and NADPH leading to synthesis of RuBP, the rate of diffusion of CO_2 to the active sites of Rubisco and CO_2 assimilation capacity of Rubisco. Electron transport *per se* does not limit CO_2 assimilation. The maximum rate of electron transport, J, between the water oxidizing process and the main acceptor for electrons, $NADP^+$, depends upon the rate of formation of ATP and on the availability of $NADP^+$, which is a function of the pool sizes of the pyridine nucleotides and their turnover, themselves dependent upon CO_2 fixation. In the discussion of chlorophyll *a* fluorescence (Sections 5.5 and 12.5)

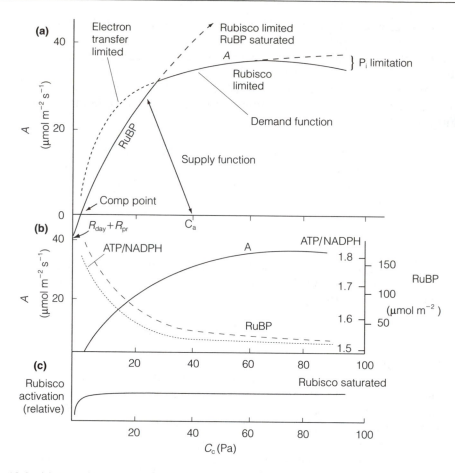

Figure 12.3. (a) Theoretical relationships between the CO_2 assimilation rate, A, and the chloroplast CO_2 partial pressure (C_c) modelled from Rubisco characteristics at high light and 25°C. The initial rise in A with increasing C_c is the result of increasing carboxylation when Rubisco is CO_2 limited but RuBP saturated. At high C_c, RuBP is limiting not Rubisco. At very large C_c inorganic phosphate may become a limitation. The curve of A is the demand function and the supply function, which relates the CO_2 partial pressure in the atmosphere, C_a, to C_c, depends on the conductances of the gas and liquid phases between ambient air and the active sites of Rubisco in the chloroplast. The CO_2 compensation point is shown and the respiration in the light, $R_{day} + R_{pr}$. (b) Generalized changes in ATP/NAPH ratio and the RuBP concentration, together with the Rubisco activation state with increasing C_c and A. (c) Rubisco activation state (relative values) with C_c. (See text: derived from literature given)

it is shown that the energy in the PSII reaction center could be utilized by the photochemical quenching processes (e.g. CO_2 and NO_3^- reduction), and by nonphotochemical quenching (e.g. by dissipation via carotenoids). With increasing PAR the amount of energy captured rises but the capacity of photochemical quenching does not rise in proportion because of limitations in other parts of the system. Therefore, nonphotochemical quenching rises and electron flow increases suggesting that electron flow is not a limiting factor in CO_2 assimilation in very bright light. Under normal atmospheric conditions in air with C_a of

36 Pa, A increases as I_o increases. However, as a consequence of small stomatal conductance, C_i rather than light becomes limiting, and eventually further increase in irradiance has no effect on A (see Chapter 11 for the dependence of internal CO_2 partial pressure on atmospheric CO_2 and on stomatal and boundary layer conductances). With g_s for CO_2 of 0.4 mol m^{-2} s^{-1} and 30 Pa CO_2 and 21 kPa O_2 in the atmosphere, the internal CO_2 in a C3 leaf becomes limiting at A of about 30 μmol m^{-2} s^{-1} when irradiance is about 1000 μmol quanta m^{-2} s^{-1}. If the PAR is increased to 1500 μmol m^{-2} s^{-1} or greater there may be progressive decrease in A. This may be due to processes such as decreased stomatal conductance resulting from induction of water stress – water loss under such conditions is large and difficult to avoid – but also to photoinhibition after prolonged exposure to bright light.

Saturation of A of leaves by high CO_2 at high I_o in 21 kPa O_2, is due to Rubisco limitation when the RuBP supply is adequate. However, in high I_o but low (2 kPa) O_2, A is limited by RuBP because of the large demand from Rubisco. With decreasing C_c, A saturates at progressively smaller I_o because less RuBP is required with low CO_2 when the Rubisco reactions are limiting but quantum yield is unaffected. At very small I_o, even when close to the CO_2 compensation point, there is ample CO_2 so the quantum yield is insensitive to I_o. Decreasing (by anti-sense to the small subunit) activity of Rubisco per unit leaf area by ca 75% had a substantial effect on A_{max} (75% reduction; *Figure 12.2c*), but decreased quantum yield by only ca 15%. Decreasing Rubisco also decreased the I_o at which the transition from RuBP to Rubisco limitation occurred. These responses are expected, as just as there is less substrate for Rubisco in the case of small C_c and there is less Rubisco in the transgenics (with the equivalent effect), so the RuBP requirements are met with a smaller photon flux. The light saturated A_{max} reflects the limited capacity of Rubisco and is of considerable importance, as it determines the maximum assimilation which a plant under a large photon flux can achieve, and ultimately the productivity achievable with large leaf area intercepting the available light. Ultimately A_{max} is the limiting factor in the efficiency of plant production when other environmental conditions are in excess.

Oxygen greatly affects A and the responses to I_o and CO_2, as *Figure 12.2b* shows. With low O_2 (2 kPa), A increases substantially compared to the rate in 20 kPa as CO_2 and I_o increase, because the CO_2 released in R_{pr} and the energy used in making RuBP for the oxygenase reaction are eliminated. Hence, with R_{pr} of 25% of gross CO_2 assimilation in 20 kPa O_2, the response of decreasing O_2 to 2 kPa is a 40% increase in A. There is a strong interaction between O_2, CO_2 and I. In low O_2, quantum yield increases to a maximum of about 0.08 with C_c between 10 and 20 Pa CO_2, but in 20 kPa O_2 it rises less, reaching about 0.05 at the same C_c and 0.7 at 50 Pa CO_2.

Electron transport and CO_2 assimilation.

Photosynthetic electron flux through PSII (J_{II}) is used for several different purposes in leaves, primarily CO_2 assimilation (J_{CO_2}) and the associated photorespiration but also reduction of nitrate (J_{NIT}) and sulfate (J_{SUL}) and O_2 in the Mehler peroxidase reaction (J_{MEH}).

$$J_{II} = J_{CO_2} + J_{NIT} + J_{SUL} + J_{MEH} \tag{12.14}$$

However, the proportion of electrons used in each process and how it changes with conditions is still not established. Fluorescence measurements are used to estimate the electron fluxes in photosynthesis (see Section 12.5), providing a comparison with rates derived from measurements of CO_2 uptake or calculated rates based on Rubisco; discrepancies between them indicate the magnitude of the alternative, that is noncarboxylation, sink for electrons such as the Mehler-peroxidase pathway. From the rate constants of energy conversion in photochemistry, and dissipation (Section 5.6), the quantum yield of photochemical quenching of PSII is $\phi_p = F_m' - F_s/F_m'$ (or $1 - F_s/F_m'\sim$) and nonphotochemical quenching $\phi_n = F_s/F_m' - F_s/F_m$, where F_s is steady-state fluorescence under actinic light, F_m' is the maximal fluorescence in the actinic light caused by a brief saturating light pulse and F_m is the flash-induced maximal fluorescence from dark adapted leaves (Genty *et al.*, 1989). These quantum yields are related to the quenching coefficients derived earlier (Section 5.6), ϕ_n is related to the nonphotochemical quenching q_N and ϕ_p to the photochemical quenching q_P. To J_{II}, the absorbed photon flux, I_o, and the fraction of photons used by PSII (a_{II}, which is about 0.45–0.5 and is commonly taken as 0.48) are required:

$$J_{II} = a_{II} I_o (1 - F_s/F_m') \tag{12.15}$$

As $4e^-$ are required per CO_2 fixed, $J_{II}/4$ is the flux of electrons required if all are used in CO_2 fixation (e.g. when photorespiration is negligible). There is a very strong relationship between quantum yield of PSII electron transport calculated from CO_2 assimilation and the F/F_m. The NADPH limited rate of electron transport may be estimated from CO_2 assimilation by $J_A = (A + R_{day}) (4C + 8 \Gamma^*)/(C - \Gamma^*)$. The rate for limiting ATP (assuming $4H^+/ATP$ and $3H^+/e^-$) is $(A + R_{day}) (4.5C + 9.33\Gamma^*)/(C - \Gamma^*)$. However, assuming $3H^+/ATP$ and $2H^+/e^-$ then $J_A = (A + R_{day}) (4.5C + 10.5\Gamma^*)/(C - \Gamma^*)$. The relation between J_A for CO_2 and J_{total} from fluorescence measurements in nonphotorespiratory conditions (2 kPa O_2) is very close with varying CO_2 supply, but the relationship is also close in photorespiratory conditions when the chloroplast CO_2 partial pressure is used, that is when the mesophyll conductance is allowed for.

Quantum yields of PSII (ϕPSII, the electron flow through PSII per unit quantum flux) decrease with increasing photon flux as fewer of the reaction centers are open the stronger the actinic light and the larger the rate of CO_2 assimilation. Although q_P remains constant or decreases with photon flux, the nonphotochemical parameters increase as more of the absorbed energy is dissipated by the xanthophyll cycle (Section 3.4.4) as heat. As irradiance increases, photorespiratory CO_2 evolution also increases, because the rate of CO_2 assimilation rises in proportion if the CO_2/O_2 ratio remains constant. However, there is considerable interaction between processes; if the CO_2/O_2 ratio in the leaf falls, as it does when stomatal conductance is small and A is rapid, the ratio of oxygenation to carboxylation of RuBP, V_o/V_c, increases and R_{pr} takes an increasing proportion of J_{II}. Thus, the proportion of electrons used by the different processes in the leaf is not constant but depends upon the magnitudes of gross CO_2 assimilation and photorespiration. The flux of electrons required for CO_2 assimilation, J_{CO_2}, may be calculated from measured CO_2 assimilation, or from the Rubisco characteristics. The difference between J_{II} and J_{CO_2} gives the electron transport to alternative non-CO_2 acceptors, $J_{ALT} = J_{NIT} + J_{SUL} + J_{MEH}$. Although there is good agreement between the estimates of J_{CO_2}, assumptions that the estimates based on fluorescence for nonphotorespiratory conditions are applicable to other rather different conditions, such as with water stress, need testing. This may lead to errors in the estimated magnitudes of the alternative sinks. Calculated J_{ALT} decreases as CO_2 assimilation rises with increasing photon flux because CO_2 provides the sink for electrons. Oxygen also provides a sink, so J_{ALT} increases as O_2 partial pressure rises from 1 up to 21 kPa, but falls thereafter, possibly because it is difficult to measure the specificity factor for calculating J_{CO_2} under these conditions.

Saturation of A has implications for the energy balance of leaves. As the plateau is reached, less and less of the captured photon flux is used for CO_2 assimilation and the energy is used for reduction of nitrate *et cetera*, which is relatively small (5–10% of the energy) and the Mehler–proxidase pathway reactions which are increasingly regarded as using only a small part of J_{II}. The excess energy captured in the light-harvesting systems is not then used for electron transport and is principally dissipated as heat by the activity of the xanthophyll cycle (see Chapter 3).

12.2.6 Photosynthesis and carbon dioxide

When a typical C3 leaf is irradiated with large photon flux (1200 μmol quanta m^{-2} s^{-1}) and A is measured as a function of the ambient partial pressures of CO_2 from 0 to 100 Pa, a response curve (*Figure 12.3a and b*) results. Without CO_2 in the gas supplied to the leaf chamber, the leaf emits CO_2; at greater pressures (*ca* 5 Pa) there is no net exchange of CO_2 and with further increase there is a linear increase in A up to *ca* 80 kPa, but little or no increase with greater pressures. Rather than relating A to C_a, the effects of stomatal conductance are removed by calculating the internal CO_2, C_i, as described in Chapter 11, and A is determined as a function of C_i giving the 'A/C_i curves'. These provide much understanding of the factors regulating photosynthesis. The CO_2 partial pressure at the Rubisco active sites, C_c (estimated as described in Section 11.4.4), is an even more important value to relate A to than C_i, because it is the CO_2 partial pressure at which Rubisco operates. The rate of A is determined by the light-driven synthesis of RuBP and the activity of Rubisco, which provide the potential demand for CO_2, also called the 'demand function' (*Figure 12.3*) which represents the requirement for CO_2 resulting from photosynthetic metabolism. With I_0 above 1200 μmol m^{-2} s^{-1} the potential electron transport for synthesis of ATP and NADPH is large and RuBP content is high, so at low CO_2 partial pressure CO_2 is limiting (*Figure 12.3b*), as expressed by the 'supply function', which is the rate at which CO_2 is available to the active sites of Rubisco, and is determined by stomatal conductance in the gas phase and by liquid phase resistances, combined with atmospheric CO_2 concentration (Chapter 11). It is important to recall that C_i is substantially smaller (*ca* 25%) than C_a and C_c is some 25% less than the C_i. The role of the variable stomatal conductance is very important under field conditions, where atmospheric humidity, water deficiencies and so on greatly modify the responses. At low C_i the availability of CO_2 limits A because Rubisco is not CO_2 saturated: this is referred to as Rubisco limitation despite the availability of active Rubisco as shown by the immediate increase in A resulting from an increase in C_i. The transition from RuBP limitation to RuBP saturation occurs close to the normal C_i, 25–27 Pa CO_2, with C_c at 18–20 Pa. Possibly there is a balance, determined genetically and therefore subject to evolutionary selection pressure, between the use of light and CO_2, the loss of water (determined by stomatal conductance) and the costs of making the leaf and its biochemical components. The photosynthetic system seems to be co-limited by many factors, all operating close to an optimum for the system, if not for the individual components (see discussion of 'control theory'; Section 7.4).

With the CO_2 partial pressure inside the leaf close to zero, obtained when CO_2-free air surrounds the leaf, $R_{pr} + R_{day}$ in mesophyll cells results in CO_2 escaping, in part, into the intercellular spaces, increasing C_i, and then into the atmosphere. The measured rate of CO_2 evolution from the leaf is smaller than the rate of CO_2 production by the leaf cells, as some of the CO_2 evolved is re-assimilated. It has been estimated that 50% or more of evolved CO_2 is re-assimilated but it is difficult to quantify the rate of dark or photorespiration under the normal atmospheric conditions, for if the CO_2 and O_2 partial pressures are decreased then the behavior of the entire system is affected. Isotope exchange provides a method of estimating respiratory fluxes (Chapter 11). At the CO_2 compensation point electron transport is slowed by lack of acceptors and probably mitochondria function in the light to provide ATP and metabolites (from the TCA cycle) and also use reductant (from chloroplasts via the malate shuttle).

CO_2 **compensation point.** With increasing CO_2, the rate of CO_2 evolution rises until $A = R_{pr} + R_{day}$ when CO_2 exchange is zero (*Figure 12.3*). This is the CO_2 compensation point (Γ) which – as with the light compensation point – is not a fixed value. It is determined by the balance between A and $R_{pr} + R_{day}$. Under these conditions, R_{day} may be largest when demands for metabolites are large but A cannot supply them, so they are derived from the TCA cycle and glycolysis. In C3 leaves at 2 kPa O_2, Γ is about 0–1 Pa CO_2 and in 21 kPa O_2 about 4–5 Pa O_2. Much progress has been made in understanding the CO_2 fluxes in relation to CO_2 partial pressures in leaves close to the compensation point. The Γ is measured by putting a leaf in a well-sealed (i.e. very small leakage rate) illuminated chamber, with different initial CO_2 and O_2 concentrations, and measuring the equilibrium (steady state) CO_2 partial pressure attained (Chapter 11). Also, Γ may be measured from A/C_i curves at different I_o (low to high) and over a range of 0–10 kPa CO_2 C_i in open gas exchange systems (*Figure 12.4*). The Γ is determined by Rubisco characteristics and respiration. The chloroplast CO_2 compensation point with R_{day} zero and respiration given by R_{pr}, is Γ^*, which is zero when α equals 2 from $A = V_c (1 - 0.5\alpha) - R_{day}$, that is, there is no net CO_2 exchange and it is independent of other CO_2 fluxes in the cell. So $\Gamma^* = V_{o_{max}} K_c O / 2 V_{c_{max}} K_o = O/2S_{c/o}$ and is linearly dependent on O_2 partial pressure. With R_{day} (the usual condition) then

$$\Gamma = (\Gamma^* + K_c (1 + O/K_o) R_{day}/V_{c_{max}}) / (1 - R_{day}/V_{c_{max}}) \tag{12.16}$$

In *Figure 12.4*, the relation between A, R_{day}, Γ, Γ^* and C^* (C_i at which $A = R_{day}$) is shown. The point at which the curves of A versus C_i at different I intersect is C^* and is given by $A = -R_{day}$. The analysis is extended by plotting A as a function of C_c, giving $-R_{day} = \Gamma^*$ and Γ is obtained from both plots with A/C_c and A/C_i zero. Thus, C^* is similar to Γ^* but is smaller by up to 1.5 Pa due to the diffusion resistance, g_i, between the internal spaces and the chloroplast, expressed by:

$$\Gamma^* = C^* + R_{day}/g_i \tag{12.17}$$

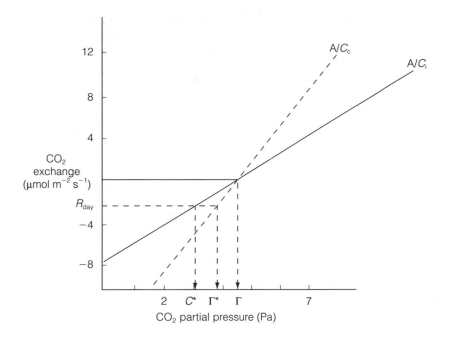

Figure 12.4. Carbon dioxide exchange related to the CO_2 partial pressure near the leaf CO_2 compensation point, Γ, at the Γ^* (without respiration), and at C^*, the intercellular CO_2 partial pressure at which $A = -R_{day}$. The lines are plots of A against the intercellular (C_i) or the chloroplast (C_c) CO_2 partial pressures

Both Γ^* and Γ increase linearly with increasing oxygen partial pressure from about 2–3 Pa at 1 kPa O_2 to *ca* 5 kPa CO_2 at 21 kPa O_2 and 10 kPa CO_2 at 40 kPa O_2 due to the Rubisco oxygenation reaction. Decreasing the amount of Rubisco in transgenic tobacco to 10% of that in the wild-type plant increased Γ substantially (from 5 to 7.5 Pa CO_2) because R_{day} was not altered but the capacity for CO_2 assimilation was decreased. The Γ also increases with higher temperatures, due to stimulation of Rubisco oxygenase relative to carboxylase, and with decreasing I_o as a consequence of inhibition of CO_2 assimilation.

Carboxylation efficiency. The initial slope of the curve relating A to CO_2 at the compensation point is called the carboxylation efficiency or mesophyll conductance, and is a measure of the ability of the system to assimilate CO_2 at maximum efficiency and independent of light limitation. Carboxylation efficiency is about 2–2.5 µmol m^{-2} s^{-2} Pa^{-1} in 2 kPa O_2 and 1–2 µmol m^{-2} s^{-2} Pa^{-1} in 21 kPa O_2. The slope of the A/C_i curve may be calculated at Γ^* from Rubisco characteristics by:

$$dA/dC = V_{cmax}/[\Gamma^* + K_c(1 + O/K_o)] \tag{12.18}$$

The calculated values of efficiency agree well with measured values, although the relation between measured carboxylation efficiency and measured activity of Rubisco is not linear, suggesting that conductance to CO_2 diffusion within the liquid phase is responsible. Correction for this improves the relationship. The effect of temperature on Rubisco

325

kinetics also explains the increase in dA/dC_i with warmer temperatures. The consequences of decreasing Rubisco amount per unit leaf on the CO_2 compensation point in transgenic plants have been mentioned. Carboxylation efficiency (*Figure 12.5a*) decreased by over 50%, and was independent of the light, showing that the transgenic plants had excess RuBP; the limitation was by inadequate Rubisco capacity to use the CO_2 becoming available as C_i increased.

Photon flux has important effects on both compensation point and carboxylation efficiency. Compensation point rises substantially at temperatures above about 25°C when

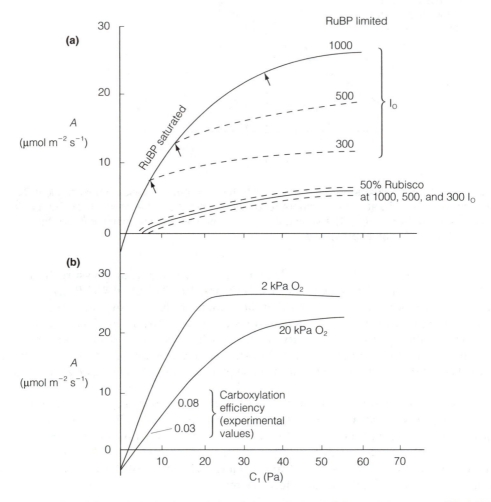

Figure 12.5. The CO_2 assimilation rate, A, in relation to internal CO_2 partial pressure (C_i). (a) The effects of photon flux, I_o (shown on the curves) on the relationship; arrows indicate the transition from Rubisco limitation at low C_i to RuBP limitation at high C_i. Also shown is the effect of approximately 50% reduction in Rubisco activity, which prevents the response to all values of I_o. (b) Effects of O_2 partial pressure on the A/C_i response. At low O_2, the carboxylation efficiency is greater than in high O_2 so A_{max} is reached much earlier, but very large CO_2 partial pressure almost overcomes the inhibition caused by oxygen (based on information in von Caemmerer, 2000)

I_o is less than 300 μmol m^{-2} s^{-2}. Possibly this is caused by increased R_{day} at low I_o or RuBP is limiting. Carboxylation efficiency is independent of I_o at photon fluxes greater than about 500–800 μmol m^{-2} s^{-1} (depending on the species), when RuBP is saturating; then the limitation is caused by Rubisco. However, with decreasing I_o below the upper threshold, carboxylation efficiency drops substantially, by 60–80% below 250 μmol m^{-2} s^{-1}. This is due to RuBP limitation. Low photon flux decreases the C_i at which A_{max} is achieved, reflecting the RuBP/Rubisco transition (*Figure 12.5a*) When Rubisco activity is decreased by anti-sense methods, A_{max} is greatly decreased and there is no response to light (*Figure 12.5a*). Using the interpretation of the A/C_i relationship given and shown (*Figure 12.3a and b*) it is seen that in high light and low C_i and C_c, CO_2 is limiting (Rubisco limitation). The RuBP concentration in leaves in bright light is large (150 μmol m^{-2}) when C_i is low (5–10 Pa) and A is slow (1–2 μmol m^{-2} s^{-1}) but electron transport potential is unaffected. As C_i rises, so A increases and RuBP concentration falls, and saturation of A occurs at about 40 μmol m^{-2} s^{-1} at RuBP pool size of 75 μmol m^{-2} and C_i of 30 Pa, although the RuBP pool may drop further without increasing A due to the increase in oxygenation and R_{pr} as C_c is decreased. As C_a, C_i and C_c increase, so Rubisco can function at maximum rate because RuBP is available. However, as A increases further, the pool of RuBP is depleted as consumption is faster than synthesis which is limited by electron transport. Synthesis of ATP is expected, from modeling studies, to decrease more than NADPH. Therefore, as CO_2 supply rises there is a transition from RuBP saturation to RuBP limitation and *vice versa* for Rubisco, when Rubisco activity is large. There is a large O_2 effect on carboxylation efficiency (*Figure 12.5b*). At 2 kPa carboxylation is very efficient, as mentioned, but it decreases from *ca* 2 to 1 μmol m^{-2} s^{-1} Pa^{-1} as O_2 increases due to the increasing competition of O_2 for CO_2 at the Rubisco active sites. Also, CO_2 is released in R_{pr} and there is decreased availability of Rubisco. The initial slope dA/dC_i is strongly dependent on Rubisco activity although it is not necessarily linearly related at large Rubisco content due to the mesophyll conductance slowing CO_2 transport.

Effects of CO_2 and O_2 on A_{max}.

The effect of CO_2 and O_2 on A_{max} is shown in *Figure 12.5b*. In the normal atmosphere with 21 kPa O_2 and 36 Pa CO_2, there is a large component of R_{pr} at A_{max}. Increasing the CO_2 partial pressure reduces competition from O_2 so R_{pr} is inhibited and the difference becomes smaller as C_i and C_c increase but the transition from RuBP limitation to Rubisco limitation is achieved at much larger C_i and C_c in high O_2. In an atmosphere of 2 kPa O_2, A_{max} of about 40 μmol m^{-2} s^{-1} is achieved at about 20–30 Pa CO_2 but 21 kPa O_2 decreases A_{max} from 40 to 30 μmol CO_2 m^{-2} s^{-1} and increases the CO_2 partial pressure at which saturation is achieved, from *ca* 20–30 to 40–60 Pa, due to photorespiration. As C_a at the leaf surface increases, C_i and C_c rise and the supply of CO_2 satisfies the demand and the activated catalytic sites become saturated with CO_2, R_{pr} is very small and A is limited by RuBP regeneration. Under these conditions, the supply of P_i to the stroma may be deficient, as the capacity of the triosephosphate translocator to export DHAP and import P_i is reached. It is important to appreciate that Rubisco in C3 plants is not saturated with CO_2 in the current atmosphere containing 21 kPa O_2: with current C_a of 36 Pa C_i is *ca* 25 Pa. The value of A is about 25% less than the CO_2 saturated A_{max} obtained at high C_i and C_c. This effect of O_2 is shown by measurements of A as a function of increasing C_i from *ca* 0 up to 60 Pa with O_2 of 2 and 21 kPa O_2 in saturating light (*Figure 12.5b*). A_{max} is eventually very similar with both O_2, but is achieved at very different CO_2, *ca* 25 Pa in 25 kPa O_2 and 40 Pa in 21 kPa O_2. In low O_2, this is due to CO_2 saturating

Rubisco, which then becomes RuBP regeneration limited. In high O_2 the CO_2 competes and, although RuBP is limited, carboxylation replaces the more energy-requiring oxygenation, thus increasing A. Also less R_{pr} is released in low O_2, also increasing A.

12.2.7 Climate change, increasing atmospheric CO_2 and A

Increasing atmospheric CO_2 will require less Rubisco activity to sustain A at the same I_o and RuBP supply. Increasing C_a from 35 to 50 Pa decreases the amount of Rubisco needed by 20% at 10°C and by 33% at 45°C. At 70 Pa (expected by the end of the twenty-first century) the decrease would be 30–50% at 10°C and 45°C, respectively. Much discussion and experimental analysis suggests that adjustment ('acclimation') of the amount and activity of Rubisco occurs in response to elevated CO_2. However, plants such as wheat, grown with adequate nutrients (especially N), do not readily re-adjust the ratio of Rubisco/electron transport capacity due to high CO_2, despite the effects of carbohydrate accumulation on some plants under some conditions; they are genetically determined. The decrease seen in Rubisco amount in experiments on a range of species are mainly due to differences caused by growth conditions, such as inadequate nitrogen supply or altered temperature, both of which change composition and photosynthesis of leaves. Temperature differs as a consequence of growth in elevated CO_2 compared to normal CO_2, for example smaller stomatal conductance results in warmer, faster-growing leaves. Selection for smaller amounts and total activity of Rubisco in leaves will probably occur with time, in response to the selective advantage of removing the inefficiency caused by over-production of Rubisco. Photosynthesis and production are expected to respond to the increased CO_2 in the future atmosphere and may have already done so over the last century. With CO_2 increase from *ca* 28 to 36 Pa over that period gross photosynthesis may have increased by 10–15% and vegetation may, therefore, have acted as a sink for some of the excess carbon generated by burning fossil fuels and destruction of forests associated with rapid industrialization and population growth (see Chapter 11).

12.2.8 Temperature and photosynthesis

Photosynthetic rate responds markedly to temperature. The magnitude of the responses and the actual temperatures at which they occur are very dependent upon the species and growth conditions. Species of a very wide range of families of plants may be very tolerant of cold, for example conifers of boreal forests, herbs of arctic-alpine regions, whilst others are very tolerant of hot conditions, for example trees and herbs of deserts. They may be relatively intolerant of warm and cold, respectively, although many have considerable ability to tolerate both extremes. Many photosynthetic organisms have very great tolerance. Cyanobacteria photosynthesize in hot springs at temperatures of 70°C and above, which kill most plants. Many lichens (combinations of algae and fungi) photosynthesize at low temperatures whilst others tolerate heat. Tolerance is not dependent on the photosynthetic processes and structures alone, but on many aspects of the plants anatomy, biochemistry and physiology. Despite this, a general response is evident (*Figure 12.6a–c*): with extreme cold and heat irreparable damage may occur to the photosynthetic system. There is an optimum (often broad) at which A is rapid, and outside it the rate is small but photosynthesis proceeds effectively. Temperatures at which particular responses occur are not only very dependent on the species but also on the adaptation of

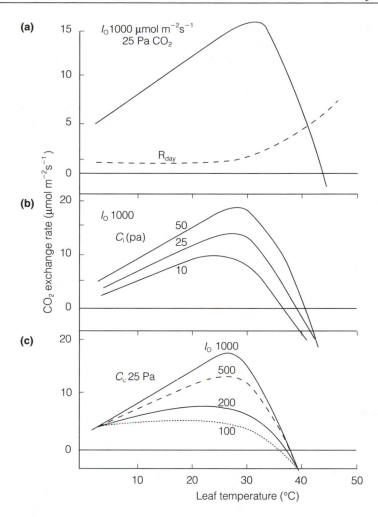

Figure 12.6. General responses of CO_2 exchange rate, A, to leaf temperature. (a) at C_i of 25 kPa, O_2 partial pressure 21 kPa and I_o of 1000 µmol quanta m^{-2} s^{-1}. The rate of respiration (R_{day}) is also indicated. (b) A as a function of different C_i at I_o of 1000 µmol quanta m^{-2} s^{-1}. (c) A related to different I_o at chloroplast compensation point (C_c) of 25 Pa

the plant: adaptations include changes to lipid and pigment composition of the thylakoid membranes and altered amounts of proteins.

Temperature affects several parts of the photosynthetic system. The rate of A at low C_i is not greatly affected by increasing temperature because CO_2 is limiting and the activation energies of the carboxylase and oxygenase reactions of Rubisco are similar. At higher CO_2 partial pressures A increases as Rubisco activity increases and also the rate of RuBP synthesis rises. However, at high temperatures decreased electron transport slows RuBP production and A falls, often greatly for a further small increase. The RuBP supply is decreased at high and low temperatures as a consequence of effects on ATP and NADPH synthesis due to the decreased electron transport, possibly caused by altered thylakoid viscosity and changed interactions between components. Temperature during growth affects the lipid composition of the membranes, and there is considerable optimization of the properties of the membranes and the components, leading to acclimation. The Q_{10} (see below) of processes changes with growth conditions, for example that for electron transport in thylakoids from cold-adapted winter rye (*Secale cereale*) plants was 1.53 compared to 1.85 for unadapted, and the photosystems differed in temperature response also (for discussion see Huner *et al.*, 1998).

The effects of temperature on electron transport are modeled by empirical relationships based on experimental data from isolated chloroplasts using the activation energies of electron transport. Changes in enzyme kinetic constants are described by an Arrhenius function:

$$\text{Parameter } (T) = \text{Parameter } (25°C) \exp [(25-T)E/(298R(273+T))] \tag{12.19}$$

where R is the universal gas constant, T is the temperature in °C and E is the activation energy. E varies between processes. Alternatively, a Q_{10} relationship may be used to represent the changes:

$$\text{Parameter } (T) = \text{Parameter } (25°C) \, Q_{10}^{[(T-25)/10]} \tag{12.20}$$

This gives the relative increase in response over a 10°C range at a standard temperature. Q_{10} values differ for different processes.

The response of Rubisco to temperature is due to differential effects on the reaction kinetics of the carboxylase and oxygenase reactions. The reaction of the enediol with O_2 has a higher activation energy than with CO_2. This increases the rate of oxygenation more than that of carboxylation, for a given temperature increase, thus decreasing the specificity factor. Also, oxygenation increases relative to carboxylation because the solubility of CO_2 is smaller, relative to O_2, at higher temperatures, thus favoring oxygenation over carboxylation. Consequently, the photorespiration/net assimilation ratio is small in cool conditions, but larger at higher temperatures, so the increase in A is offset, resulting in a small increase in A and a marked drop in φ. Other processes may influence Rubisco as temperature changes, for example inactivation of Rubisco activase. The balance between electron transport and Rubisco activity is important. In some cases RuBP pools may increase as temperature rises; in others they may fall, probably depending on the change of temperature relative to the optima of different processes. Production of CO_2 by R_{day} increases with warmer temperatures (Q_{10} of about 2.5), so the A/C_i compensation point rises substantially. However, the carboxylation efficiency is not greatly changed as activation energies for the processes are similar. In contrast, A_{max} decreases at high photon flux and temperature, due to RuBP limitation. The relationship is greatly affected by CO_2 availability (*Figure 12.6b*), which greatly increases the temperature at which the carbon balance is positive. Also, with larger I_o at the C_c experienced in the current atmospheric CO_2, A is stimulated, particularly at higher temperatures (*Figure 12.6c*). Although R_{day} increases and is important, it is a relatively small fraction of photosynthesis. Photorespiration is a much larger component of total CO_2 evolution and it becomes increasingly larger at higher temperatures. Its importance may be seen in the response of quantum yield of C3 plants to temperature. In small O_2 partial pressures they have quantum yields of about 0.08 mol CO_2 mol photon^{-1} over the range 10–40°C. However, in 21 kPa O_2 the value decreases progressively to half that at 40°C.

Adaptation to temperature. Adaptation to temperature is complex and strongly species-dependent. Temperature affects processes such as utilization of assimilates, and transport and diffusion of metabolites; at low temperatures the rates of these processes are generally slower than at high. Supply of phosphate ions to the chloroplast stroma at low temperatures probably limits A and prevents a response to O_2 in C3 plants. In plants able to tolerate cold, growth at low temperatures increases the contents of chlorophylls, carotenoids, plastoquinones and cytochromes, but generally the proportions remain very similar (e.g. the ratio of light-harvesting complexes to photosystems). Lipids of thylakoids may also alter with the temperature at which they develop. Some evidence for a decreased MGDG/DGDG ratio and increased unsaturated fatty acids in cold-grown leaves of different species suggests that this may be a general response, although thylakoids seem to conserve their structure and composition rather more than other plant membranes. Enzymes of photosynthesis may be modified in leaves grown at different temperatures. Rubisco of rye, for example, shows changes in kinetic constants related to alterations in the exposed SH groups on proteins; at low temperature these effects result in greater efficiency *in vitro*. Cold seems to alter the capacity for photosynthesis, increasing the content of 'machinery' per unit leaf area and perhaps the ratio of some key components, for example plastoquinone or other electron transport components, and improving some enzyme characteristics. Components of protective metabolic sequences increases to dissipate energy from the photosynthetic apparatus in cool conditions to avoid photoinhibition, but this appears not to have been examined. Respiration of plants, both R_{day} and R_{pr}, is small in cool conditions and often large accumulations of carbohydrate result from continued photosynthesis in the cold, partly as a consequence of the low respiration and partly because of reduction in 'sink demand'. This may affect the use of metabolites and result in feedback inhibition. Regulation of the gene expression and development of the photosynthetic system is such that over a very wide range its characteristics are very tightly controlled to give a stable, efficient system, well protected against many aspects of the environment. Higher temperatures generally lead to smaller production of photosynthetic components per unit area: because the rates of processes are increased, less of each component is required. Generally, there is a decrease in the proportion of electron transport components compared to enzymes but, as at the other temperature extreme, ratios of components do not change drastically and A remains relatively constant over a wide range before being inhibited.

Increased respiration of leaves and of non-photosynthetic organs as temperature rises (*Figure 12.6a*), places a substantial burden on the photosynthetic production of assimilates. As a result plants not acclimated to high temperature may eventually die of assimilate starvation. How these processes are regulated in the short and long term is unknown. Tropical and subtropical plants and C4s have much higher temperature optima than many temperate plants, which are predominantly C3s and are more tolerant of heat than cold. Many factors contribute to their differences in response to extremes of temperature, including growth of leaves and of organs which consume photosynthate. The photosynthetic system of plants, which cannot adjust to cold, may be inhibited because specific processes are damaged. Membrane instability may be a cause of chilling damage; the mechanisms by which this is overcome in some species but not others are not yet understood. Chilling results in phase changes in the lipids of membranes and disruption of membrane stability and functions (e.g. PSII). This may be at the level of accumulation of high energy intermediates in the electron transport chain and photoinhibition, which is characteristic of chilling injury. Slower metabolism at low temperatures decreases the use of light reaction products and thus increases the possibility of photochemical damage (particularly in bright light) which occurs as the energy load on the photosystems increases and the energy dissipating mechanisms, for example carotenoid 'safety-valve' or SOD and peroxidase detoxification mechanisms, do not have the capacity or are too slow to remove the energy or products. Perhaps the plant is genetically unable to produce the required intermediates or they are not made in adequate amounts because of the growth conditions. Other plants may suffer lesions during development of organs under low temperatures, reducing expression of their photosynthetic capacity, for example, chlorophyll synthesis is prevented in maize by temperatures below 8–10°C.

High temperatures also have marked effects on photosynthesis and associated processes, but what constitutes a high temperature depends on species and growth conditions. Instability of membranes and denaturation of enzymes change photosynthesis and increase respiratory demands on the assimilate supply. Prevention of proper development of organs with high temperatures may also contribute to the inability of the plant to form effective photosynthetic structures. Gene expression, protein synthesis and assembly of proteins into the correct forms may be affected and heat shock responses, with accumulation of chaperonin proteins (see Chapter 10) may occur. These changes reflect the inability of tissues to make and assemble protein complexes. Possibly chaperonins have a role as protein protective agents or in synthesizing protective metabolites by which plants avoid heat damage. These factors are outside the realm of the present discussions of photosynthesis but are vitally important from an ecological perspective. Adaptations to the growth conditions may involve overcoming several limitations in growth, development and metabolism in order to form mechanisms by which plants adjust to particular ecological conditions.

12.3 **C4 photosynthesis**

The photosynthetic mechanisms of C4 leaves have been discussed in Chapter 9. The differences between C4 and C3 biochemistry result in the very different response of photosynthesis to light, CO_2 *et cetera*. There are different mechanisms of C4 photosynthesis, so the physiological responses to the environment are rather heterogeneous, however the general responses are outlined here.

12.3.1 *The characteristics of C4 photosynthesis*

Before discussing C4 photosynthetic responses to the environment, brief repetition of the important features of the C4 mechanism may help understanding (*Figure 12.7a and b*). Analysis of C4 processes is based on light capture, electron transport and Rubisco kinetics,

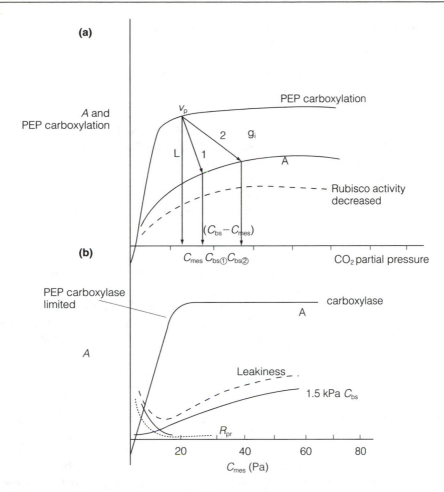

Figure 12.7. Assimilation of CO_2 in C4 plants. (a) The rate of CO_2 uptake by the PEP carboxylase reaction in the mesophyll is saturated by a small increase in CO_2 partial pressure above zero. The assimilation rate, A, rises as the CO_2 partial pressure in the bundle sheath increases as the flux of assimilates from the mesophyll increases. However, the CO_2 partial pressure in the bundle sheath is also determined by the leakiness of the bundle sheath cell walls. If the conductance (g_i) is large (1) then the cells are leaky so the C_{bs} is smaller than if the g_i is small (2). (b) General response of A to CO_2 in the mesophyll, showing the effect of the PEP carboxylase concentrating mechanism. Leakiness may increase as the CO_2 partial pressure rises. The decrease in photorespiration is indicated.

with the response of Rubisco determined by the CO_2 partial pressure in the bundle sheath. The deciding feature, and major difference between modeling C4 and C3 photosynthesis, is the CO_2-concentrating mechanism in C4 leaves. This involves the PEP carboxylase reaction in the mesophyll, transport of metabolites to the bundle sheath and release of CO_2 there, which increases C_c, the CO_2 partial pressure at the active sites of Rubisco. However, there is leakage of CO_2 from the bundle sheath, which depends on the conductance of the bundle sheath cell walls to CO_2. Although CO_2 is re-assimilated by PEPc in the mesophyll it is a futile cycle which wastes energy. Models of C4 photosynthesis thus combine models of Rubisco activity in the bundle sheath, with PEPc in the mesophyll and assessment of the CO_2

partial pressure from the concentrating mechanism, together with leakage. Here the analysis by von Caemmerer (2000) is followed.

For the entire leaf, A is given by the same relation as in C3 plants:

$$A = V_c - 0.5V_o - R_{day} \tag{12.21}$$

with A determined by the V_c and V_o in the bundle sheath PCR cycle activity and R_{day}, which is the sum of R_{day} from mesophyll and bundle sheath. The CO_2 supply to the bundle sheath is given by the rate of the PEPc carboxylation, V_p, minus the leakage rate, L, which depends on the gradient of CO_2 partial pressure, C, between the bundle sheath (bs) and mesophyll (mes), and the conductance for CO_2, $[L = g_{bs}(C_{bs} - C_{mes})]$, and the R_{day} from the mesophyll, R_{daymes}. The rate of supply of CO_2 to the bundle sheath determines A, so:

$$A = V_p - L - R_{daymes} \tag{12.22}$$

The CO_2 partial pressure in the bundle sheath is controlled by the CO_2 in the mesophyll, the rate of PEP carboxylation minus A and R_{daymes} and conductance to CO_2 leakage. The O_2 in the bundle sheath is derived from estimates of the PSII activity in bundle sheath and mesophyll, as it is an important variable in different types of C4 plants, in contrast to C3s, in which O_2 partial pressures are virtually constant. Efficiency of use of the CO_2 concentrating mechanism is given by $\varphi = L/V_p$. From this analysis, the rate of PEP carboxylation is greater than net A because part of the CO_2 fixed is recycled. This is expressed as 'overcycling', which is the ratio of L to A. The rate of A in the bundle sheath is given by the same equations for carboxylation and oxygenation as in C3 assimilation, with kinetic constants for C4 Rubisco. These are not the same as those of C3 plants (*Table 12.1*): K_c is 1.5–3 times greater and K_o is also larger, with k_{cat} (the rate of enzyme site turn over) 1.2 times larger, but the specificity factor is similar. In the analysis, Γ^* in the mesophyll and bundle sheath differs as the O_2 pressure is not the same in both, so it is related to the O_2 concentration by $\gamma^* = 0.5 \times S_{c/o}$, where $S_{c/o}$ is the specificity factor. So, for A in the bundle sheath, the Rubisco limited rate (RuBP saturated) is

$$A = (C_{bs} - \gamma^*: O_{bs}) V_{cmax}/ C_s + K_c (1 + O_{bs}/K_o) - R_{day} \tag{12.23}$$

Assuming that PEP carboxylation equals decarboxylation, and that it is the rate-determining step in the mesophyll under limiting CO_2 in the atmosphere,

$$V_p = (C_{mes}V_{pmax})/(C_{mes} + K_p) \tag{12.24}$$

where V_{pmax} is the maximum rate of PEP carboxylation and K_p is the Michaelis–Menten constant for PEPc with respect to CO_2. Combining the models of the different processes leads to a relationship:

$$A = [(C_{mes} V_{pmax})/(C_{mes})] + K_p - R_{daymes} + g_s C_{mes} \tag{12.25}$$

in which A is determined by the CO_2 partial pressure in the mesophyll, the Michaelis–Menten constant for PEPc with respect to CO_2, the respiratory release of CO_2 and the conductance to CO_2, gs, of CO_2 diffusion into the bundle sheath.

Calculation of the rates of synthesis of ATP and NADPH is complex, and differs from that for C3 photosynthesis because of the different demands of photosynthesis. Also the electron transport and the distribution of electrons depends on the proportion of PSII activity in the two tissue types, which differs between the groups of C4s. In C4 photosynthesis, it is possible that both Rubisco and PEPc can limit A due to deficient electron transport and PEP and RuBP supply, or because of limited CO_2 supply or enzyme activity. Because of the many features of C4 models they are less developed than those of C3s.

12.3.2 Light response of C4 photosynthesis

The response of A to I_o, the light response curve, of a C4 leaf (*Figure 12.8*) shows respiration in darkness, a linear increase in A up to about 500 μmol m^{-2} s^{-1} and a slower rate of A to very large photon flux. Some species are less responsive to I_o than this, with a more pronounced plateau. The absence of a marked plateau is, however, characteristic, and is one reason for the very high productivity of C4 plants in tropical conditions. The curve is little affected by CO_2 or O_2 partial pressure. C4 leaves respire in darkness, R_{dark}, and the rate varies with conditions and duration of the dark period. With increasing irradiance and thus increased gross photosynthesis, CO_2 evolution decreases and the light compensation point is reached. At low I_o, the bundle sheath CO_2 partial pressure is almost the same as that of the mesophyll and electron transport is needed for recycling R_{pr}. Optimal partitioning of electrons decreases compared to high light. Quantum yield of C4 plants is smaller than that of C3 plants in dim light (*Figure 12.9*) because of the greater requirement for ATP in C4 than in C3 metabolism. The proportion of electrons going to the PEPc reactions is about 0.4 over a wide range of I_o but decreases in very low light. Thus, C4 plants are less efficient in dim light than C3 plants, even under photorespiratory

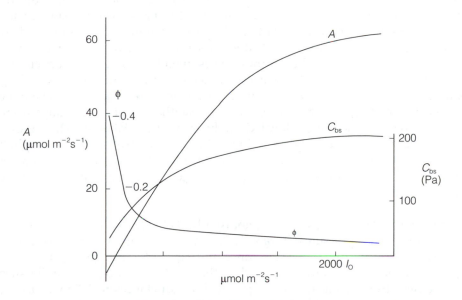

Figure 12.8. The response of CO_2 assimilation (A), the CO_2 partial pressure in the bundle sheath (C_{bs}) and the leakiness (φ) of the bundle sheath to CO_2 in relation to photon flux, I_o, in C4 photosynthesis

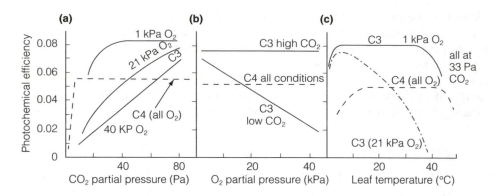

Figure 12.9. Photochemical efficiency of C3 and C4 leaves (a) as a function of CO_2 and O_2 partial pressure. (b) Photochemical efficiency of C3 and C4 leaves as a function of oxygen partial pressure and CO_2. (c) Photochemical efficiency of C3 and C4 species in relation to temperature at air partial pressure of O_2 and CO_2 or at low CO_2 and normal oxygen; generalized responses based on information in the literature

conditions. Also there is wide variation between species of C4s, with largest difference in quantum yield between NAD-ME dicots (mean 0.05) and the rest (mean 0.063). This difference is probably not related to leakiness of the bundle sheath to CO_2. As I_o increases bundle sheath CO_2 increases as PEPc activity rises and leakiness probably decreases. When irradiance reaches 200–400 µmol m^{-2} s^{-1} the advantage of the C4 mode of photosynthesis exceeds the disadvantages and photosynthesis continues to increase up to very high irradiance. There is an effect of CO_2 partial pressure in the mesophyll up to about 10 Pa but not above it. The increase in A per unit of radiation is often smaller above *ca* 1000 µmol m^{-2} s^{-1} PAR than below, but at this I_o assimilation of C3s is saturated, even in elevated CO_2. Only with very large I_o does A saturate, but often not reaching a true plateau. If Rubisco is decreased as in transgenics, then A falls because capacity for CO_2 assimilation is limited. If the PEPc reaction is decreased it results in a low CO_2 partial pressure in the bundle sheath so A saturates at lower I_o. Leakiness is a very important parameter determining the efficiency of C4 photosynthesis and it may range from 10 to 40% of CO_2 fixed in mesophyll; at low CO_2 in the mesophyll leakiness, which is of course zero at zero CO_2, rises, probably becoming stable at about 0.3 above a mesophyll concentration of some 150 Pa CO_2 when the C_{bs} is 1000 Pa. Leakiness may depend on the age of tissues, I_o and temperature, and perhaps differences in activities of C3 and C4 cycles affect leakiness.

CO_2 **response.** The response of C4 photosynthesis to CO_2 is insensitive over a wide range of CO_2 partial pressures (*Figure 12.7*). C4s have very low CO_2 compensation points, less than 0.5 Pa, due to the CO_2 concentrating mechanism greatly decreasing R_{pr}. Compensation points are linearly related to the conductance of the bundle sheath to CO_2, increasing from about 0.2 Pa at very small conductances to 1 Pa CO_2 as conductance increased to 0.1, but this depends on R_{day}, the O_2 partial pressure, the K_m for PEPc and the maximum rate of the PEPc reaction. Decreasing PEPc would have large effects on A, judging from modeling studies.

Oxygen response. PEPc is insensitive to O_2, and the large CO_2 partial pressure in the bundle sheath keeps the Rubisco oxygenase activity very small, so there is a negligible effect of O_2 on A over a wide range of partial pressures (10% decrease as O_2 increases from 0 to 50 kPa). There is probably increased photorespiration but it is only a very small proportion of A and the CO_2 is refixed. Consequently, efficiency remains constant (0.05) over a range of CO_2 concentrations from about 10 to 90 Pa (*Figure 12.9*). Thus, although the photochemical efficiency of C3 plants exceeds that of C4 in low O_2, as oxygen increases so more CO_2 is required to overcome the oxygenase induced photorespiration whereas C4 plants are insensitive above CO_2 of 5–10 Pa (*Figure 12.9*).

Response to temperature. C4 plants are relatively inefficient at low temperatures as the majority are of warm, bright, dry habitats. Some C4 species are adapted to cooler, low irradiance environments. One is *Spartina townsendii* a temperate salt marsh grass of temperate regions; the features which allow it to occupy its ecological niche are not understood. Other C4s, such as species of *Miscanthus*, which grow in climates with pronounced hot and cold seasons, largely photosynthesize at high temperatures and become dormant in the cold. The low-temperature intolerance of C4 plants, shown by the decrease in photochemical efficiency (*Figure 12.9*) involves the processes already discussed, for example, damage to membranes and inhibition of chlorophyll synthesis. Possibly, the greater demand for ATP by C4 metabolism places restrictions at low temperature when electron transport and ATP synthesis are slowed. In hot, dry environments the major advantage of the C4 syndrome is clearly shown. Whereas C3 photosynthesis decreases beyond about 30–35°C, C4s have an optimum generally substantially above 25°C and may not be damaged until 45–50°C. Also their photochemical efficiency remains constant over a wide range.

Response to water. Another advantage of the C4 syndrome compared with C3, is the ability to achieve large A despite the low CO_2/O_2 ratio in mesophyll cells and chloroplasts, resulting from the small stomata conductances (0.2–0.4 mmol m^{-2} s^{-1}). These are generally much smaller than those of C3s (0.4–1 mmol m^{-2} s^{-1}), and can decrease without substantial effects on carboxylation, in contrast to C3s. This decreases water loss in hot, dry conditions compared to C3 but with relatively little CO_2 depletion and stimulation of Rubisco oxygenase activity relative to carboxylase. The C4 mechanism also avoids the detrimental effects of decreased CO_2/O_2 solubility ratio with increasing temperature. The advantage of C4 metabolism, its independence of O_2 and CO_2 in the atmosphere and to warmer temperatures, is due to the CO_2 'pump' based on the PEP carboxylase reaction which results in C_i and C_c in the bundle sheath where Rubisco functions under CO_2 saturating conditions, thus avoiding photorespiration (Chapter 9). Hence, C4s form a larger proportion of vegetation under hot, dry conditions with water deficits. This is shown by the increase in the proportion of C4 species in the grasslands of the North American great plains in hot compared to cooler seasons and their dominance of the herbaceous flora of hot, dry ecosystems such as desert fringes in southern Africa. However, the C4 syndrome does not automatically confer drought and heat tolerance; some C4 species, such as maize and other NADPH-ME types, are sensitive to dry conditions compared to NAD-ME types (e.g. sorghum; Chapter 9). The C4 mechanism explains many of the differences between C3 and C4 plants, although the long-term ecological adaptations involve other metabolic and physiological mechanisms. The ability of C3 plants to adapt to very high temperatures should not be overlooked, for example the C3 *Larrea divaricata* is able to grow in the extreme heat of Death Valley, California.

12.4 Photosynthetic response to environmental stresses

A plant species growing and reproducing successfully over a long period in a particular environment may be regarded as broadly adapted to the physical and chemical characteristics of that environment. Such species are genetically capable of forming the structures required for photosynthesis and effective growth. However, the plant's genetic ability to adapt has limits, although in the evolutionary time scale selection pressure may result in modifications to the genotype, giving an ecotype, variety or sub-species with a new balance with the environment. In the short term, environmental conditions outside the plant's capacity to maintain its functions at the normal – and presumably relatively large – efficiency required to remain competitive, constitute 'stresses' which result in 'strain' to the plant's physiological and metabolic system. If an environmental factor exceeds the upper or lower limit for regulation of photosynthesis or other process, it constitutes a stress. Thus, 'stress' is not a characteristic of the environment but of the plant's ability to deal with environment. If the photosynthetic mechanisms adapt to new conditions without loss of efficiency then A is effectively not subject to a stress. The physical/chemical features of the environment may be regarded as neutral. It is the plant's ability to respond which effectively determines the stress–strain relationship.

The main stresses encountered by plants are associated with extremes of temperature, light (both large and small energy flux or changes in spectral characteristics such as increased UV-B radiation), water supply and chemical composition of the environment. In the latter are included nutrients required for formation of the photosynthetic system (e.g. nitrogen, phosphorus, potassium), and those which damage it, for example sodium ions in saline conditions, SO_2^{2-} and heavy metals, which may be from natural sources or human activity (pollutants).

12.4.1 Temperature stress

Temperature has been discussed in relation to the photosynthetic rates of C3 and C4 plants, including brief mention of its role in determining tissue composition and the relative rates of carboxylation and oxygenation of Rubisco. Extreme cold and heat damage the photosynthetic system at many different levels of organization and, because of the interacting processes in metabolism, cause and effect are difficult to disentangle. Detailed discussions of the effects of temperature stresses on photosynthetic metabolism may be found in the references (Alscher and Cumming, 1990).

12.4.2 Light stress

When the capture of energy exceeds the capacity of the sinks for electrons, the excess energy must be dissipated, otherwise photoinhibition results from damage to the light-harvesting apparatus, particularly to the D1 protein of the PSII reaction center. Excess energy may result from the limited capacity of the non photochemical quenching processes in the thylakoids (involving the xanthophyll cycle) to dissipate energy. The capacity to utilize and dissipate energy captured depends on the adaptation of the plant to particular conditions of growth. Some plants are particularly susceptible to photoinhibition in very bright light as a consequence of genetic limitations (they are said to be 'obligate'), for example shade plants have rather limited capacity for electron transport and for utilizing electrons but have a large antenna for photon capture; hence the system may be limited by electron transport in large I_0 and damaged by it: they are obligate shade species.

Light-generated toxic oxygen radicals cause damage to the photosynthetic system (photoinhibition) by destruction of reaction center proteins and by attacking thylakoid lipids, thus forming lipid peroxides which are themselves very destructive of membrane components. This type of response to adverse environments is common and prevents normal electron transport and CO_2 fixation. The mechanisms of adaptation to radiation stress may involve increased formation of protective compounds, for example, carotenoids of the chloroplast zeaxanthin cycle (Section 3.4.4). In the case of UV-B radiation, which is damaging to many plants and may increase as a consequence of global climate change, plants form large quantities of flavonoid compounds which absorb and dissipate the energy and are thus protective. Adaptation to such adverse radiation is very species-specific and may depend on the genetic ability to produce flavonoids and other protectants (Teramura and Sullivan, 1991).

Dim light is a stress as well as bright light but for rather different reasons, because the available energy is insufficient for normal growth of nonadapted plants which suffer from deficient assimilation and inadequate energy and substrates for growth. Sun plants which adapted to dim light may be severely photoinhibited on exposure to bright light for even relatively short periods, unless allowed to acclimate. Dim light may also cause photomorphogenetic responses, aspects of which are considered in Chapter 10.

Sun and shade plants.

Metabolism and characteristics of C3 plants, their light-harvesting systems, chemistry of CO_2 fixation and rate of photosynthesis have been presented as if without variation. Even within the distinct types of C4 and CAM plants, uniformity of response has perhaps been implied. However, this is not the case. C3 and also C4 plants show great diversity of response to the environment, particularly to the major determining factor in plant growth – light. Distinction may be made between plants with high rates of photosynthesis and growth in very intense light, so-called 'sun plants', which are inefficient in dim light, with poor photosynthesis and survival, and 'shade plants' which photosynthesize and survive in dim light but are unable to function efficiently in bright light (low maximum rates of photosynthesis and photochemical damage). Although many species are obligate sun or shade plants, many species show flexibility in response to varying light intensity; they are facultative sun or shade species but may lack ability to adapt to extremes. Sun plants, which include many crops and plants of tropical regions, achieve maximum rates of photosynthesis greater than 30 µmol CO_2 m^{-2} s^{-1} and respiration rates in darkness of 2 µmol CO_2 m^{-2} s^{-1} and are not damaged by very large photon fluxes. Shade plants may have photosynthetic rates less than 10 µmol CO_2 m^{-2} s^{-1} at an I_0 of perhaps one-tenth of that of sun species and may be damaged by light intensities above half that of full sunlight. Extreme adaptation to low light is shown by plants from the floor of tropical forests, where the PAR may be less than 3% of the radiation at the top of the forest canopy and greatly enriched in wavelengths of green, far red and infrared, which are not absorbed by the foliage above. Sunflecks, small spots of intense light which pass between the leaves in the canopy, contain over half the PAR in rainforests, and plants may be adapted to use the energy quickly. Many taxonomically different plants (e.g. *Alocasia* and *Cordyline*) are able to grow in poorly illuminated habitats. Even in temperate conditions, dim light within the canopy of a single tree is also related to the development of shade leaves which contrast with leaves exposed to large photon flux.

Anatomically, sun plants have thick palisade and spongy mesophyll tissue so that there is a 2–3-fold increase in the number of cell layers. Mesophyll cells are large and thick-walled, and there is less intercellular air space and a greater (up to five-fold) surface area of cells to total leaf than shade plants. Leaf thickness appears to be an important variable in many plants and allows flexibility in the use of light and CO_2, *et cetera*, without changing the other physiological properties of the leaf. Shade and sun plants and leaves of the same plant from different illuminations differ in numbers of chloroplasts, in thylakoids (granal stacking) and ratio of membranes, light-harvesting and electron transport mechanisms to PCR cycle enzymes. Chloroplasts are usually more numerous in mesophyll cells of shade plants, and arranged near the upper leaf surface, whilst cells of the lower mesophyll have few chloroplasts. However, as the number of cell layers is smaller, shade plants often have less chlorophyll per unit leaf area. Thylakoid membranes are stacked into many grana with many lamellae and less stromal lamella. *Alocasia* grown in dim light has four times as many granal partitions per stack as spinach grown in bright light, thus increasing the area of light-capturing membrane. Grana are often irregularly orientated which may increase capture of diffuse or variably orientated light.

Shade plants may contain four to five times more chlorophyll *a* and *b* per unit volume of chloroplast and have a higher *b/a* ratio than sun plants because the LHC increases. Shade enhances the capacity for light capture and energy transfer to the reaction centers. However, the capacity of the electron transport chain in shade plants is not increased, as there is relatively much less (one-fifth) cytochrome *f*, plastoquinone, ferredoxin and carotenoid per unit of chlorophyll than in sun plants. Shade plants have, therefore, more light-collecting apparatus, but a smaller complement of electron carriers than sun plants. In dim light the rate of electron transport is limited by the number of photons falling on the leaf; it would be no advantage for shade plants to possess a large-capacity electron transport chain. With a pool of plastoquinone receiving electrons from several PSII reaction centers, this rate-limiting step is minimized in the shade plants, except when the plant is exposed to light of a brightness outside its normal range. Sun plants have less developed thylakoid systems, fewer granal stacks and partitions and less LHC, so they are less efficient at absorbing light energy at low photon flux than shade plants and so have lower quantum yield. Electron transport, or photophosphorylation, may be limiting at very large photon flux for many species, hence the larger capacity of electron transport acceptors, particular plastoquinone. Electron transport in sun plants may be 15–30 times faster than in shade species (uncoupled rate). However, the quantum yields are very similar in very dim light, so there is no difference in photochemical efficiency, that is, the reaction centers are able to transfer electrons at the same rate. However, the uncoupled electron transport rates are saturated at 600–800 µmol quanta m^{-2} s^{-1} for sun plants and 50–100 µmol quanta m^{-2} s^{-1} for shade species. Clearly the main differences between the two groups of plants is the capacity of the light-harvesting system and of electron transport. The light-absorbing system in shade plants makes them very effective at gathering the light available and passing it to the reaction centers, especially in dim light, but they are limited in bright light by the rate of electron transfer; sun plants in contrast are very efficient at transporting electrons but not at gathering weak light.

Of course it is not electron transport *per se* which determines CO_2 assimilation but NADPH and ATP synthesis. Shade plant photosynthesis may be limited by NADPH synthesis; extensive granal stacking may be essential to obtain sufficient rates of electron flow to reduce $NADP^+$. Cyclic electron flow could drive ATP synthesis to match NADPH production and the extensive grana may restrict the coupling factor to a small area of thylakoid to increase the ΔpH gradient near CF1. Sun plants probably have adequate $NADP^+$ reduction, but may be limited by ATP supply; this also applies to C4 plants. Cyclic electron transport then may produce the required ATP without $NADP^+$ reduction. Photosynthetic systems of strongly illuminated shade plants may be irreversibly damaged by very intense light, whereas sun plants are apparently insensitive. Slow movement of electrons through plastoquinone at high rates of electron flow in bright illumination causes 'backing up' of electrons and reaction centers cannot use excitation energy so that the high energy states of chlorophyll accumulate and damage increases. Photosystem II is more sensitive to photoinhibition than PSI. The structure and balance of activities of shade plant thylakoids makes them more easily damaged. Possibly the structure of the photosystems, or carotenoid complement, which reduces the energy load and provides a 'safety valve', is inadequate in shade plants. Sun plants have relatively smaller light-harvesting systems so are inefficient in weak light, but the greater number of cell layers and the greater capacity of the electron transport chain for electron flow and excess energy-dissipating systems contribute to their greater efficiency and capacity to assimilate CO_2 in intense illumination.

Assimilation is expected to be limited by CO_2 supply when photosynthesis is rapid; sun plants have more stomata per unit area of leaf and larger stomatal conductances than shade species. Internally the cell surface area is greater, increasing the conductance of the CO_2 supply pathway. Sun plants also have more enzyme capacity in a greater stromal volume. There is more Rubisco per unit of chlorophyll; as this enzyme is a major protein of leaves, the larger protein content of sun leaves compared with shade plants is expected. In very brightly illuminated sun and shade plants, with saturated rates of electron transport, the capacity of the enzyme system might be insufficient to exploit all the NADPH and ATP synthesis. However, this remains to be tested; there is a parallel between rate of photosynthesis and the amount of Rubisco under such circumstances.

Dark respiration in shade plants is much less than in sun plants, *ca* 0.15 compared with 2 µmol m^{-2} s^{-1}. Also, photorespiration is probably less because, although the RuBP oxygenase to carboxylase ratio of the protein is similar, the CO_2/O_2 ratio in the tissue is more favorable and dimly lit environments are cooler. Light compensation is very much smaller in shade than sun plants. Of course, the ratio of respiration to CO_2 assimilation at saturating light intensities is smaller in sun plants compared with shade plants and so the growth rate of sun plants is much greater than shade species. These features may be genetically determined for maximum efficiency in use of resources, such as nitrogen for protein production, with the capacity to adapt very dependent on species.

12.4.3 Water stress

The rate of water loss and onset of stress depends on many features of the soil–plant–atmosphere continuum. If water loss from plants exceeds water uptake then water content, turgor pressure and water and osmotic potentials of cells in leaves and other tissues decrease. One of the earliest responses is a decrease in g_s resulting from decreased turgor, which reduces the rate of transpiration, slowing dehydration of the tissue and also decreasing growth of organs so that less leaf area is formed under drought. Also, mild desiccation of root systems stimulates production of abscisic acid (ABA) by an unknown mechanism, which causes stomatal closure, and other longer-term adaptations such as altering gene expression, although the significance of many of the changes observed remains to be established. Decreased leaf area together with smaller g_s are the major ways of slowing loss of water but, as a consequence, A also falls due to smaller leaf area per plant and decreased g_s and CO_2 entry. Photosynthesis is slowed by tissue dehydration initially by stomatal conductance. However, there is much evidence for increasing metabolic inhibition at larger deficits. In the first phase of water loss from a leaf, with reduced stomatal conductance, the internal CO_2 partial pressure decreases because assimilation is not substantially inhibited so the major site of regulation is the stomata. There is disagreement about the effects of stress on C_i: evidence from analysis of CO_2 exchange and fluorescence suggests that C_i decreases with relatively mild stress but increases under more severe stress. With longer exposure to, and greater degree of, dehydration, particularly with decreased relative water content (RWC), low turgor and decreased osmotic potential which are internal to the plasma membrane and therefore likely to be the components of changed water balance perceived by the cellular contents, mesophyll cells lose the ability to photosynthesize.

Some studies suggest that capacity for CO_2 assimilation is unimpaired by 20–30% decrease in RWC, others that the capacity decreases progressively as RWC decreases, as a consequence of metabolic inhibition not because of decreased CO_2 partial pressures, that is the demand function decreases with stress, so despite stomatal closure C_i remains large or even increases. Increasing C_a does not restore A to the control rate in some studies but does in others. Analysis of the relative stomatal and mesophyll limitations in studies where elevated C_a fails to restore A to the control value indicates that, despite stomatal closure, control of CO_2 assimilation resides in the mesophyll. Incorrect estimation of C_i in stressed leaves may occur because stomatal closure occurs irregularly across the leaf surface ('patchy stomata'), invalidating the calculation, but patchiness is probably not as important as once thought. Decreased A has been ascribed to the effects of water stress on several photosynthetic processes. Rubisco, for example, is affected by the decrease in CO_2 partial pressure and the proportion of photorespiration relative to gross photosynthesis increases under stress, although with severe stress the increased compensation point is probably due to TCA cycle respiration rather than photorespiration. Water stress may decrease Rubisco activity but it is not the primary site of action. Lest it be thought that the effect of stress is only through photorespiration, C4 plants suffer decreased photosynthesis when water stressed despite their smaller g_s, lower c_i and virtual absence of photorespiration.

One process which may be sensitive to decreasing RWC, or more likely increasing osmotic concentration in the chloroplast, is photophosphorylation. The ATP content of stressed leaves is sometimes decreased at more severe stress, and this is probably the cause of the decreased content (slow synthesis?) of RuBP in the chloroplast correlating with decreased photosynthesis. Decreased ATP synthesis by isolated chloroplasts, and smaller ATP content and also less CF1 in stressed leaves provide evidence that inhibition of photosynthesis is related to impaired ATP production. When CO_2 assimilation is impaired electron transport and reduction of $NADP^+$ continue at a substantial rate and a large ratio of NADPH to ATP results. Oxygen evolution continues at stresses where A is completely inhibited. Fluorescence increases substantially as A is strongly inhibited with progressively increased nonphotochemical energy quenching in thylakoids and decreased photochemical quenching at more severe stress. Clearly the photosynthetic system cannot tolerate such metabolic imbalances for long and leaves become damaged and senesce rapidly under stress conditions. At very low turgor, leaves in the stressed, highly reduced state accumulate large amounts of characteristic secondary metabolites, for example, amino acids (particularly proline), betaine glycine and ABA. These may protect the photosynthetic mechanism under stress conditions. However, such changes may not have a functional role, but represent accumulation of end products of disturbed metabolism. Proline synthesis requires NADPH which may stimulate its accumulation and ABA may be a product of the breakdown of carotenoids (Young and Britton, 1990). The accumulation of neutral osmotica such as proline may minimize exposure of the photosynthetic system to decreased osmotic potential and protect cells and membranes against increased concentration of ions.

However, there is controversy about the cause of decreased photosynthetic rate, with spectrophotometric studies indicating that CF1–CF0 is not impaired, and oxygen isotope exchange and fluorescence studies indicating that mild loss of water does not impair A, because increasing the CO_2 partial pressure overcomes stomatal closure. Despite many studies of the effects of water stress on basic photosynthetic metabolism, there is no generally accepted model of the processes occurring in dehydrating leaves.

12.4.4 Nutrient deficiency

Plants require many different chemical elements from their environment in order to form fully functional organs, including leaves which can photosynthesize effectively. The response of A depends on the particular nutrient, the extent of the deficiency, and on other environmental factors. Several general points should be noted. The effects of deficiency will depend on the supply relative to the demand by the plant. If development of the photosynthetic system is a genetic process controlled by, for example, temperature, then the rate at which a nutrient must be supplied is determined by the rate of development. With supply slower than demand, the formation of photosynthetic components will be restricted. The time at which the nutrient is required and the rate also impose demands upon the supply from the storage capacity of the plant as well as from the environment. Effects of a particular nutrient will depend on the step in the photosynthetic system at which it operates, for example nitrogen (needed for proteins, chlorophyll *et cetera*) will affect the organ growth as well as its composition whereas iron (e.g. in Fe–S centers) may have specific effects on electron transport (Terry and Rao, 1991).

Nitrogen. Nitrogen is a major nutrient and photosynthesis requires considerable investment of this element because it is a very active metabolic process requiring many components. In a C3 leaf, the proportion (very approximate) of the leaf N in the thylakoids is about 25%, with components in the light harvesting and photosystem components about 18% (PSII 5%, LHC 7% and PSI 6%); and bioenergetics 7% (CF1–CF0 5% and electron transport 2%). Soluble proteins comprise 31% of total leaf N, with 23% in Rubisco, 7% in other PCR enzymes and 1% in carbonic anhydrase. Biosynthetic proteins form 22%, nucleic acids and ribosomes (9% and 7%, respectively), amino acids (3%) and the envelope proteins (3%). The remainder of 23% is 13% of other soluble proteins and 10% in structural components. Thus, Rubisco is the dominant protein in leaves. In addition, N is needed for production of leaf area which determines, together with the rate of photosynthesis per unit area, total plant production (Chapter 13). The response of A_{max} and changes in some proteins and of chlorophyll to the N-content of leaves is shown in *Figure 12.10*. Decreasing the N supply results in nitrogen deficiency, with smaller content of proteins (e.g. Rubisco) and of chlorophyll per unit area of leaf. Consequently, A_{max} and carboxylation efficiency decrease (*Figure 12.10a–c*, compare with *Figure 12.2c*) as a consequence of decreased electron flux and Rubisco activity. However, the proportion of components may remain relatively constant (e.g. chlorophyll *a/b* ratio and protein to chlorophyll, and CF1 to Rubisco) even with severe deficiency (*Figure 12.10*). So the effect on quantum yield is small, because photon capture is adequate to maintain *A*, but as I_o increases saturation occurs at smaller I_o in N-deficient than in N-sufficient leaves. At large I_o, deficiency of Rubisco slows *A*. Carboxylation efficiency and A_{max} decrease substantially with N deficiency as Rubisco is limiting and also RuBP is deficient due to inadequate electron transport. Leaves grown with very large nitrogen supply often contain a large amount of Rubisco, which does not result in significantly increased photosynthesis at the average photon fluxes experienced normally. However, given greater I_o and CO_2, this additional enzyme increases *A* compared to leaves with deficient Rubisco, with potential advantage to the plant. This is also shown by comparing transgenic plants, with small Rubisco content, to the wild-type with normal Rubisco; the transgenics respond poorly to increased I_o or to CO_2. A decrease of up to 50% did not reduce *A* in light of 340 µmol m^{-2} s^{-1} but did at 1000 µmol m^{-2} s^{-1}. Reduction in the amount of Rubisco had no effect at 30 or 100 Pa CO_2 but at 10 Pa CO_2 *A* was restricted. Many processes in the plant place demands on nitrogen supply: deficiency often induces remobilization of nitrogenous compounds from older leaves for use in developing organs, accelerating senescence and slowing *A* of leaves in crop canopies (see Chapter 13). Rubisco has been regarded as having a storage function because it is used, for example, during grain development in cereals; this may be of long-term benefit for plant survival, even if photosynthetic carbon assimilation capacity is lost.

Phosphate. Phosphate supply affects many aspects of photosynthetic function and other plant processes, so it is difficult to ascribe specific metabolic lesions caused by deficiency. Deficiency during growth decreases the number and sizes of leaves and their content of proteins and pigments but, as with nitrogen, the proportion of photosynthetic

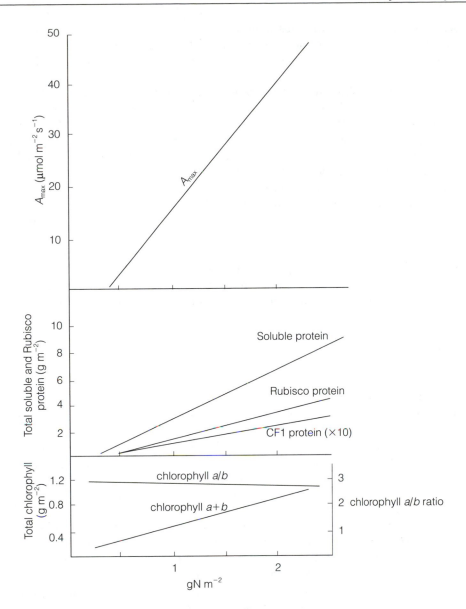

Figure 12.10. The dependence of maximum CO_2 assimilation rate, A_{max}, the amounts of total soluble, Rubisco and CF1 protein, chlorophyll and the chlorophyll *a/b* ratio on the N content of wheat leaves of plants grown with different N supply

components remains rather constant over a wide range of supply of the nutrient. One of the main effects of phosphate deficiency is to decrease the ATP content of tissues and thus to reduce the regeneration of RuBP. Under such conditions electron transport may not decrease as much as CO_2 assimilation so the pyridine nucleotides are highly reduced, and qN increases with energy dissipation by the xanthophyll cycle. The content of xanthophyll pigments often increases with P deficiency. However, the regulation of metabolism with phosphate deficiency also involves changes to enzyme activity (e.g. low P decreases

Rubisco activity). Several processes may limit photosynthesis, depending on the conditions. Under large I_o for example, synthesis of ATP is particularly important and plants exhibit symptoms of deficiency at larger P concentrations in the tissue than in small I_o.

These brief examples show how the photosynthetic system responds to different forms of environmental stress depending on the nature of the stress, the point in the system at which it operates and how the severity depends on other environmental factors. Photosynthesis is not generally a linear function of nutrient supply; deficient plants may grow much less but maintain those leaves which are formed with sufficient and balanced nutrient to allow a viable system to develop and therefore permit assimilation to take place at quite high capacity. However, such leaves may senesce rapidly as materials are remobilized to younger organs. This shortens the assimilation life of the leaf and, with severe deficiencies, impairs assimilation. Long-term effects of nutrition and other conditions on development of the photosynthetic apparatus and its function are not well understood, despite the agronomic importance. All environmental conditions have the potential to affect some aspect of photosynthesis; future analysis will focus on quantifying the responses to particular stresses and the interaction with environment so that a proper assessment of controls is obtained. This may allow the stress effects to be minimized.

12.5 Leaf fluorescence and photosynthesis

The relation between the changes in chlorophyll a fluorescence emission from the antenna complexes and the use of excitation energy in electron transport through PSII coupled with biochemical processes is discussed in Chapter 5. Application of fluorescence measurements to intact leaves has greatly advanced understanding of the interactions between these processes, under a range of environmental conditions. Analysis of the transient changes with time in fluorescence (Kautsky curves, shown in *Figure 5.10*), measured experimentally on dark-adapted intact leaves, has proved a very valuable way of understanding the relation between light harvesting, electron transport, thylakoid energetics and the processes of CO_2 fixation and oxygen evolution in the intact leaf (Krause and Weis, 1991). The transients observed are related to the rate constants of the electron transfer steps, to the rates of formation of O_2 and pH in the thylakoids, to enzyme activation and to the onset of CO_2 assimilation. In dark-adapted leaves, a very powerful, short light pulse (20 mmol quanta m^{-2} s^{-1}, 10 times greater than the brightest daylight, for 1 or 2 s, rapidly (in microseconds) induces basal fluorescence, F_o, and the excitation energy from the antenna causes all reaction centers to transfer electrons to Q_A, so closing them and giving the maximum fluorescence, F_m. After the flash, Q_A is again oxidized and the photosystem returns to the dark state. The method is reversible, and can be repeated and used on leaves without damage.

The Kautsky fluorescence induction curve gives much valuable information on thylakoid energetics, and on the efficiency of the reaction centers. With the simple Kautsky curve, the analysis is limited to the induction period because the leaf has to be adapted to darkness. Analysis in the light is difficult, as wavelengths of the actinic light which interfere with the fluorescence detection, particularly with strong background illumination, must be removed, and changing the intensity alters the fluorescence produced. Development of modulated fluorescence techniques allows measurement of chlorophyll *a* fluorescence under normal actinic light intensities during steady-state and variable I_o with changing *A* (*Figure 12.11*). The method uses a weak, pulsed (100 Hz) measuring beam (685 nm) from light-emitting diodes to induce fluorescence which is measured, with suitable electronic circuitry and amplifiers, in phase with the induction light. Thus, fluorescence is measured independently of the actinic light source and can be analyzed under varying environmental conditions. In *Figure 12.11* the course of modulated fluorescence during induction and the measurements of fluorescence parameters under steady-state assimilation is shown. The F_o and F_m values are the basal and maximal fluorescence yields for the dark-adapted leaf, as in the Kautsky curve. The steady-state fluorescence in actinic light is F_s. By applying the brief saturating light pulse to the leaf to fully reduce Q_A and close all PSII reaction centers the value of F_m under fully activated photosynthetic conditions (F_m') is obtained. The value of F_o changes as the thylakoids adapt to light, so a value F_o' applicable to the conditions is obtained immediately afterwards by stopping the actinic light by exposing the leaf to weak, far-red light, which allows electrons to flow from Q to PSI in the fully oxidized state, so removing electrons from PSII.

As with the Kautsky analysis, $F_v = F_m - F_o$ and F_v/F_m expresses the efficiency of excitation energy capture (photon yield) of 'open' PSII reaction centers (ϕ) and is almost constant (0.83) for intact, unstressed leaves of many species: it is equivalent to 9–10 quanta

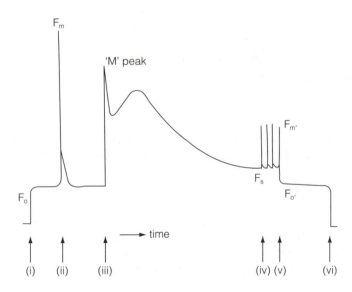

Figure 12.11. Fluorescence transient following illumination of a leaf which has been in darkness. A typical modulated fluorescence induction curve, showing: (i) modulated light on after darkness, (ii) a pulse of saturating light which induces maximal fluorescence, F_m; (iii) actinic light on; (iv) four pulses of saturating light at 15 s intervals, inducing maximal fluorescence in the light (F_m') at steady-state fluorescence (F_s); (v) actinic light off and far-red light on, giving F_o'; and (vi) all illumination off. The fluorescence parameters shown are discussed in the text together with their use in describing thylakoid energetics and quenching coefficients

per O_2 absorbed and to 0.85 electrons transported through PSII if the two photosystems receive equal proportions of excitons. A decrease in the ratio F_v/F_m indicates damage to PSII; it may be affected by environmental factors, such as very bright light during a period of cold when photosynthesis is slow, or by application of herbicides which block electron consumption. Both prevent the use of excitation energy in the PSII complex and may cause photoinhibition and damage to PSII.

Inverse relationships are observed between photochemical reactions (e.g. CO_2 assimilation) and fluorescence. However, because of the complex nature of electron flow, its coupling to photophosphorylation and the different uses made of the electrons (only a proportion are consumed in CO_2 reduction), fluorescence is not directly related to any one process but expresses the state of the PSII electron acceptor. The efficiency of electron flow through PSII per unit quantum flux, $\phi PSII = (F_m' - F_s)/F_m'$ (Genty *et al.*, 1989), is a very important value, as the supply of electrons ultimately determines A. If A is not adequate to use the electrons, then energy must be dissipated. As previously discussed, whole chain electron transport, J_{total}, may be calculated from the incident light, I_o, the leaf's absorbtance, a (about 0.85), the fraction of the light absorbed by PSII, b (values between 0.45 and 0.5, with average 0.48), and $\phi PSII$:

$$J_{total} = I_o (\phi PSII).a.b \tag{12.26}$$

There are several routes of electron use. The largest, J_{CO_2}, is for CO_2 assimilation, including photorespiration, and is related to Rubisco carboxylation by: $J_{CO_2} = (4 + 4a) V_c$. Nitrate reduction, J_{nit}, is probably 5–10% of J_A. Reduction of oxygen in the Mehler–ascorbate peroxidase reaction, J_{MEH}, is probably small and does not increase substantially with decreasing A or photorespiration, for example in transgenic tobacco with decreased Rubisco activity (Ruuska *et al.*, 2000):

$$J_{total} = J_{CO_2} + J_{Nitrate} + J_{Meh} \tag{12.27}$$

The rate of J_A determines how much excitation energy is used by photochemistry, that is electron consumption from the Q_A acceptor of PSII. This is described by photochemical quenching, qP:

$$qP = (F_m' - F_s)/(F_m' - F_o') = \phi PSII/\phi PSI_{max} \tag{12.28}$$

which approaches one with fully open PSII centers, the upper limit to the amount of photochemical work that the photosystems can do. Excitation energy not used in electron transport is dissipated in the antenna complexes of PSII by nonphotochemical quenching, qN:

$$qN = (F_m - F_m')/(F_m - F_o') \tag{12.29}$$

qN increases greatly when Q_A is reduced and is the main route for energy dissipation. It has several components, including energy-dependent quenching, qE. The qE requires a large proton gradient across the thylakoid membrane and a large $NADPH/NADP^+$ ratio (indicating a reduced state of the chloroplast stroma). This leads to the operation of the xanthophyll cycle, with epoxidation of violaxanthin to antheraxanthin and zeaxanthin, which dissipates the excitons in the chlorophyll matrix as heat (increased thermal de-excitation). Some 80% of variable fluorescence is quenched by qE when the PSII reaction centers are closed in the fully light-adapted leaf. Other components of qN are slower mechanisms of quenching fluorescence. State transition, qT, involves energy transfer ('spill-over') from PSII to PSI associated with phosphorylation of the light-harvesting complexes. Photoinhibitory quenching, qI, is related to damage at the reaction centers. The two components are related because both are determined at the same time by the rate constants of all the processes using the excitation energy (Havaux *et al.*, 1991).

Estimation of electron fluxes. Relation of electron fluxes in leaves to the processes consuming electrons has been touched on in Section 12.2.5. It may be considered from

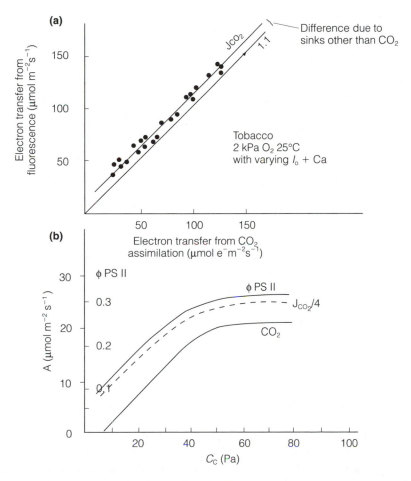

Figure 12.12. Measurement of electron flow by fluorescence and CO_2 assimilation. (a) Rates of electron transport through PSII (J_{CO_2}) calculated from fluorescence, compared with electron transfer calculated from CO_2 assimilation. (b) The relationship between the electron flow through PSII per unit quantum flux (ϕ PSII), electron flow (scaled to same units) and the measured rate of CO_2 assimilation, A, related to chloroplast CO_2 partial pressure (C_c)

two points: calculating e-transport from fluorescence and from the characteristics of Rubisco. The flux of electrons required for CO_2 assimilation and photorespiration is given by:

$$J_{CO_2} = (A + R_{day})\,(4C + 8\Gamma^*)/(C + \Gamma^*) \tag{12.30}$$

using standard abbreviations. This is the rate limited by NADPH. The rate of O_2 evolution is $J_{CO_2}/4$. The rate of electron transport limited by ATP depends on the stoichiometry of the processes, as discussed earlier. Assuming that four protons are required per ATP synthesized and that three protons are transported into the thylakoid per electron passing along the electron transport chain, then $8\Gamma^*$ increases to $9.33\Gamma^*$, so J_{CO_2} increases. If $3H^+/ATP$ and $2H^+/e^-$ is the true requirement of the system, then J_{CO_2} must increase further $(4.5C + 10.5\Gamma^*)$ in order to 'drive' the processes. The rates of electron transport estimated from fluorescence and from Rubisco characteristics, agree very well in C3 leaves under nonphotorespiratory conditions (*Figure 12.12a and b*). When CO_2 assimilation reaches a plateau at high C_i, so electron transport also saturates. When the rate of assimilation is decreased by low C_i and Rubisco is limiting, so electron transport follows. In transgenic plants with altered Rubisco contents down to 10% of the wild type, there is good agreement. Thus, the characteristics of Rubisco largely account for the use of electrons in leaves, under a range of conditions. As further support for the use of fluorescence to estimate electron fluxes *in vivo*, oxygen isotope evolution from water splitting correlates very closely with it.

Fluorescence and measurement of CO_2 exchange also provide estimates of C_c, from:

$$C_c = \Gamma^*[(J_{CO_2} + 8(A + R_{day})]/[J_{CO_2} - 4(A + R_{day})] \tag{12.31}$$

but it is necessary to assume a value for Γ^* and also that J_{total} equals J_{CO_2}, using a calibration curve for non-photorespiratory conditions (2 kPa O_2) to estimate J_{CO_2}. This is a large assumption when applied to normal O_2, and also careful determination of the number of photons resulting in photochemistry is essential. From the methods, and estimate of C_c, the conductance of the path between the intercellular spaces and the active sites of Rubisco has been estimated from:

$$g_i = A/(C_i - C_c) \tag{12.32}$$

giving values similar to those estimated from O_2 isotope exchange. The methods have been used to assess the rates of CO_2 assimilation and photorespiration in leaves under different environmental conditions. For example, in water-stressed leaves with stomatal closure it has been concluded that the ratio of photorespiration/A increases substantially and C_c approaches the compensation point. However, these conclusions depend on the assumption that J_{total} and J_{CO_2} are equal. The role of the Mehler–peroxidase reaction is clearly, from the complexity and ubiquity, very important in removing the toxic products of electron transfer to O_2, but the total flux of electrons is perhaps small. Hence, the similarity of the estimates of electron flux estimated from gas exchange and fluorescence. Quantifying the fluxes of electrons to the different potential sinks in leaves under different conditions is a major research focus.

References and further reading

Alscher, R.G. and Cumming, J.R. (eds) (1990) *Stress Responses in Plants: Adaptation and Acclimation Mechanisms.* Wiley-Liss, New York.

Baker, N.R. (1991) The relationship between photosystem 2 activity and CO_2 assimilation in leaves. In *Impact of Global Climatic Changes on Photosynthesis and Plant Productivity* (eds Y.P. Abrol, P.M. Wattal, Govindjee, D.R. Ort, A. Gnanam and A.H. Teramura). Oxford and IBH, New Delhi, pp. 379–389.

Bowes, G. (1991) Growth at elevated CO_2: photosynthetic responses mediated through Rubisco. *Plant Cell Environ.* **14**: 795–806.

Coombs, J., Hall, D.O., Long, S.P. and Scurlock, J.M.O. (eds) (1985) *Techniques in Bioproductivity and Photosynthesis* (2nd edn). Pergamon Press, Oxford.

Farquhar, G.D. and Sharkey, T.D. (1982) Stomatal conductance and photosynthesis. *Annu. Rev. Plant Physiol.* **33**: 317–45.

Farquhar, G.D. and von Caemmerer, S. (1982) Modelling of photosynthetic responses to environmental conditions. In *Physiological Plant Ecology II* (eds O.L. Lang, P.S. Nobel, C.B. Osmond and H. Ziegler). Encyclopedia of Plant Physiology, New Series, Vol. 12B. Springer, Heidelberg, pp. 550–587.

Genty, B., Briantais, J.M. and Baker, N. (1989) The relationship between quantum yield of photosynthetic electron transport and quenching of chlorophyll fluorescence. *Biochim. Biophys. Acta* **990**: 87–92.

Havaux, M., Strasser, R.J. and Greppin, H. (1991) A theoretical and experimental analysis of the qp and qN coefficients of chlorophyll fluorescence quenching and their relation to photochemical and non-photochemical events. *Photosynth. Res.* **27**: 41–55.

Heath, O.V.S. (1969) *The Physiological Aspects of Photosynthesis.* Stanford University Press, Stanford, CA.

Huner, N.P., Öquist, G. and Sarhan, F. (1998) Energy balance and acclimation to light and cold. *Trends Plant Sci.* **3**: 224–230.

Krause, G.H. and Weis, E. (1991) Chlorophyll fluorescence and photosynthesis: the basics. *Annu. Rev. Plant Physiol. Plant Mol. Biol.* **42**: 313–349.

Laisk, A. and Oja, V. (1998) *Dynamics of Leaf Photosynthesis. Rapid-Response Measurements and their Interpretations.* Techniques in Plant Sciences No. 1. CSIRO Publishing, Melbourne.

Lawlor, D.W. (1991) Concepts of nutrition in relation to cellular processes and environment. In *Plant Growth: Interactions with Nutrition and Environment* (eds J.R. Porter and

D.W. Lawlor). Society for Experimental Biology Seminar Series 43. Cambridge University Press, Cambridge, pp. 1–28.

Makino, A., Shidama, T., Takumi, S. *et al.* (1997) Does a decrease in ribulose-1,5-bisphosphate carboxylase by antisense rbsc lead to higher N-use efficiency of photosynthesis under conditions of saturating CO_2 and light in rice plants? *Plant Physiol.* **114**: 483–491.

Maxwell, K., Björkman, O. and Leegood, R.C. (1997) Too many photons: photorespiration, photoinhibition and photoxidation. *Trends Plant Sci.* **2**: 119–121.

Nishida, I. and Murata, N. (1996) Chilling sensitivity in plants and cyanobacteria: the crucial contribution of membrane lipids. *Annu. Rev. Plant Physiol. Plant Mol. Biol.* **47**: 541–568.

Osmond, C.B. and Grace, S.C. (1995) Perspectives on photoinhibition and photorespiration in the field: quintessential inefficiencies of the light and dark reactions of photosynthesis? *J. Exp. Bot.* **46**: 1351–1362.

Pearcy, R.W. (1990) Sunflecks and photosynthesis in plant canopies. *Annu. Rev. Plant Physiol. Plant Mol. Biol.* **41**: 421–453.

Prasil, O., Adir, N. and Ohad, I. (1992) Dynamics of photosystem II: mechanism of photoinhibition and recovery processes. In *Topics in Photosynthesis*, Vol. II, *The Photosystems: Structure, Function and Molecular Biology* (ed. J. Barber). Elsevier, Amsterdam, pp. 295–348.

Ruuska, S.A., Badger, M.R., Andrews, T.J. and von Caemmerer, S. (2000) Photosynthetic electron sinks in transgenic tobacco with reduced amounts of Rubisco: little evidence for significant Mehler reaction. *J. Exp. Bot.* **51**: 357–368 (GMP special issue).

Šesták Z., Čatský, J. and Jarvis, P.G. (eds) (1971) *Plant Photosynthetic Production, Manual of Methods.* Junk, The Hague.

Stitt, M. (1991) Rising CO_2 levels and their potential significance for carbon flow in photosynthetic cells: Commissioned review. *Plant Cell Environ.* **14**: 741–762.

Teramura, A.H. and Sullivan, J.H. (1991) Field studies of UV-B radiation effects on plants: case histories of soybean and loblolly pine. In *Impact of Global Climatic Changes on Photosynthesis and Plant Productivity* (eds Y.P. Abrol, P.N. Wattal, Govindjee, D.R. Ort, A. Gnanam and A.H. Teramura). Oxford and IBH Publishing, New Delhi, pp. 147–161.

Terry, N. and Rao, I.M. (1991) Nutrients and photosynthesis: Iron and phosphorus as case studies. In *Plant Growth* (eds J.R. Porter and D.W. Lawlor), Society for Experimental Biology Seminar Series, vol. 43. Oxford University Press, Oxford, pp. 55–79.

Tezara, W., Mitchell, V.J., Driscoll, S.P. and Lawlor, D.W. (1999) Water stress inhibits plant photosynthesis by decreasing coupling factor and ATP. *Nature* **401**: 914–917.

Theobald, J.C., Mitchell, R.A.C. Parry, M.A.J. and Lawlor, D.W. (1998) Estimating the excess investment in Rubisco-1,5-bisphosphate carboxylase/oxygenase in leaves of Spring wheat grown under elevated CO_2. *Plant Physiol.* **118**: 945–955.

von Caemmerer, S. (2000) *Biochemical Models of Leaf Photosynthesis.* Techniques Plant Science, No. 2. CSIRO Publishing, Collingwood.

von Caemmerer, S. and Edmondson, D.L. (1986) Relationship between steady-state gas exchange, in vivo ribulose bisphosphate carboxylase activity and some carbon reduction cycle intermediates in *Raphanus sativus*. *Aust. J. Plant Physiol.* **13**: 669–688.

Young, A. and Britton, G. (1990) Carotenoids and stress. In *Stress Responses in Plants: Adaptation and Acclimation Mechanisms* (eds R.G. Alscher and J.R. Gumming). Wiley-Liss, New York.

Chapter
13

Photosynthesis, plant production and environment

13.1 Introduction

The end product of the photosynthetic process may be considered the production of the assimilates necessary for growth. However, it is the plant's capacity to use them in growth, development and formation of organs of reproduction which assures the long-term survival of the species. Photosynthesis is, thus, not an end in itself, but a means of continuing the viability of the species, by providing the resources for growth, dispersal and establishment. In terms of the growth and reproduction of non-photosynthetic organisms (including a large human population – currently 6 billion globally and likely to increase to 10 billion by 2050) in the biosphere, it is the production of plant biomass and its components which is important, rather than the primary processes of photosynthesis. In this chapter, the way in which photosynthesis interacts with other processes at the scale of vegetation, in the natural environment, is considered.

Photosynthetic responses to the complex diurnal and seasonal changes in environment – weather conditions such as temperature, radiation and rainfall, plus variation in other environmental conditions (e.g. nutrient supply) – determine the gross assimilation of carbon and nitrogen by plants. However, although the primary photosynthetic processes underpin plant production, it is the way that they are integrated with all the other processes in development, growth and reproduction that is critical. Simple addition of basic processes, for example, of chloroplast CO_2 assimilation with dark respiration, does not describe production by the leaf under a range of conditions. Extrapolation of photosynthetic rate per unit of leaf area to the leaf area of a stand of vegetation does not describe the production of the whole system. There is great 'nonlinearity' in how processes relate, in proceeding from chloroplast through the leaf to vegetation and eventually the ecosystem, which often is not appreciated in extrapolating analysis of processes at the scale of photosystems and enzymes to the whole system. The balance between photosynthesis and respiration determines the production of dry matter by crops or natural vegetation. Accumulation of dry matter involves formation of the photosynthetic surface (leaf area) and also formation of reproductive and storage organs: this includes synthesis of the metabolic machinery, and requires assimilates to generate the 'energy' as well as the material for growth. All these processes occur in very variable environments, where temperature, light, water and nutrients as well as biotic factors often change rapidly, greatly, and in different directions. Plants are adapted to different climates and ecosystems, with large differences in means and ranges of conditions, for example tropical temperatures may not vary by more than a few degrees throughout the year, compared with 40–50°C variation in subtropical deserts. Future increase in atmospheric CO_2 and temperature, together with other aspects of climate change, represent yet further fluctuations in the environment, similar to those experienced and successfully exploited by the ancestors of current plants. No doubt the 'accumulated wisdom' in the plant's genome will allow adaptation, survival and evolution. However, the extent to which plants can adapt, and the ecological consequences, will depend on what environmental factors change, how much, and their interactions, such as the particular combinations of increasing temperature and CO_2, and their effects on photosynthesis and respiration, which are very hard to predict.

13.2 'Source' and 'sink' processes

The ecological behavior of plants is determined by photosynthetic productivity, coupled to use of the photosynthetic products in growth and development. Although the mechanisms are poorly quantified, they are clearly very sophisticated, with interrelated metabolic processes, which were the subject of the earlier chapter. Terms such as 'source' and 'sink' for the organs and processes which produce assimilates and consume them are inadequate, although they are a useful short hand. The environments of the sinks are often very different to those of the sources for the majority of plants; consider roots in soil and leaves in air to appreciate the problems faced in achieving a long-term balance between supply and demand. It is apparent that assimilate production is regulated in relation to the requirements of the rest of the plant, demanding considerable flexibility in source and sink characteristics within plants. Photosynthetic mechanisms are regulated to function within a wide range of inputs and outputs, and 'sinks' are also. There are considerable differences between species, with genetically determined characteristics evolved under different conditions. The mechanisms for achieving this balance have hardly been explored. Clearly, evolutionary pressures have led to the selection of very flexible source–sink systems, capable of functioning under varying conditions and maintaining the species in the long-term. Manipulation of this system by genetic engineering, using molecular biological modification of sources and sinks, will require sound understanding of the mechanisms by which whole plants produce assimilates and use them for growth and reproduction. By examining the way in which photosynthetic and other processes are related in plant communities, and how they respond to their environment, the mechanisms responsible for production in different current and future climates may be assessed, and the ways in which they may be altered evaluated.

13.3 Global plant productivity

Global carbon balances and the role of photosynthesis and respiration are considered in Sections 8.9 and 11.7. There is great variation in the mass of standing vegetation in different climates and habitats within them, from almost nothing in some deserts to hundreds of tonnes of dry matter per hectare in tropical forests. Rates of dry matter production by photosynthesis in the different conditions also differ greatly. Large standing biomass (e.g. in forests) does not necessarily imply rapid rates of assimilation and growth: often such ecosystems are poor in the nutrients needed for photosynthesis. Indeed, such mature ecosystems do not grow in dry matter but are in balance with respect to carbon, with CO_2 assimilation equal to respiration. When considering biomass production globally, differences in three environmental factors dominate production: temperature, precipitation and evaporation. The Holdridge Life-Zone Classification relates the distribution of the main terrestrial ecosystems of the globe to biologically relevant temperature (above 0°C), mean annual precipitation and the ratio of potential evapotranspiration (determined by the environment – humidity, temperature wind speed) to precipitation. By plotting the logarithm of temperature, precipitation and the potential evaporation ratio on the sides of an equilateral triangle, 36 biomes are defined (and can be plotted as hexagons within the triangle) which occur under the same climate in different parts of the world. Photosynthetic assimilation by plants within the biomes can be predicted from soil carbon and nitrogen contents (see Woodward and Smith, 1994). The areas of the different biomes differ greatly from 4×10^4 km² of warm temperate rainforest to 1512×10^4 km² of subtropical moist forest. The maximum rate of terrestrial photosynthesis (A_{max}) measured in different biomes differs from about 1 µmol m^{-2} s^{-1} in polar dry tundra to 25 µmol m^{-2} s^{-1} in tropical dry forest, and this can be accounted for by a model based on the supply of N, because photosynthesis has a strong linear dependence on leaf N content (about 8 µmol m^{-2} s^{-1} for each 100 mmol N m^{-2} leaf), and the environmental conditions. The A_{max} is least in very dry regions, for example hot deserts such as the Sahara, and in cold dry regions such as high mountains and tundra regions. Photosynthesis is greatest in the equatorial rainforests and cool temperate moist forests, although plants in dryer environments such as subtropical thorn steppe and boreal dry bush have large assimilation rates.

13.4 Crop productivity

Understanding of the factors determining the productivity of vegetation in natural environments is progressing rapidly, based on analysis of mechanisms at all levels of organization. The processes may be analyzed by considering an agricultural crop; natural vegetation functions in the same way but it is often very heterogeneous in terms of species, age of plants, structure of the canopy *et cetera*, and is more difficult to conceptualize and analyze. Crops are generally (in modern industrial agriculture) of genetically uniform plants, have defined growth periods, and environmental conditions may be manipulated, for example by irrigation, fertilization, control of pests and diseases. This allows the effects of light, temperature and atmospheric composition to be evaluated. Crops adequately supplied with such resources may then approach the genetic potential in that environment, with the maximum rates of photosynthesis per unit leaf area, optimum leaf area *et cetera*. For example, the C3 cereal wheat (*Triticum aestivum*) in temperate zones such as the cool, moist, long-day conditions of northwestern Europe, has the potential to produce more than 25 t dry matter ha^{-1} in a 9 month growing season. In the tropics, sugar cane (C4) forms over 80 t dry matter ha^{-1} year^{-1}. Average United States yields of grain in wheat and soybean are 2.5 t ha^{-1} and of rice 5.8 t ha^{-1}, whilst those of maize and sorghum are 7.8 and 4. 5 t ha^{-1}, respectively. Inorganic material (nutrients) constitutes less than 10% of dry matter so virtually all dry matter is organic matter derived from photosynthesis.

13.4.1 Crop growth

Growth of a wheat crop, sown as seed at uniform spacing in monoculture in homogeneous soil, is shown in *Figure 13.1*. After germination, the individual leaves of the plants grow. At full maturity the leaves provide assimilates for their own metabolism and also generate excess assimilates which are transported to growing organs. As the plant

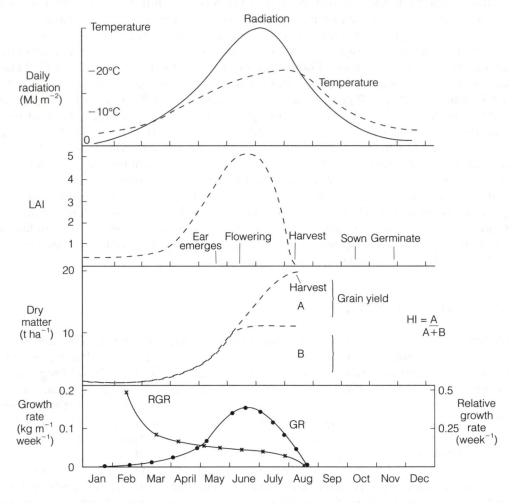

Figure 13.1. Growth of a wheat crop in a temperate climate, southern England. The crop is sown in early autumn, germinates and grows, slowing as winter advances. With the increased radiation and temperatures (a) in early spring, growth accelerates, increasing the leaf area index (LAI) (b) and dry matter (biomass) of the crop (c). With the onset of reproductive development, leaf production ceases and LAI reaches its maximum. After flowering the grain growth is rapid, and LAI decreases to zero at grain maturity. Relative growth rate is rapid for the young crop but decreases as the crop matures

develops a succession of leaves is produced, forming the canopy. Roots grow into new regions of the soil and provide nutrients and water to the plant. As leaves age, they become less efficient and slowly senesce, the constituents (e.g. proteins) being remobilized and used for new growth, and eventually die. As the leaf canopy grows, and thus leaf area index (LAI=plan area of leaf/unit plan area of soil surface) increases, more of the solar radiation incident on the surface is intercepted, providing the potential energy for photosynthesis (*Figure 13.2*). Eventually, the leaf canopy intercepts most of the incident light and growth rate is rapid, for example 15–25 g dry matter m^{-2} day^{-1} for a wheat crop in early summer in north-western Europe. Formation of buds and the growth of shoots (called tillers in the case of cereals) result in even faster formation of the leaf area and canopy. In a crop such as wheat, the determinate developmental sequence leads to the cessation of leaf production and formation of ears, with flowers which are then fertilized, thus starting the process of grain development. After the last leaf is fully expanded, the continued senescence of leaves decreases the LAI, light interception falls and dry matter accumulation ceases as the crop matures (*Figure 13.1*). The period over which the leaf area is maintained is called leaf area duration (LAD, units time).

13.4.2 Leaf growth and LAI

Development and growth of leaves are essential components of crop production, and are briefly mentioned, although outside the main theme of this book. Leaf canopy size and structure and rates of development and growth during the season are determined by the times when leaves appear and the rates and durations of their growth: these processes are greatly influenced by the temperature of the meristems and growing leaves. Below a particular temperature growth does not occur; above this threshold, leaf appearance and growth rate are related to accumulated temperature, the thermal time (°C days) or possibly to thermal time and photoperiod (photothermal time). The potential size of leaves, determined by genetic factors and expressed under nonlimiting conditions, sets the maximum LAI attainable, but the actual size depends on environmental conditions. Adequate water and nutrients allow large laminae to grow, and stress conditions result in very small leaves. After the maximum LAI is formed (in a cereal crop this is when the last leaf has grown, as a consequence of the transition to reproduction), the senescence of old leaves, which occurs throughout the formation of the canopy, leads to a decrease in LAI, eventually to zero when the crop is mature. The rate of leaf senescence is, as with leaf growth, very dependent on environmental conditions.

13.4.3 Dry matter growth

The dry matter accumulation, W, is the difference between the total net photosynthetic CO_2 assimilate production and consumption of it by respiration associated with maintenance of the crop structures and growth:

Dry matter m^{-2} $season^{-1}$ = (total net assimilate produced m^{-2} day^{-1} − total respiration m^{-2} day^{-1}) × day $season^{-1}$

$$(13.1)$$

Respiration is about 50% of net daily assimilate production, and a major component of the carbon balance. For the period from the start of rapid growth in spring to attainment of maximum dry mass, the rate of change of W with time, the growth rate ($R_w = \mathrm{d}W/\mathrm{d}t$), is relatively constant at about 20 g m^{-2} day^{-1} for a crop such as wheat. However, the relative growth rate (RGR = $\mathrm{d}W/\mathrm{d}t.W$) decreases as crop dry matter increases (*Figure 13.1*), because more nonphotosynthesizing stems, old leaves *et cetera* accumulate. They also consume assimilates in respiration, so productive leaves are a smaller proportion of dry matter as the crop develops. Although relative growth rate decreases, absolute growth rate increases in the very early stages of growth, until it becomes resource-limited due to competition. When the transition occurs depends on resources. If an environmental factor limits the growth of leaves and photosynthesis then the rate of dry matter accumulation is slowed, and productivity decreased.

13.4.4 Dry matter and radiation

Substantial experimental evidence shows that accumulation of W over a period of a few days to a whole season is very closely related to the amount of radiation energy intercepted by a crop during that period. Indeed, the relationship is very similar for well-fertilized and irrigated C3 plants in the same environment. Also, C4 plants have similar W for a given amount of intercepted radiation but produce more dry matter per unit of radiation intercepted than C3s. The relative constancy of the relationships between W and radiation, the radiation use efficiency (RUE = dry matter produced/radiation absorbed both over the same area and period), shows similar efficiency of plants and crops within the main photosynthetic types, when grown with similar conditions. This important generalization (*Figure 13.3*), emerged from detailed analysis of crops and is of great practical importance in agronomy (see Monteith, 1977, Monteith *et al.*, 1994). The relationship between the increase in W of a crop over time (*t*, e.g. growing season) per unit ground area (m^2) and light energy absorbed, *I*, is given by:

$$W = \int \mathrm{RUE}.I.\mathrm{d}t \tag{13.2}$$

Constancy of RUE shows that the efficiency of the photosynthetic and respiratory processes cannot vary greatly between different seasons, or even between crops. Differences in crop production, therefore, will depend on the amount of radiation intercepted, either because the incident radiation varies, for example between periods within a season or between seasons, or the LAI differs, due to the effects of environmental conditions on plant, and specifically leaf, growth. RUE is, however, not constant and changes with environmental conditions such as water stress and nutrient deficiency. Nevertheless, the RUE of crops grown under optimal conditions in similar environments may be considered the 'genetic potential RUE', providing a baseline for comparison.

Radiation-use efficiency is assessed from dry matter production and the energy intercepted by the crop over the same period. Dry matter is generally determined by destructive harvesting over time; it is the reference method, as calculation (modeling) of net productivity is less accurate. The efficiency depends not only on the relation of gross photosynthesis to light, on CO_2 (but this is relatively constant from day to day and over the life of an annual crop) and on temperature but also on the response of photo- and dark-respiration of both leaves and respiration of other organs (e.g. roots and stems) to conditions.

13.5 Crops and radiation capture

Radiation 'drives' photosynthesis and thus plant growth. Interception of radiation in the canopy of a crop is illustrated in *Figure 13.3*. About 10 % of incident PAR (400–700 nm) is reflected and 10% transmitted by leaves (*Figure 13.4*). Leaf absorbtivity decreases greatly at about 700 nm. At longer wavelengths (800–1500 nm) much more is reflected and transmitted, which is important for the heating effects. Solar radiation not intercepted by upper leaves in the canopy is transmitted and may pass into the lower canopy and may be intercepted there. The greater the LAI, the more of the incident radiation is intercepted. With an LAI greater than 5, the overall maximum efficiency of light capture is *ca* 90% of incident PAR. Considering the relation between the interception of the full spectrum of solar radiation or of PAR for photosynthesis (PAR is close to 50% of total when the light is predominantly diffuse) and the incident flux above the canopy, Beer's law states that:

$$1 = I_o \exp^{-KL} \tag{13.3}$$

where I is the light flux at a point in the canopy below a LAI of L, I_o the incident flux above the canopy and K is the canopy extinction coefficient. This describes an exponential decrease in light (net radiation in the day, *Figure 13.2*) from the top to the base of the canopy (*Figure 13.2*) and assumes an homogeneous canopy; leaves are effectively opaque to radiation (absorbtivity 90% to PAR wavelengths) so Beer's law is applicable. A plot of $\log_n (I/I_o)$ against LAI gives a straight line of slope K, which has values of about 0.4–0.7

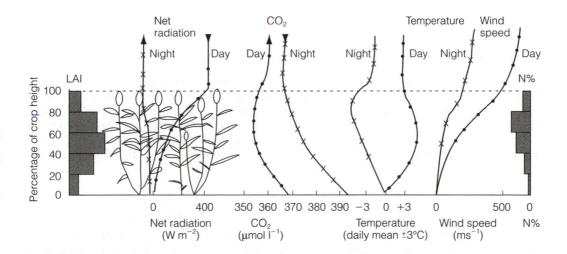

Figure 13.2. Depiction of generalized profiles of leaf area index (LAI) and percentage of above-ground nitrogen in dry matter (%N) in different leaf strata of a wheat crop at maximum LAI. Interception of incident net radiation (approximately equivalent to photosynthetically radiation) during the day in relation to LAI is indicated. The profiles of carbon dioxide, temperature and wind speed show the variation within the canopy. For comparison, the night and day profiles are given, showing the changes in direction of flux for radiation and CO_2 and large differences for all components in different strata

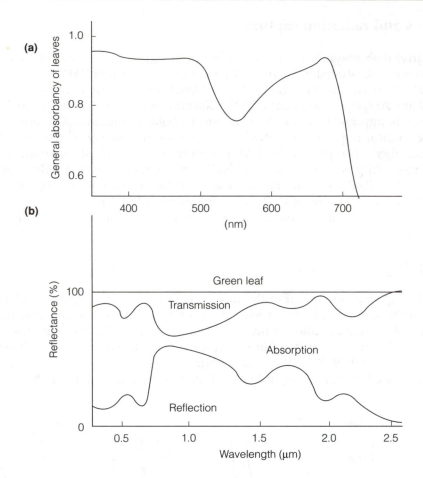

Figure 13.3. Light absorbing characteristics of leaves of crops: general for a range of species. (a) Absorbtivity of leaves as a function of wavelength. The high efficiency in the 400–500 nm and *ca* 650–700 nm regions is related to the characteristics of photosystems discussed in Chapter 3. Radiation outside these wavebands is largely reflected and transmitted from about 0.7 to 1.5 μm but longer wavelengths are increasingly absorbed (Redrawn from Monteith and Unsworth (1990), *Principles of Environmental Physics*, with permission from Butterworth Heinemann)

in wheat. It varies depending on crop architecture, for example narrow, erect leaves in a cereal crop result in light penetrating deep into the canopy and *K* is smaller for a given LAI than in crops such as cotton with broad, horizontal leaves. Also, *K* changes with growth stage and presence of ears or flowers, but is usually very constant for a given crop during the period of vegetative growth. From a knowledge of LAI, *K* and the incident radiation, an estimate of the intercepted radiation may be obtained. To measure the energy intercepted by a crop, the incident flux of radiation (or the net radiation) above the crop is measured and that penetrating to the base of the crop is also determined by radiometers which integrate over sufficiently large areas. The total energy captured by the canopy is obtained by integrating the short-term fluxes over the period of interest, for

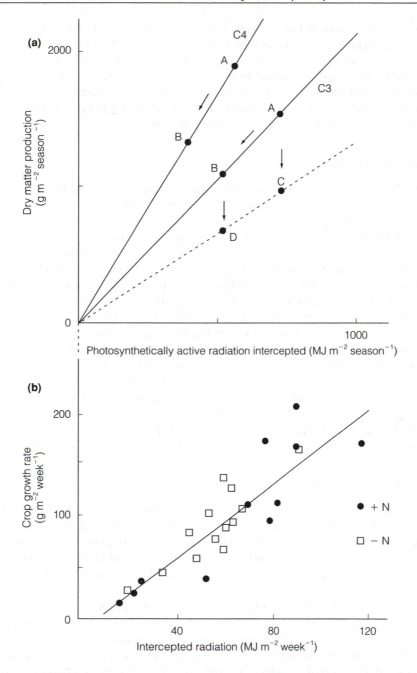

Figure 13.4. Dry matter production as a function of photosynthetically active radiation. (a) Generalized response for C4 and C3 crops with adequate water, nutrients and temperature (solid lines). Changes in leaf area decrease light interception and move production from A to B. Impaired metabolism caused by stress decreases production at a given light interception (broken line; A–C, illustrated for C3 only). Stresses which decrease both area and efficiency decrease production via A–B–D. (b) Observed dry matter production with amount of intercepted radiation for sugar beet. (After Scott *et al.*, 1994)

example the whole season. Values of from 1500 to 2000 MJ m^{-2} are typical for the whole duration of crop growth in north-western Europe. Total absorbed energy (0.4–3 μm) can be partitioned into photosynthetically active radiation (PAR, 400–700 nm waveband) which is effectively absorbed (absorbtivity *ca* 0.8), and the near-infrared (0.7–3 μm), which is not efficiently absorbed (absorbtivity *ca* 0.2). As mentioned earlier, *ca* 50% of total radiation is PAR under a wide range of conditions; environmental physics texts such as Monteith and Unsworth (1990) consider the concepts and methods.

13.5.1 Crop architecture

Canopy structure ('architecture') is important for efficiency of energy capture and, thus, productivity. To calculate and model the distribution of radiation and photosynthesis in crops, knowledge of canopy structure is required. Architecture is determined by the size and arrangement of leaves and stems, and there are a number of methods of describing such complex structures, from direct techniques such as stratified cutting of the canopy, removing and describing the size, shape and position of each part, to semi-direct methods such as 'point quadrates'. This method originally used thin rods, manually inserted into the canopy, to determine organ position; by applying mathematical methods the structure could be described. Now descriptions of canopy structure are based on electronic methods of determining and recording positions of organs, and for simulating the canopy. Indirect ways of measuring LAI, leaf inclination *et cetera* are based on light interception, for example the 'gap-faction' method, and have the advantage of integrating the canopy and are potentially much faster in application. Such technology will become more important for modeling crop processes, including photosynthesis (Pearcy *et al.*, 1990). Generally, at small LAI, horizontal leaves are most efficient at intercepting radiation, but at large LAI erect leaves are more efficient. There is a large gain in dry matter for a LAI of up to 3–4 and smaller gain at greater LAI when interception may reach 80–90% of the incident radiation.

13.5.2 Canopies and radiation capture

Canopies reflect and transmit equal amounts of PAR and intercept 80–90% of total solar radiation. Direct beam radiation is absorbed less efficiently than diffuse radiation. The ratio of direct to diffuse depends on the cloud cover and atmospheric factors such as dust or particulates, which affect the scattering of light beams. In the UK with a moist atmosphere and small incident radiation, the ratio of diffuse/direct is about 0.5, but in very dry, high irradiance conditions it is 0.2. In addition, in the canopy, reflection of light from surfaces, leaves, stems or reproductive organs affects the energy distribution. Canopies are not uniform in intercepting radiation, and gaps, often of short duration caused by motion of the vegetation due to wind, allow the penetration of patches of bright light for short periods, sunflecks, into the lower part of the canopy. In natural vegetation this may an important source of energy for understory and terrestrial plants (see Pearcy *et al.*, 1990).

Given a constant rate of photosynthesis m^{-2} leaf and increasing LAI, the total rate of assimilate production should increase exponentially in the early phase of increasing W, and if the 'fixed costs' of the crop do not increase respiration in proportion, then W should increase exponentially. It may do so for a very short period, but soon the rate of dry matter production, dW/dt, does not increase. There are several reasons for this. Many of the leaves at larger LAI are in the lower canopy, and so receive only small photon flux; thus increasing leaf area does not increase total energy capture and photosynthesis in proportion. Leaves become less efficient with age and Senesce and LAI decreases. The spectrum of the light changes due to absorption of the blue and red light by chlorophyll so there is an increase in the green, which is less efficiently absorbed. Leaves also become less efficient photosynthetically, because the efficiency of older leaves lower in the canopy is decreased, which is related to a smaller concentration m^{-2} of photosynthetic components, as shown by the small N content (*Figure 13.2*). This may be related to development of the system and to an increased rate of senescence of older leaves. Proteins, pigments, *et cetera*, are broken down to basic components (e.g. amino acids) and transported to growing leaves and grain. Thus, there is approximate balance maintained between the capacity for photosynthesis and the light environment within the canopy. All these factors and subtle interactions have important consequences for the function of the whole canopy and its production.

13.5.3 Plant factors determining radiation use efficiency

Accumulation of dry matter is strongly related to intercepted radiation and therefore, with a constant photosynthetic efficiency, productivity is a function of LAI and LAD (Section 13.4.1), LAI and LAD are determined by the rate of leaf appearance and growth, and by the rate of senescence. Conditions which increase leaf number, size and longevity, such as ample nitrogen fertilizers and water, increase LAI, LAD and dry matter production. If these resources are inadequate, then fewer, smaller short-lived leaves are formed and less radiation is intercepted. In addition, such conditions decrease the formation of the photosynthetic system (e.g. inadequate N results in less Rubisco and chlorophyll per unit leaf area), thus decreasing photosynthetic capacity and efficiency. Thus, the decrease in rate of dry matter accumulation with limiting uptake of nitrogen (*Figure 13.5*) is explained by reduced LAI and decreased efficiency. Under most conditions in the field, except with relatively extreme nutrient deficiency or water stress, the principal effect is on leaf area and light interception and not on photosynthetic efficiency or altering the

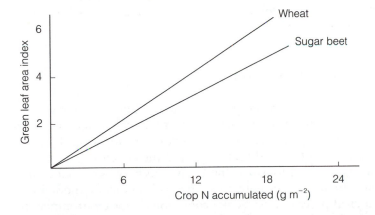

Figure 13.5. The maximum green leaf area index in relation to the amount of nitrogen taken up by crops of wheat and sugar beet. Radiation interception is strongly related to LAI, and dry matter production also. Hence, the linear dependence of dry matter on intercepted radiation shown in *Figure 13.4*.

photosynthesis/respiration ratio. Over a day/night cycle leaf carbon balance is given by $A-(R_{pr}+R_{day}+R_{dark})$ and the assimilate is respired by other parts of the crop (stems, roots, reproductive organs or 'nonleaf'), so the daily net carbon balance, and hence the increment of growth, depends on the difference between them. For constancy of RUE in a range of species and over different condition the balance between the production and use of assimilates must be relatively closely coupled. However, RUE is not fixed, probably because the coupling between processes is flexible and depends on conditions. The RUE of different C3 plants under optimal conditions ranges from about 1.3 to 2 g MJ^{-1} for total radiation and 2.6–4 g MJ^{-1} for PAR. Equivalent values for C4 crops are 2.4–2.8 g MJ^{-1} for total radiation and double for PAR. RUE is decreased by stresses such as water and nutrient shortage. In wheat, for example, in experiments with maximum RUE of about 3 g dry matter/MJ of PAR intercepted, with 300 kg ha^{-1} of nitrogen applied, and all other factors not limiting, reducing N input to zero (N only from the rather depleted soil) reduced RUE by about 30%, but total dry matter production by 70% due to the effect on LAI and LAD. The effects of conditions on RUE depend on the extent to which particular processes are affected.

Increasing crop production per unit of radiation available requires increases in both RUE and LAD. In advantageous environments (adequate water, temperature etc.) application of fertilizers has resulted in both attaining maxima which are close to the genetic potential of the crop. The increase in LAD is possible by breeding crops with longer-lived leaf area; the large increases in yield of wheat crops over the last 50 years in Europe is related to increasing use of winter wheat with a much longer growing period. Some increase in efficiency may come from altering crop architecture to improve light capture by leaves lower in the canopy. However, RUE has not increased with breeding, as the many processes contributing to RUE form a very stable integrated system which presumably changes only by small increments. Selection for higher rates of A has not increased either total production (as A has generally decreased as a result of breeding compensated by larger LAI) or RUE. Increasing the sinks for assimilates may increase the dry matter gain, as occurred with the introduction of dwarfing genes in wheat, which allowed the harvest index to be increased although total dry matter production has either not increased or only marginally (Reynolds *et al.*, 2000).

13.6 **Modeling crop production**

To combine knowledge of the partial processes in plant growth in such a way that the total productivity and efficiency of the vegetation can be calculated is still extremely difficult to achieve with very great accuracy. This is desirable for several reasons – the need to predict the production of agricultural crops and the effects of adverse environments on them and the urgent need to assess the likely effects of potential changes in global climate are but two. To describe growth, mathematical simulation modeling is employed. The response of a plant process to particular conditions in the environment is determined experimentally and the relation expressed as a mathematical function. A group of these functions expressing the responses of the most important plant processes is combined into a plant or crop (because most of this work has considered the simplified situation in homogeneous stands and has direct links to profitability of agriculture) or vegetation model. Then, using estimates of environmental factors the productivity may be calculated. An advantage of this approach is that the effects on plant production of combinations of unusual conditions may be assessed, even if they are experimentally very difficult to achieve, relatively rapidly. A brief example is given.

Assimilation may be calculated (although difficult to achieve in practice) for an entire growing season by integration of the photosynthetic rate of individual leaves over light, temperature, CO_2 supply and water stress, and then over the entire leaf area for each day. The daily assimilation may then be summed over the season to give total assimilation. Net photosynthesis is a function of light intensity (photon flux), CO_2 concentration in the atmosphere within the vegetation, temperature, water supply, nutrition and the physiological state of the plant, for example, leaf age and reproductive state. Net photosynthesis of a crop may be calculated if the response of CO_2 assimilation, A, of a leaf to light flux is known. This may be expressed in several ways; one of the simpler, frequently used functions is the rectangular hyperbola [equation (13.4)]:

$$A = \alpha I g_1 C / (\alpha I + g_1 C) - R_{dark} \tag{13.4}$$

where α is light utilization efficiency (quantum efficiency) given by the initial slope of the CO_2/light response curve, g_1 is leaf conductance to the flux of CO_2, R_{dark} is the dark respiration of the leaf and C is the CO_2 concentration. Values may be assigned to these parameters based on experiments. Assimilation by leaves may still increase at light intensity where the rectangular hyperbola is 'saturated' so in many studies nonrectangular hyperbolae are used; these include a 'curvature factor' which is selected to fit the observed data better.

13.6.1 *Photosynthesis, respiration and production*

Quantitative models of biochemistry and physiology in relation to environment are now well developed (Chapter 12), and provide understanding of the way that productivity of vegetation is determined.

Photosynthesis. Assimilation of CO_2 depends on the photon flux incident upon the canopy and its absorption by leaves in different layers, as described above, and on the efficiency of conversion to assimilate. The ratio of the direct to diffuse components in radiation are important in determining the light environment of leaves in the canopy and thus efficiency. As A of C3 leaves is saturated at PAR of half or less full sunlight (maximum flux approximately 2000 μmol m^{-2} s^{-1} in some dry, middle latitudes and at high altitude), photosynthetic radiation use efficiency decreases markedly at intensity above saturation of A. This occurs despite leaves adapting, in part, to bright light at the top of canopies. Conditions during development of the photosynthetic system modify the structure of light-harvesting, energy-transducing and enzyme systems and thereby alter the efficiency of assimilation (Section 12.4). In the lower canopy, the photon flux is low, assimilation is a linear function of intensity and photosynthetic efficiency is relatively large. C3 plants are more efficient in low to intermediate light intensities than high; C4 plants, in contrast, have photosynthetic rates which do not saturate with very large photon fluxes, and so are very efficient in bright light, using the radiation efficiently even at the top of the canopy. This is in addition to the greater CO_2 assimilation in C4 compared to C3 plants.

In natural vegetation, plants with large photochemical efficiency in dim light will have an advantage in photosynthesis and growth rates for a given light absorption and may be more successful, for example, in dense vegetation or other shady habitats, where competition is very strong. Natural vegetation canopies are stratified, with the success of different species in occupying niches mainly dependent on their ability to capture light. Plants of very dim light may be unable to adjust to bright light without photo-chemical damage and some have developed structural and other characteristics, such as the ability to change leaf orientation to light, for protection. Vegetation does not receive bright radiation at all times. Many habitats are dimly lit for long periods, with clouds and twilight, when the sun's elevation is low. Efficiency in dim light is then of great importance and C3 species would have an advantage compared with C4 species. Strat-ification of leaves of different efficiency within a canopy allows very effective light absorption and high productivity. Canopy architecture influences the efficiency of light utilization; thus C3 crops suffering nutrient or water stress produce less leaf than unstressed crops and so intercept less light overall. However, the light intercepted is of higher intensity where assimilation may be higher but the conversion is less efficient. In contrast, C4 crops would absorb less total light with less leaf area but absorb it with higher efficiency.

Respiration. Respiration consumes assimilates and productivity is, therefore, depend-ent on respiratory losses throughout the period of crop growth, the growing season [equation (13.1)]:

dry matter production season^{-1} = total net photosynthesis m^{-2} ground day^{-1} × number of days season^{-1} – total respiration m^{-2} ground day^{-1} × number of days season^{-1} (13.5)

As with photosynthesis, the rate of respiration per unit dry mass of different organs must be correctly integrated over total mass, age and conditions throughout the life of the crop if the net crop production is to be modeled. Respiration is important as it proceeds throughout the day and night in nonphotosynthetic tissue (Section 8.9 and Chapter 12 for the situation in photosynthesizing tissues) and is about 50% of the net production of assimilates. Respiration provides energy and substrates for all biochemical processes, including turnover of cell structure and also formation of new growth, although part of the requirement may come directly from photosynthesis. Relatively small changes in res-piration therefore have large effects on production and RUE (Amthor, 1989, 2000). Respiration is often divided into 'maintenance' and 'growth' corresponding to the demands of the existing system and to the requirements for synthesis of new cell com-ponents: the relative magnitude of these will depend on growth stage and conditions.

$$R = k_1 P + cW \qquad (13.6)$$

where R is daily respiration and P is daily 'gross' photosynthesis (both grams of CO_2 m^{-2}

ground day^{-1}) and W is living dry mass in grams of CO_2 equivalents m^{-2} ground day^{-1}: k_1 is a dimensionless value and c is a rate (per day). The term k_1P is taken as the growth respiration and cW the maintenance respiration.

The rate of maintenance respiration per unit dry matter is large in slowly growing crops, and growth respiration is large, providing the energy and substrates for vigorous growth. However, although this distinction is a useful concept to analyze the interaction between respiration and plant functions, the type of respiration is identical; only the use to which the products are put differs. It is difficult, in practice, to distinguish between the two components of respiration. Respiratory release of CO_2 is not tightly linked to cellular metabolism under all conditions, so that electron transport in the mitochondria and ATP production vary according to demands of metabolism. In one pathway, the normal electron transport is coupled to three ATP synthesis and in the other electrons are consumed by an alternative oxidase pathway. Electron flow in the alternative oxidase pathway gives only one ATP, 'short circuits' the normal process and is less energetically efficient, but may provide a 'spill-over' mechanism to regulate carbohydrate and reductant concentrations in plants (Section 8.9.1).

Despite the close relation between growth and respiration rate, and the apparent constancy in the ratio of assimilation to respiration implied by constant RUE, there are differences between plant species and under different conditions that are ecologically important. Small rates of respiration, as in shade-adapted plants, minimize loss of energy and carbon and thus increase the efficiency of energy use under limited light. Plants adapted to bright light have a relatively large respiration rate; presumably energy and carbon are not limiting factors in growth, which is often rapid. Species adapted to adverse conditions often have low metabolic activity. There is evidence that plants such as rye grass (*Lolium*) and tobacco (*Nicotiana tabaccum*) may be selected by breeding for small rates of respiration, resulting in increased biomass production; this appears easier than selection for enhanced photosynthetic capacity. Combination of low respiration with high capacity for photosynthesis gives the greatest dry matter production in a range of environments. Both processes may be optimized to different conditions so their role in determining total crop production may vary with environment. However, plants are integrated organisms so that the net carbon balance is related to growth processes ('sink demand') in the long term; highly productive crops are optimized with respect to photosynthesis, respiration and organ growth, and to the environment. Selection of single factors (e.g. by manipulating single processes in metabolism with molecular and genetic engineering techniques) may unbalance processes, making the plant less efficient and susceptible to changes in the environment.

13.6.2 Leaf growth and LAI

Modeling development and growth of leaves is not possible from basic understanding of cellular processes, so they are modeled empirically, from measurements of leaf appearance, and rates and durations of growth, plus knowledge of size attained, related to temperature. Leaves develop and grow only above a base temperature, which is very dependent on the species (e.g. wheat is about 2°C, maize 10°C). Above the base temperature development and growth depend on the accumulated temperature, the thermal time (°C days). In some cases day length may interact with thermal time, and photothermal time is used. Leaf area decreases as leaves senesce, but senescence is very difficult to assess experimentally and to model. With such complexity, LAI and LAD of crops are difficult to simulate accurately, and consequently the calculation of crop photosynthesis and water use is affected. Often, experimental data for the particular conditions are used.

From knowledge of the incident and transmitted radiation, or of K and LAI and the incident radiation, the radiation intercepted by the canopy can be determined. Using this as input to [equation (13.2)] and with the parameters of the model established (allowing the effects of temperature, nitrogen content etc. to be incorporated), the net photosynthesis for the leaves is calculated. Photosynthesis is summed over time, giving the net carbon assimilation. To obtain the net increment of dry matter over time, total respiratory consumption of carbon by the crop is determined from a temperature response function for the process and subtracted from the assimilate production. Assimilate is partitioned into vegetative (leaves, stems) or storage and reproductive (tubers, grains) organs. Crop models use basic biological information obtained on the small scale and short term (e.g. from experiments) to explain what controls crop production, and are increasingly used as management tools in agriculture.

13.6.3 Relation between photosynthetic rate and production

Experiments on many crops do not show a very close correlation between rates of A, measured on individual leaves (or often only small areas of selected leaves) over short periods (often seconds, some times minutes, rarely hours) at high irradiance, and dry matter production or yield. This apparent anomaly is readily understood as a consequence of the neglect of factors which contribute to production. Measurements of CO_2 exchange over longer periods and of larger areas of crop have generally agreed better with production. Short-term estimates of A from younger, active leaves, particularly in bright light, with neglect of the contribution of older, shaded leaves, only give a part of the total assimilation. Leaf area changes in response to conditions rather more than assimilation per unit area, and variation in leaf area may dominate production. Also, respiration of dry matter of different organs and crops is not usually measured but is a major component of dry matter accumulation. As both assimilation and respiration may change to different extents, depending on conditions, and are not included in short-term measurements, the lack of correlation is perhaps not surprising. Whole crop measurements over longer periods are therefore expected to correlate better with production. Although photosynthesis 'drives' biomass production, it is clear that it is not the sole determining factor, as discussed above. Production and efficiency depend on many genetically determined plant factors which cannot, currently, be described from first principles. Therefore, the overall efficiency of production must be determined empirically by relating radiation interception to dry matter yield over the same interval.

13.6.4 Photosynthesis and yield

Yield (i.e. a particular part of a crop required for human use) is part of the biomass and so they are related, but are not constant. Yield/total of above ground dry matter is called the harvest index (HI). Dry matter is distributed ('partitioned)') between harvestable organs (e.g. cereal grains) and organs (e.g. cereal straw) which are not consumed. Distribution of assimilates depends on the number of potential storage sites, their capacity and supply of assimilate. In cereals, for example, yield is determined by the number of ears produced per unit ground area, the number of grains per ear and the mass per grain. Conditions which prevent grain or ear formation (nutrient or water stress) decrease yield irrespective of photosynthesis although there are generally some related effects of stress as the parts of the system are very closely integrated. Crop yield is then limited by the storage capacity for assimilate (often-called 'sink' capacity) under some conditions and the rate of production of assimilate ('source' of supply) in others. The relation of assimilation to yield and dry matter production depends on environment and plant characteristics and many processes, which occur at different times and respond in different ways to conditions, are involved. Deficiency of resources (nutrition, water supply *et cetera*) may decrease biomass but often results in relatively high HI, whereas improving resources may increase photosynthesis and biomass but not the sink capacity, so HI is low. Despite many years of research into the 'source–sink' problem and the great social and economic importance of plant yield, surprisingly little is understood of the factors which determine the growth of organs, particularly those bearing yield, so that prediction of the effects of environment on partitioning are largely based on extrapolations from observations and are not understood mechanistically. It is important that the factors regulating yield, biomass and HI are better understood.

Present crop yields, worldwide, are much smaller than potential because of poor nutrition, drought *et cetera*, and also because varieties may not allow the potential of the environment to be exploited. Better husbandry would increase the food supply for a rapidly increasing world population without increasing intrinsic (i.e. genetic), or biochemically determined, photosynthetic efficiency. Plant breeding and selection of high-yielding crops such as modern cereals have increased yields largely by improving harvest index. Paradoxically, selection has decreased photosynthetic efficiency m^{-2} leaf but increased LAI and sink capacity compared to older varieties, so that total productivity of old and new varieties are similar under comparable conditions. Further improvements in yield may come from selection for yet larger harvest index but it may become increasingly difficult to reduce support and assimilatory organs (e.g. stems and leaves, respectively). In the longer term, greater photosynthetic efficiency is needed. Genetic manipulation of enzymes (e.g. Rubisco) or of processes which limit assimilation or use of assimilates may increase efficiency and therefore yield. However, this is unlikely to result from changes in a single metabolic process in photosynthesis and alterations to other, distantly related processes will be necessary.

13.7 Transgenic plants and increasing production

Genetic engineering of plants to modify their characteristics has been discussed earlier. Trangenics provides a valuable way of analyzing processes in photosynthesis (Stitt and Sonnewald, 1995). It also offers benefits for improving crop production in the field. Currently, there is controversy about the use of genetically modified plants for this purpose, which is not dealt with here; some aspects and details are considered by Dunwell (2000). Many of the modifications so far made and introduced into agriculture, particularly in North America, address processes such as herbicide resistance. Resistance to herbicides frequently arises from single amino acid changes in the active site of a process. This change can be introduced into plants which are sensitive to the herbicide, thus allowing the herbicide to be used as a selective weed-killer when applied to the resistant crop. No specific photosynthetic modifications appear to have yet been used in agriculture, although much research and commercial activity is considering ways of improving assimilate production and accumulation in harvested components, and some are in advanced stages of testing. Photorespiration is regarded as an inefficiency in C3 plants and many attempts have been made to decrease the process. Introduction of the gene for PEP carboxylase from the C4 plant maize into the C3 species rice, using *Agrobacterium* transformation, increased PEPc 2–3-fold and decreased O_2 sensitivity of rice photosynthesis without affecting the rate. Also, altering starch to sucrose ratios has proved feasible, for example by changing sucrose phosphate synthase. The results are often unexpected with pleiotropic effects on other metabolic and developmental processes, as expected for very tightly integrated systems. Slowing senescence may improve photosynthetic production, although, as with all such changes, the effects on other processes must be considered, for example availability of N (from remobilization of leaf proteins and pigments) for grain filling in cereals. Improving stress resistance is a reason for manipulating many biochemical processes, for example increasing accumulation of mannitol to detoxify active oxygen compounds. Trehalose has also been made in tobacco, by incorporating two bacterial enzymes for its synthesis, with increased dry matter production, together with improved photosynthesis under drought stress (Pilon-Smits *et al.*, 1998). However, there is still uncertainty about the best targets for modification which will be successful in increasing basic production in variable environments, and indeed it may be a difficult problem, because of the characteristics of photosynthetic systems, requiring greater understanding for application. In the medium term, it is likely that altering secondary processes using photosynthetic products will result in improved yield (by altering harvest index) and perhaps quality of products.

13.7.1 Composition of dry matter

Conversion of CO_2 assimilated into crop dry mass depends on the type of organic molecules synthesized and the relative proportions. One gram of CO_2 gives 0.4 g fats, 0.62 g starch and 0.5 g protein; an average value (allowing for the proportions of fats, proteins, *et cetera*, in dry matter) is 0.58 g dry matter. Plants producing much oil or protein produce less dry matter than those forming carbohydrates, but the energy yield may be similar. Each gram of CO_2 assimilated is equivalent to an energy content of 38 kJ g^{-1} in fats, 12.6 kJ g^{-1} in proteins and 17 kJ g^{-1} in starch, with an approximate value of 15–20 kJ g^{-1} in dry matter from the leaves of a range of species. From the known total energy requirement for a given dry matter production and the energy in the crop, the efficiency of energy conversion of crops and natural vegetation can be calculated. For a C3 crop requiring 0.5 MJ g^{-1} dry matter and producing 25 tonnes dry matter per hectare, the energy input is 1250 MJ m^{-2} and the energy content (taking the upper value of 20 kJ g^{-1}) is 50 MJ m^{-2}, an efficiency of about 4%. Alternatively, assuming that the average energy of a mole of PAR photons is 0.2 MJ and that 1 g dry matter is equivalent to 1.7 g CO_2 (0.039 mol CO_2), then the quantum yield is less than 0.02 (mol CO_2/mol photons), an efficiency much smaller than the theoretical conversion efficiencies or those obtained under ideal conditions. The conversion efficiency of C4 crops is 50% greater than C3, but is still very inefficient in use of light energy.

13.8 Production and environmental stress

Vegetation is subject to environmental conditions which decrease growth and are regarded as 'stresses'. Each combination, e.g. of temperature and water deficit, may influence different plant processes, such as photosynthesis or leaf growth, with different quantitative and qualitative effects, although largely modifying the same plant mechanisms. The processes are most efficient under a particular range of conditions, which will depend on the plant species. Differences caused by the environment and between species are explicable in terms of the interactions between environment and plant biochemistry and physiology. Understanding of environmental control of net photosynthesis is developing rapidly, particularly at the level of organization of the leaf, as discussed in Chapter 12. Variation in total production and RUE of a particular crop of the same genotype grown at different places, in different seasons and with different agronomic practices is due to the environment interacting with plant metabolism. Environmental impacts may be divided into those which affect light interception via leaf area, and those that affect RUE. However, these are often not separate processes; for example increasing nitrogen supply increases both cell number and sizes and hence gives larger leaves, but it also increases the content of Rubisco and chlorophyll and so increases photosynthetic capacity and thus RUE.

13.8.1 Temperature

Crop temperature is very important as it determines the rate of leaf formation, rate of growth, duration and final size as well as longevity of the leaves, that is the components of LAI and LAD (Section 13.6.2). In the same way, it determines the formation of reproductive organs. If low temperature (or any other stress) slows early growth of LAI, light interception is decreased when radiation is most available, so biomass production may suffer. If the conditions do not alter RUE then productivity moves along the curve relating W to I (radiation absorbed) (*Figure 13.4*); an example is sugar beet in the cool climate of Britain. However, if low temperature decreases efficiency then the slope of the curve relating W to I decreases. Hot conditions compared to the optimum for the crop also affect crop production, by speeding up development and ripening so that LAI and LAD are small, energy capture is reduced and biomass yield falls. Also, photosynthetic efficiency may drop due to increased respiration and inadequate storage capacity and thus RUE (*Figure 13.3*). Crops growing in areas to which they are not adapted are affected by such temperature conditions. Economic production may depend on the formation and growth of organs, e.g. ears and grains of wheat, at particular times; if the temperature is too high, for example, then growth is too fast and small, shrunken grain may develop and yield suffers. The efficiency of conversion of light energy to dry matter is relatively insensitive, however respiration increases with a Q_{10} of about two so heat stimulates respiratory losses and so decreases RUE. As metabolic processes vary less than plant morphology with temperature, much variation in production comes from change in photosynthetic area.

13.8.2 Water stress

Water use of natural vegetation and crops depends on the temperature of the leaf surfaces (which determines the saturated vapor pressure inside the leaves) on the atmospheric vapor pressure and stomatal and boundary layer conductances (the latter a function of wind speed): this is discussed in Chapter 11. It also depends on the LAI: low LAI increase water loss per unit leaf area, because leaves are more exposed to radiation, wind speed is greater and vapor pressure smaller in the crop than at large LAI (*Figure 13.2*). However, total water use may be smaller. If the LAD is large then water use increases, but if photosynthesis is also maintained then water-use efficiency (WUE = dry matter produced/unit of water used) may be little altered. Given the large number of factors involved in water loss and in production it is therefore not surprising that there is great variation between environments and seasons in WUE; very approximately it ranges, for C3 crops from 1 to 3 g dry matter kg^{-1} water transpired, and for C4 crops from 3 to 8 g dry matter kg^{-1}. These differences are related to the larger stomatal conductances and smaller assimilation rates, particularly under light-saturating conditions, of C3 compared to C4 plants. In temperate climates, the advantages of C4 plants are minimized but in bright but dry environments, their advantages increase. Water deficit, which develops in plants as the soil water is used by the crop and is not replaced by rain or irrigation, rapidly and progressively slows the growth of leaves, reducing their size and number and accelerating senescence. LAI and LAD decrease more if water is limiting early in growth, rather than later (although mature leaves may roll or drop, thus decreasing area); light interception is thus smaller in stressed than unstressed crops. RUE also decreases because stress induces stomatal closure and thus CO$_2$ assimilation, whilst not decreasing respiration. The more severe the stress the greater the reduction in RUE. Generally, water stress decreases LAI and radiation interception more than RUE. Because stomatal closure affects water loss and CO$_2$ assimilation approximately equally, WUE may not change greatly with deficits. In natural vegetation of very dry environments, leaf area is generally small and stomatal closure ensures that plants do not suffer water deficits, but crops planted in such areas may produce large LAI and develop severe water deficits.

13.8.3 Nutrient stress

Adequate nutrition is of greatest importance if the genetic potential for maximum production is to be expressed. Brief consideration of the effects of nitrogen shows why it is essential, other nutrients have similar effects. The formation of organs and their metabolic components depends on synthesis of adequate amounts of proteins, both structural and enzymatic, and a wide range of other components discussed in earlier chapters. N deficiency reduces the formation of proteins and thus the efficiency of organs. This applies particularly to leaves: limited N slows the growth of leaves and reduces LAI, LAD and light interception. Leaves have the largest proportion of nitrogen of all organs in the whole plant, because of the large content of Rubisco necessary for active CO$_2$ fixation, the light-harvesting system *et cetera*. There is a strong relation between *A* and the N and protein contents of leaves. Much of the high productivity of modern intensive farming comes from the use of industrially fixed N to supply plant needs. In N-deficient soils, growth is decreased, with loss of production; constant removal of nutrients (e.g. by crop-ping or grazing) requires that they are replaced, otherwise plant production falls. In natural vegetation, where competition for N is strong, production may be limited by N supply, and there are mechanisms for absorbing and retaining N in the plant and ecosystem, for example perennial plants absorb, store and recycle N from year to year. Plants adapted to such environments often have small N content, slow rates of assimilation and correspondingly slow biomass production, but accumulate large total biomass over long periods. In addition, such vegetation may have enhanced ability to survive poor conditions compared to plants of N-rich environments, including many crops. Similar constraints on production come from deficiencies in all other nutrients if supply does not match demand, for example phosphorus and micronutrients are often particularly limiting in many tropical soils.

Responses of photosynthesis and plant growth to nutrients are not necessarily linearly related to the amount or rate of supply of a particular nutrient. In the case of N, photosynthetic rate generally increases as the N content of the leaf rises (related to the increased amount of photosynthetic machinery) but at larger content the rate of increase slows; similar responses are seen for organ (e.g. leaf area) growth, although growth seems to be capable of greater response than the composition of the leaf. Thus, the plant processes become 'saturated' and with abundant nutrition the efficiency of conversion of light energy to dry matter may increase little and total light capture likewise. The supply of reduced N to the plant depends on nitrate reduction, and hence on the supply of reductant, carbon skeletons from photosynthesis and on the amount and turnover of enzymes, as well as on NO_3^- supply.

References and further reading

Amthor, J.S. (1989) *Respiration and Crop Productivity*. Springer-Verlag, Berlin.

Amthor, J.S. (2000) The McCree-deWit-Penning de Vries-Thornley respiration paradigm: 30 years later. *Annal. Bot.* **86**: 1–20.

Araus, J.L. (1996) Integrative physiological criteria associated with yield potential. In *Increasing Yield Potential in Wheat: Breaking the Barriers* (eds M.P. Reynolds, S. Rajaram and A. McNab). CIMMYT, Mexico, pp. 150–160.

Basra, A.S. (ed.) (1994) *Mechanisms of Plant Growth and Improved Productivity. Modern Approaches*. Marcel Dekker, New York.

Boote, K.J. and Loomis, R.S. (eds) (1991) *Modeling Crop Photosynthesis from Biochemistry to Canopy*. CSSA Special Publication Number 19. Crop Science Society of America, Madison.

Boote, K.J., Bennett, J.M., Sinclair, T.R. and Paulsen, G.M (eds) (1994) *Physiology and Determination of Crop Yield*. American Society of Agronomy, Madison.

Campbell, G.S. and van Evert, F.K. (1994) Light interception by plant canopies:

efficiency and architecture. In *Resource Capture by Crops* (eds J.L. Monteith, R.K. Scot and M.H. Unsworth). Nottingham University Press, Nottingham.

Coombs, J., Hall, D.O., Long, S.P. and Scurlock, J.M.O. (eds) (1985) *Techniques in Bio-production and Photosynthesis* (2nd edn). Pergamon Press, Oxford.

Dunwell, J.M. (2000) Transgenic approaches to crop improvement. *J. Exp. Bot.* **51** (GMP special issue): 487–496.

Goudriaan, J. and van Laar, H.H. (1994) *Modelling Potential Crop Growth Processes: a Textbook with Exercises. Current Issues in Production Ecology*, Vol. 2. The Netherlands: Kluwer Academic, p. 238.

Hall, D.O., Scurlock, J.M.O., Bolkàr-Nordenkampf, H.R., Leegood, R.C. and Long, S.P. (eds) (1993) *Photosynthesis and Production in a Changing Environment: a field and laboratory manual*. Chapman and Hall, London.

Lawlor, D.W. (1995) Photosynthesis, productivity and environment. *J. Exp. Bot.* **46**: 1449–1461.

Lambers, H., Chapin, S.F. and Pons, T. (1998) *Plant Physiological Ecology*. Springer, Berlin.

Loomis, R.S. and Amthor, J.S. (1996) Limits to yield revisited. In *Increasing Yield Potential in Wheat: Breaking the Barriers* (eds M.P. Reynolds, S. Rajaram and A. McNab). CIMMYT, Mexico, pp. 150–160.

Monteith, J.L. (1977) Climate and the efficiency of crop production in Britain. *Phil. Trans. R. Soc. Lond. B*. **281**: 277–294.

Monteith, J.L. and Unsworth, M.H. (1990) *Principles of Environmental Physics*, 2nd edn. Edward Arnold, London.

Monteith, J.L., Scott, R.K and Unsworth, M.H. (eds) (1994) *Resource Capture by Crops*. Nottingham University Press, Nottingham.

Nobel, P.S. (1990) *Physicochemical and Environmental Plant Physiology*. Academic Press, San Diego, CA.

Ong, C.K. and Monteith, J.L. (1992) Canopy establishment: light capture and loss by crop canopies. In *Crop Photosynthesis: Spatial and Temporal Determinants* (eds N.R. Baker and H. Thomas). Elsevier Science, Amsterdam.

Pearcy, R.W., Ehleringer, J.R., Mooney, H.A. and Rundel, P.W. (eds) (1990) *Plant Physiological Ecology, Field Methods and Instrumentation*. Chapman and Hall, London.

Pilon-Smits, E.A.H., Terry, N., Sears, T., Kim, H., Zayad, A., Hwang, S.B., Van Dun, K., Voogd, E., Verwoerd, T.C., Krutwagen, R.W.H.H. and Goddijn, O.J.M. (1998) Trehalose producing transgenic tobacco plants show improved growth performance under drought stress. *J. Plant Physiol.* **152**: 525–532.

Reddy, K.R. and Hodges, H.F. (eds) (2000) *Climate Change and Global Crop Productivity.* CABI Publishing, Wallingford.

Reynolds, M.P., van Ginkel, M. and Ribaut, J.-M. (2000) Avenues for genetic modification of radiation use efficiency in wheat. *J. Exp. Bot.* **51** (GMP special issue): 459–473.

Richards, R.A. (1996) Defining selection criteria to improve yield under drought. *Plant Growth Regulation* **20**: 57–166.

Schulze, E.-D. and Caldwell, M.M. (eds) (1994) *Ecophysiology of Photosynthesis.* Ecological Studies 100. Springer, Berlin.

Scott, R.K., Jaggard, K.W. and Sylvester-Bradley, R. (1994) Resource capture by arable crops. In *Resource Capture by Crops* (eds J.L. Monteith, R.K. Scot and M.H. Unsworth). Nottingham University Press, Nottingham.

Stapleton, A.E. (1992) Ultraviolet radiation and plants. Burning questions. *Plant Cell* **4**: 1353–1358.

Stitt, M. and Sonnewald, U. (1995) Regulation of metabolism in transgenic plants. *Annu. Rev. Physiol. Plant. Mol.* Biol. **46**: 341–368.

Thornley, J.M. and Johnson, D.N. (1990) *Plant and Crop Modeling.* Clarendon Press, Oxford.

Tollenaar, M. and Wu, J. (1999) Yield improvement in temperate maize is attributable to greater stress tolerance. *Crop Sci.* **39**: 1597–1604.

Woodward, I.F. and Smith, T.M. (1991) Predictions and measurements of the maximum photosynthetic rate, A_{max}, at the global scale. In *Ecophysiology of Photosynthesis* (eds E.-D. Schulze and M.M. Caldwell). Ecological Studies 100. Springer, Berlin, pp. 491–509.

Zamski, E. and Schaffer, A.A. (eds) (1996) *Photoassimilate Distribution in Plants and Crops. Source–Sink Relationships.* Marcel Dekker, New York.

Index